Remarkable Physicists
From Galileo to Yukawa

The 250 years from the second half of the seventeenth century saw the birth of modern physics and its growth into one of the most successful of the sciences. The reader will find here the lives of fifty of the most remarkable physicists from that era described in brief biographies. All the characters profiled have made important contributions to physics, through their ideas, through their teaching, or in other ways. The emphasis is on their varied life-stories, not on the details of their achievements, but, when read in sequence, the biographies, which are organized chronologically, convey in human terms something of the way in which physics was created. Scientific and mathematical detail is kept to a minimum, so the reader who is interested in physics, but perhaps lacks the background to follow technical accounts, will find this collection an inviting and easy path through the subject's modern development.

Remarkable Physicists

From Galileo to Yukawa

Ioan James

Mathematical Institute, Oxford

CAMBRIDGE
UNIVERSITY PRESS

PUBLISHED BY THE PRESS SYNDICATE OF THE UNIVERSITY OF CAMBRIDGE
The Pitt Building, Trumpington Street, Cambridge, United Kingdom

CAMBRIDGE UNIVERSITY PRESS
The Edinburgh Building, Cambridge, CB2 2RU, UK
40 West 20th Street, New York, NY 10011–4211, USA
477 Williamstown Road, Port Melbourne, VIC 3207, Australia
Ruiz de Alarcón 13, 28014 Madrid, Spain
Dock House, The Waterfront, Cape Town 8001, South Africa

http://www.cambridge.org

First published 2004

Printed in the United Kingdom at the University Press, Cambridge

Typeface Trump Mediaeval 9.25/13.5 pt.   *System* LaTeX 2_ε [TB]

A catalogue record for this book is available from the British Library

Library of Congress Cataloguing in Publication data
James, I. M. (Ioan Mackenzie), 1928–
Remarkable physicists: from Galileo to Yukawa / by Ioan James.
   p. cm.
Includes bibliographical references and index.
ISBN 0 521 81687 4 – ISBN 0 521 01706 8 (pb.)
1. Physicists – Biography. 2. Physics – History. I. Title.
QC15.J36 2003
530′.092′2 – dc21   [B]   2003055423

ISBN 0 521 81687 4 hardback
ISBN 0 521 01706 8 paperback

Contents

Preface

This book is intended for those who would like to read something, but not too much, about the life-stories of some of the most remarkable physicists born between the middle of the sixteenth century and the first decade of the twentieth, a period of just over 350 years. There are five subjects in each of the ten chapters, making fifty profiles altogether. The subjects have all made an important contribution to physics, through their ideas, through their teaching, or in other ways. The emphasis is mainly on their varied life-stories, not on the details of their achievements. By minimizing technical detail, I have been able to concentrate on a representative selection of physicists whose lives seem to me of special interest. The reader who wishes for more detail about the technicalities can so easily find it elsewhere that only the briefest of indications are given here.

In writing this book I have had in mind the reader who is interested in physics but is not necessarily familiar with the history of the subject. The biographies are arranged chronologically by date of birth, so that when read in sequence they convey in human terms something of the way in which physics developed. Each of the profiles is illustrated by a portrait of the subject, except for one case where none is known. As we shall see, the remarkable physicists of our period were a surprisingly diverse collection of people. One thing that emerges clearly is that there is no such thing as a typical physicist. Any student of physics who might be looking for a role model will find some interesting possibilities. At the end I have tried to draw some general conclusions. I have also provided some suggestions for further reading.

My thanks are due to the many people who have helped me either by reading parts of the text in draft and commenting or by dealing with particular questions. Among them are Blemis Bleaney, David Brink, Sir Roger Elliott, Dominic Flament, Robert Fox, John Roche, Paolo Salvatore, Rosemary Stewart, David Thomson, David Tranah, and John Tyrer. As far as possible the sources of the illustrations and longer quotations are given at the end of the book.

Mathematical Institute,
Oxford
April 2003

Prologue

All of us, as children, have a strong desire to learn about the natural world. What we are taught about it, at home and at school, is the result of centuries of enquiry and thought. To make it easy for us we are not taken through all the stages of the historical process of discovery, and may not realize the epic struggle which went on in order to establish the basic facts of physics. What we are taught about heat, light and sound may seem rather obvious, but it was not always so. We may be knowledgeable about the universe but much of what we know was discovered within living memory. If we are at all scientifically inclined we will be fascinated by electricity and magnetism and by many other mysterious phenomena that were poorly understood until recently and perhaps are not fully understood even now.

I have chosen to begin with Galileo and Kepler, key figures in the Renaissance of science. The scientific revolution which followed fifty years later is associated primarily with the ideas of Newton but of course others were involved, notably Huygens. In the eighteenth and nineteenth centuries there were enormous advances in the understanding of heat, light, sound, electricity and magnetism, to name just a few of the fundamental concepts. At the end of the nineteenth century it was possible to find scientists who believed that there were no more major discoveries in physics to be made. However, the twentieth century saw the birth of quantum theory and the theory of relativity. Although modern physics arose out of classical physics, there was such a profound and far-reaching discontinuity that use of the term revolution is again justified. Although its implications are still being worked out, a natural place to finish my story seems to be with the period sometimes referred to as the golden age. I begin, therefore, with physicists born in the middle of the eighteenth century, and end with some of those born in the early twentieth. To have included subjects born later in the twentieth century, when the invisible college of physics was growing so rapidly in size, would have unduly extended a book that is already long enough.

Although the subjects of these profiles are of many different nationalities, I would have preferred to have achieved a wider geographical spread. Including Russia, ten different European countries are represented; Britain, France and Germany are particularly strongly represented, with justification, I believe. However, only four countries outside Europe are represented, the USA, New Zealand, India and Japan. To a large extent this is a reflection of the way physics has developed. In many countries it is only relatively recently that remarkable physicists have begun to appear. I would also have liked to include more women, but until quite recently it was so difficult for a woman to become a physicist that it is surprising that so many succeeded, rather than so few. Even today it is quite normal for a woman to abandon a promising career on marriage, in order to concentrate on raising a family.

Biographies of the men and women who contributed something important to physics in this period do not all make interesting reading; careful selection is necessary. With an eye to variety I have chosen those which seemed to me the most remarkable. There were many other subjects I should like to have included, but not enough is on record to allow a satisfactory profile to be written. It is not sufficient just to rely on an obituary notice or eulogistic memorial address. All too often personal papers have been lost and no biography has been written because not much survives for a biographer to work on. For example, take the case of Rudolf Clausius, one of the greatest German physicists of the nineteenth century. We know that he was severely wounded during service as a non-combatant in the Franco-Prussian war. We know that he was married and had six children, that his wife died in childbirth and that he married again. However, the only aspect of his personality that can be inferred from comments of his contemporaries is his contentious nature. We read in letters of 'that grouch Clausius'; in portraits we see a strong, unforgiving face. That is about all there is on record about his life, apart from listing the successive stages in his career.

The period from the birth of Galileo Galilei in 1564 to the death of Louis de Broglie in 1987 spans over four centuries, during which there were substantial changes in scientific terminology. The term physics, in anything like the sense we use it today, had not come into use at the start of our period; the term natural philosophy was often used instead, and physicists were referred to as philosophers. Of course men like Descartes, Leibniz and Kant were philosophers in the modern sense, but they were deeply interested in physics as well and so the former usage is not inappropriate. In the eighteenth century the Paris Academy distinguished between the mathematical sciences, which included physics, and the physical sciences, which did not. In fact experimental physics was in its infancy, and it was natural to group theoretical physics with mathematics. University students who later became physicists normally started out as mathematicians. Nowadays mathematical physics is usually regarded as part of mathematics and theoretical physics as part of physics but in many respects the distinction is an artificial one and serves no purpose in what follows.

Mediaeval universities had much in common, with curricula based on the *quadrivium* and *trivium*. After the Reformation, however, they developed in different ways in different parts of Europe, although Latin remained the academic language. Throughout the eighteenth century and even later, they were almost exclusively concerned with education, especially preparation for entry into the professions. Divinity, law and medicine were taught, but the physical sciences were largely ignored. Until relatively recently universities did not regard research as part of their mission. That was left to academies, especially those of Berlin, Paris and St Petersburg. Such academies were in the nature of research institutes, under control of the state.

British scientists, above all Newton, played a leading role in the scientific revolution of the seventeenth century, but the ascendancy of Britain did not last. Towards the end of the eighteenth century Britain was being left far behind in the field of scientific research after more than a century of steady progress on the continent, particularly in France. 'It is a source of wonder and regret to many that this island, having astonished Europe by the most glorious display of talents in mathematics and the sciences dependent upon them, should have suddenly suffered its ardour to cool and almost entirely to neglect those studies in which it infinitely excelled other nations', wrote one of the few British scientists who tried to do something about it. In France science was becoming increasingly professionalized; in other countries this process occurred much later. As a result France came to dominate most aspects of early-nineteenth-century science. The foundations of theoretical physics were laid in Paris and transmitted in various ways to other countries. Laplace's physical astronomy was followed by Poisson's theory of electricity, Ampère's theory of electromagnetism, Fresnel's theory of light and Fourier's theory of heat.

In Britain, the Royal Society of London did not function like the continental academies but nevertheless served as a focus for research activity. 'Men of science', to use the phrase in vogue, might well become fellows of the Royal Society but were not usually attached to any other institution. Apart from a few wealthy amateurs, scientific training was still largely an apprenticeship entered into for love of the subject. Only a few scientists made a living through teaching or other scholarly professions; a few scattered practitioners found posts at the Royal Institution, the British Museum or similar establishments, but no-one embraced science as he might the church or law or medicine to support himself and a family. In the informal apprenticeship that produced a scientific practitioner, a master guided the novice into full participation in his speciality through advice or example. Discussion of scientific principles and findings, observation of scientific activities and criticism of scientific efforts were the chief tools of instruction. The master directed the reading of his apprentice, showed him how to use apparatus and how to design experiments and instruments, and introduced him to the scientific community.

Although Britain had no precise equivalent of the continental academies, the combination of the Royal Society and the Royal Institution served just as well, if not better. Moreover, there was hardly a town of any consequence that could not boast a Philosophical Society, where the progress of science could be reported upon, and the annual meetings of the British Association for the Advancement of Science performed a similar function at a national level. In nineteenth-century Britain, as we shall see, it was the north, rather than the south, which took the lead in scientific education and research, partly because the Scottish universities had always been strong in science. In the second half of the century reform of the ancient universities of Oxford and Cambridge, and the foundation of a number of new institutions of higher education, began to transform the situation in England.

Thus a distinctive school of physics developed in Britain, and the same was true in other countries, although at all times the subject tended to transcend national boundaries. While the international character of the subject was maintained, a particularly strong rivalry developed between the French school and the German school of physics. From about 1830 science in Germany became increasingly strong; towards the end of the nineteenth century Germany's reputation in chemistry, physics, biology and medicine was rivalled only by Britain. In the twentieth century, if scientific success can be measured by the award of Nobel prizes, Germany's record far outshone that of any other country. Of all the 100 Nobel prizes in science awarded between 1901, when the awards were founded, and 1932, the year before Hitler came to power, no less than 33 were awarded to Germans or scientists working in Germany. Britain had 18 laureates; the USA had six. Of the German laureates about a quarter of the scientists were of Jewish extraction, although the Jewish population made up no more than one per cent of the German people at the time. It might be added that Austria-Hungary supplied a considerable proportion of the physicists who contributed most to German leadership in scientific research.

Until the nineteenth century scientific research was usually published in book form. This was the age of the treatise, of which Newton's

Principia is a prime example. However, correspondence between the leading researchers also played an important role, as we shall see. At the same time individuals moved around a surprising amount, considering how difficult travelling was until quite recently, and they disseminated new ideas in the process. The earliest scientific journals were *Le journal des sçavans* and the *Philosophical Transactions* of the Royal Society of London. Both first appeared in 1665, the French journal a few months before the British. The former was clearly intended to serve the interests of the European educated public generally; after the French Revolution it was renamed the *Journal des savants*, and became more of a literary and less of a scientific journal. The latter was always more focused on science but even so was originally designed to 'give some accompt of the present undertakings, studies and labours of the ingenious in many considerable parts of the world'. Similar publications soon began to appear in other countries. It has been estimated that, out of 755 titles of serials of some scientific interest that had appeared up to the end of the eighteenth century, 401 were published in Germany, 96 in France, 50 in Great Britain, 43 in the Netherlands and 37 in Switzerland. The first specialized journal in physics is generally considered to have been the *Journal der Physik*, issued at Halle and Leipzig from 1790. The *Philosophical Magazine* in England, which is still extant, began to appear in 1798.

In what follows, expressions in foreign languages will usually be translated into English, with or without the original as seems appropriate. Literal translation is sometimes unsatisfactory, for example solar system seems preferable to world system for the French *système du monde* and counsellor or excellency to privy councillor for the German title *Geheimrat*. Expressions such as *Lycée* and *Grande Ecole* in French and *Gymnasium* and *Technische Hochschule* in German seem better left untranslated. It is important to remember that the meaning of a term may vary a good deal according to time and place. The term professor might often be interpreted as lecturer, otherwise it might seem strange that almost all university posts were professorships and that they could be held in plurality: they were often ill-paid. It seems best to elucidate any other points that might cause difficulty, such as the special features of the educational systems in different countries, when they first arise. Regarding place-names, I prefer the old name Breslau rather than the new name Wrocław, for example, but Dubrovnik rather than Ragusa and Regensburg rather than Ratisbon, as being more likely to be familiar to the reader: at first I write Leyden, later Leiden – consistency in such matters seems unnecessary.

1   From Galileo to Daniel Bernoulli

Our first five remarkable physicists were born in the 137 years from 1564 to 1700. They came from Italy, Germany, the Netherlands, England and Switzerland.

GALILEO GALILEI (1564–1642)

The great scientist we know as Galileo was born in Pisa on February 15, 1564. He was the eldest son of the notable composer, lutenist and musical theorist Vincenzio Galilei, of a long-established Florentine family, and his wife Giulia (née Ammananti), a native of Pisa who considered herself socially superior to her husband. They had five or six other children. Like Dante, Leonardo, Michelangelo and other great Italians of that period he is universally known by his first name rather than by his family name. In 1574 the family moved to Florence. After four years of education in the Camaldolese abbey of Vallombrosa on the upper Arno, Galileo was expecting to make his career in the church. However, his father decided otherwise and arranged for his son to live in Pisa with a cousin, who would train him as a wool merchant. Before long it became clear that the young man was unusually able and so, at the age of seventeen, he entered the University of Pisa, training to become a doctor in accordance with his father's wishes. However, he was dissatisfied with the lectures provided, left after four years without taking a degree, and when he returned home it was to work on mathematics. Apparently he was introduced to the subject by Ostilio Ricci, said to have been a student of Tartaglia's, who was mathematician at the court of the Grand Duke of Tuscany. Galileo took pupils and gave some public lectures on mathematics in Siena and Florence. In 1587 he visited the leading Jesuit astronomer and mathematician Father Christopher Clavius at the Gregorian University in Rome, who was interested in his first research papers, one on the determination of centres of gravity of parabolic conoids and another on an ingenious balance (*La bilancetta*) he had designed for determining specific weights with precision.

In 1589 Galileo was appointed to a teaching post in mathematics at the priest-dominated University of Pisa, at that time something of an intellectual backwater. During his time there he wrote, but did not publish, a

paper called *De motu*, about the flight of projectiles and other dynamical problems. Three years later he moved to a professorship in Padua, one of the leading universities of Europe, where Copernicus had taught and Dante had studied. Padua, in the Venetian Republic, offered a far more congenial atmosphere than Pisa. Professors were not well-paid; they were expected to supplement their modest salaries by private tuition. Galileo was an excellent teacher, whose students were devoted to him. He presided over a lively household of young men to whom he taught practical subjects such as military architecture, elementary astronomy and perspective. He also ran a small workshop to manufacture scientific instruments, and, as a result of his entrepreneurship, he became a man of means. Even so, after the death of his father in 1591, he found it difficult to meet his responsibilities towards his improvident brother, who frequently came to him for money, also his sisters needed dowries if they were to marry, so that at times he ran the risk of being arrested for debt.

  In the nearby city of Venice Galileo found friends in the nobility. The most important of these was Gianfrancesco Sagredo, a confirmed bachelor, who seemed to have been exhausted by dissipation in his youth. However, as he grew older he turned to tamer pursuits, including wild parties at his country estate on the River Brenta. Sagredo was interested in science and he formed a lasting friendship with Galileo, which they continued by correspondence when the Doge sent Sagredo to Aleppo for three years in a diplomatic capacity in 1608. Galileo never married but he formed a lasting relationship with a twenty-one-year-old Venetian serving-woman named Marina Gamba, said to be beautiful, hot-tempered, lusty and probably illiterate. Galileo's shrewish mother thoroughly disapproved of her and caused trouble. Marina had three children by Galileo: Vincenzo, Virginia and Livia. Galileo took an interest in their son's education; once they were old enough he placed their daughters in the convent of San Matteo in Arcetri, on the outskirts of Florence.

Galileo was already coming round to the view that the heliocentric system of Copernicus was much more plausible than the geocentric system of Aristotle and Ptolemy. In this he was influenced by the German astronomer Johannes Kepler, whose profile comes next. Among other things, Galileo invented a machine for raising large amounts of water from aquifers, an air thermoscope and a computing device for geometrical and ballistic purposes described in his first printed work *Le operazioni del compasso geometrico e militare* (Padua, 1606), which described the operation of a lightweight military compass he had designed in collaboration with a Venetian toolmaker. In pure science his research led him about 1602 to the discovery of the isochronicity of the pendulum and to the preliminary but wrong discussion of the law of falling bodies. In 1609 he was the first to apply the newly invented telescope to astronomical observations, revealing the mountains on the moon, numerous stars invisible to the naked eye, the nature of the Milky Way and four of Jupiter's satellites (named the Medicean stars). These sensational discoveries were described in his *Sidereus nuncius* (*The Sidereal Messenger*) (Venice, 1610), one of the most important scientific books of the seventeenth century, which at once made Galileo famous all over Europe. The popular excitement was overwhelming.

The Pope had declared the first year of the new century to be a Jubilee year. It was to be a year of celebration but also of renewed determination to stem the tide of reform. The greatest intellectual of the church of Rome, Cardinal Robert Bellamine, led the drive to stamp out heresy. One of the first victims was the Dominican friar Giordano Bruno, who was imprisoned, tortured and burnt at the stake for his beliefs. He conjectured that 'There are countless constellations, suns and planets; we see only the suns because they give light; the planets remain invisible, for they are small and dark. There are also numberless earths, circling round their suns.'

Students came from many parts of Europe to sit at Galileo's feet, including French, English, German and Polish nobility, also the Swedish King Gustavus Adolphus. The new Grand Duke of Tuscany, young Cosimo

de Medici, was one of his former pupils. In 1610, feeling he was not sufficiently appreciated in the Venetian Republic, Galileo relinquished his chair
at the University of Padua after eighteen years of great creative activity and
accepted an appointment as chief mathematician and philosopher at the
court of the Medicis. Back in Florence, he devoted his entire energy to scientific research under the benevolent protection of the Grand Duke. In his
social circle, the place of Sagredo was taken by a wealthy and accomplished
young patrician named Filippo Salviati. His country retreat Le Selve in the
hills above the lower Arno became a centre for philosophical discussions,
in which Galileo was surrounded by young disciples.

Galileo decided it was time for another visit to Rome, this time as a
kind of scientific ambassador, sponsored by the Grand Duke. Galileo was
received by Pope Paul V, the successor of Clement VIII, and generally lionized. He set about promoting the new cosmology by demonstrating the latest
discoveries. These included the phases of the planet Venus, the composite
structure of Saturn and the existence of sun-spots, all described in his *Istoria
e dimostrazioni intorno alle macchie solari* (Treatise on Sunspots) (Rome,
1613). However, the Jesuits at the Gregorian University continued to cling
to the old cosmology. One of them was Father Clavius, on whom he had
called twenty-four years earlier; the German mathematician took note of
Galileo's discoveries, but refused to embrace Copernicanism.

Galileo found an important new patron in Federico Cesi, an influential
young nobleman who possessed an enormous curiosity and the courage to
break the confines of his aristocratic upbringing. When he was only eighteen
Cesi had established the Accademia dei Lincei, arguably the first successful scientific society to be founded in the seventeenth century. The stated
aim of the Lincei was to bring together 'philosophers who are eager for real
knowledge, and who will give themselves to the study of nature, and especially to mathematics. At the same time, it will not neglect the ornaments
of elegant literature and philology, which like graceful garments, adorn the
whole body of science.' Initially the society had only four members, all non-
scientists and all under thirty years of age; there hung about the Lincei a
certain air of the occult and of pseudo-science, even the taint of scandal.
The society held its meetings in Cesi's palace, which contained a splendid library, including many proscribed books, and a collection of scientific
instruments, specimens and curiosities. Its early fame rested mainly on
Galileo's participation; later it gave him much-needed support, stimulus
and encouragement. Cesi was too powerful to have to worry about what the
Jesuits at the Gregorian University thought.

Galileo was becoming more and more audacious in pointing to the incompatibility of the new celestial phenomena with traditional astronomy. He openly confessed his Copernican conviction, already stated in a letter to Kepler, at the same time as he successfully attacked current views on hydrostatics in his *Discorso intorno alle cose che stanno in su l'acqua* (*Bodies in Water*). Increasingly Galileo had to defend his discoveries and opinions against numerous attacks from scientific opponents and jealous academic enemies. A conspiracy among the latter aiming at Galileo's downfall led first to an abusive sermon against him in Florence in 1614. There were signs of paranoia in his reaction, although the enemies were real enough. The most powerful of these was Cardinal Bellarmine, the persecutor of Giordano Bruno, who warned him not to defend the Copernican system in public. As a result, he wrote a letter to Christina, the mother of the Grand Duke, giving his carefully considered opinions about the proper relation between science and religion; this *Letter to the Grand Duchess Christina* was not published until 1636.

Galileo was already fifty years of age. He was suffering from arthritis, a condition of long standing, and from pains in the chest and kidneys. On his return from Rome he first took advantage of Salviati's villa Le Selve to recuperate before settling down in a modest villa of his own, at Bellosguardo, overlooking Florence and not too far from Arcetri where his daughters lived inside their convent. However, it is hardly surprising that the following years saw some decline in his scientific activity. He mainly occupied himself with computing tables of the motion and eclipses of the moons of Jupiter, which could be used to determine longitude at sea. He tried in vain to sell this idea to the Spanish and Dutch governments. In 1618 he was involved in a bitter argument over the nature of comets, which lost him the sympathy of his former supporters among the Roman Jesuits. A result of this controversy was the polemical work *Il saggiatore* (*The Assayer*) (Rome, 1623) in which Galileo expressed his thoughts on epistemological and methodological questions, stressing the necessity of quantitative experiments and observations and the strength of hypothetical–deductive reasoning.

In 1623 one of his former supporters, Cardinal Maffeo Barberini, became Pope Urban VII, and, after a fourth visit to Rome, Galileo felt himself encouraged to begin with the composition of a major work on astronomy, planned many years before and finally published under the title *Dialogo sopra i due massimi sistemi del mondo* (*Dialogue Concerning the Two Chief World Systems*). This was a technical account in the form of a dialogue among a supporter of the Aristotelian–Ptolemaic tradition named

Simplicio, a youthful enquiring mind named Sagredo and an advocate of the new astronomy named Salviato. Galileo had tried to safeguard himself by letting Simplicio prevail, and the book was published with the imprimatur of the ecclesiastical authorities. Nevertheless the strength of Salviato's arguments was evident.

The initial reception of the book was generally favourable, but it gave Galileo's enemies the opportunity they had been waiting for. The Pope thought the imprimatur should never have been granted and tried to have the book suppressed, but it was too late. He decided that Galileo must stand trial and summoned him to Rome. The Grand Duke was powerless to shield Galileo from the wrath of the Pope. For reasons of health Galileo asked for the proceedings to be held in Florence. This was refused but, as a concession, when he arrived in Rome, instead of being committed to prison while awaiting trial, he was allowed to live in the Tuscan embassy. Formally the charge against him was one of disobedience. His accusers maintained that Bellamine in 1616 had formally admonished Galileo not to promote Copernicanism in public; Galileo denied this, documentation was lacking and Bellamine was no longer alive to give evidence. During the trial the ailing Galileo was imprisoned in the Vatican, until eventually, more dead than alive and under threat of torture, he was forced solemnly to abjure his Copernican convictions before the Congregation of the Holy Office, before being sentenced to life imprisonment and punished in other ways.

Initially he was confined to the palace of the Archbishop Ascanio Piccolomini of Siena, a man of broad cultural interests, where he was treated as an honoured guest. Before long, however, the Pope's agents reported that the episcopal palace did not keep him sufficiently isolated and he was allowed to move to his villa in Bellosguardo. In 1631, finding the journey from there to see his daughters too much, Galileo proposed to move to a house in Arcetri itself. The Pope agreed that he could do so, although still effectively under house arrest, since he was not even allowed to go to nearby Florence without permission, which was sometimes withheld. His younger daughter Livia suffered from depression but Galileo was to find great comfort in the company of his elder daughter Virginia in his declining years. In the simple beauty of the weekly letters she sent him, as 'Suor Maria Celeste', we can follow her efforts to comfort him and lift his spirits; unfortunately his side of the correspondence has not survived. Sadly, she died from dysentery not long after he had arrived in Arcetri, at the age of thirty-three.

Galileo's own health was seriously threatened; there was a trouble-some hernia and palpitations of the heart, and he also suffered from insomnia and melancholia. He continually heard his beloved daughter calling him. The Florentine Inquisitor was right to believe that the aged Galileo would never again attempt to promote Copernicanism. In fact, Galileo went further by stating that the falsity of Copernicanism must not on any account be called into doubt, especially by Catholics. All Copernican conjectures, he wrote, are removed by the most solid arguments from God's omnipotence. He had resigned himself to the fact that his own part in the campaign to establish Copernicanism was over, although his personal convictions remained the same and many were protesting against the injustice of his condemnation and sentence.

Galileo engaged in new research, although hampered both by cataracts and by glaucoma, ending in complete blindness, and by the constant supervision of the Inquisition. He succeeded in finishing his final and most important work, the *Discorsi e dimostrazioni matematiche intorno a due nuove scienze (Discourses on Two New Sciences)* (Leyden, 1638), which was, significantly, published beyond the reach of the Inquisition. This work, containing among other things the proof of the laws governing the fall of a body in a vacuum, the principle of the independence of forces and the complete theory of parabolic ballistics, was destined to become one of the cornerstones upon which Huygens and Newton one generation later built classical mechanics. The laws of fall made it possible to study accelerated motion. Simplicio, Sagredo and Salviati reappear to debate the arguments in another dialogue like the one he had used in 1632; Galileo's fondness for this manner of presentation may have come from his father, who in 1581 had published a *Dialogo della musica antica e moderna*. Galileo died at Arcetri during the night of January 8, 1642. He was buried privately in Santa Croce, the great church where so many famous Tuscans lie, but not in the Galilei family tomb, for fear of Papal disapproval. No monument to his memory was erected until 1737, when he was re-interred and the skeleton of a young woman was found beneath his in the original grave; it is thought that this could have been his beloved daughter.

Galileo had a versatile mind. He was an accomplished amateur musician and a master of the vernacular language; his polemical work *Il saggiatore* is one of the Italian classics. He occupied himself with almost every branch of physics, but is chiefly remembered for the example he gave of the efficacy of the hypotheco-deductive method combined with quantitative experiments. In general history too he occupies an important place because of his personal fate, which was an important factor in the widening fissure between natural science and the spirituality of the counter-Reformation. The last traces of official anti-Copernicanism were not removed until 1822. While geocentrism was the official doctrine, there was some latitude for teaching heliocentrism as a working hypothesis in schools and universities where Jesuits were in control.

JOHANNES KEPLER (1571–1630)

Kepler was a near-contemporary of Galileo but his life-story was very different, as was his family background. He was born in the small Lutheran town of Weil der Stadt, near Stuttgart, on December 27, 1571. Judging by the account Kepler wrote of his early life, he seems to have had a most miserable childhood. He described his father Heinrich as 'criminally inclined, quarrelsome, liable to a bad end' and his mother Catharina (née Guldenmann), as 'small, thin, swarthy, gossiping and quarrelsome', adding that 'treated shabbily, she could not overcome the brutality of her husband'. When he was three years old, his father joined a group of mercenary soldiers to fight the Protestant uprising in Holland. His mother followed her husband to Flanders. The children were abandoned to the care of grandparents who treated them roughly. When their parents returned in 1576 the family, in disgrace because of Heinrich's part in the persecution of Protestants, had

to leave Weil for nearby Leonberg, in the Grand Duchy of Württemberg. Heinrich rejoined the infamous Duke of Alba's military service for a few more years; by 1588 he had abandoned his family forever.

The future astronomer was a sickly child, with thin limbs and a large pasty face surrounded by dark curly hair. He was born with defective eyesight – short-sighted in one eye, multiple vision in the other. His stomach and gall bladder gave constant trouble; and he nearly died from smallpox. He began his education at the German Schreibschule in Leonberg but soon moved to the Latin school, there laying the foundation for the complex Latin style displayed in his later writings. After a period of 'hard work in the country', during which he did not attend school at all, he entered the Adelberg monastery school at thirteen; and two years later enrolled at the more senior Maulbronn, one of the preparatory schools for the Protestant University of Tübingen. In October 1587 Kepler formally matriculated at the university; but because no room was available at the Stift, the seminary where, as a student supported by the enlightened Duke of Württemberg, he was expected to lodge, he continued at Maulbronn for another two years. In September 1588 he passed the baccalaureate examination at the university, although he did not actually take up residence there until the following year. He was unpopular with his fellow-students, who gave him a hard time.

At Tübingen, Kepler's thought was profoundly influenced by Michael Maestlin, the professor of mathematics and astronomy. Although Maestlin was at best a very cautious Copernican, the 1543 *De revolutionibus* he owned is probably the most thoroughly annotated copy extant; he edited the 1571 edition of the *Prutenicae tabulae* and used them to compute his own *Ephemerides*. Kepler was an exemplary student; and, when he applied for a renewal of his scholarship, the university senate noted that he had 'such a superior and magnificent mind that something special may be expected of him'. Nevertheless, although Kepler himself wrote concerning his university education that 'nothing indicated to me a particular bent for astronomy', in student disputations he often defended Copernicanism.

In August 1591 the twenty-year-old Kepler received his master's degree from Tübingen and thereupon entered the theological course. Halfway through his third and last year, however, there occurred an event that completely altered the direction of his life. The teacher of mathematics and astronomy at the Lutheran school in the Styrian capital of Graz had died, and Tübingen was asked to nominate a replacement. Kepler was chosen, and, although he was reluctant to abandon his intent of becoming a Lutheran pastor, at the age of twenty-two he embarked on the career destined to immortalize his name.

Kepler arrived in southern Austria in April 1594 to take up his duties as teacher and as provincial 'mathematicus'. In the first year he had few pupils in mathematical astronomy and in the second year none, so he was asked to teach Virgil and rhetoric as well as arithmetic. However, the young Kepler made his mark in another way; one of the duties of the mathematicus was to produce an annual calendar of astrological forecasts. His first calendar, for 1595, contained predictions of bitter cold, peasant uprisings and Turkish invasions. All were fulfilled, to the great enhancement of his local reputation. Five more calendars followed in annual succession, and later, when he had moved to Prague, he issued prognostications for the years 1602 to 1606. Later still Kepler produced a series of calendars from 1618 to 1624, excusing himself with the remark that, when his salary was in arrears, writing calendars was better than begging.

Kepler's attitude to astrology was mixed. He rejected most of the commonly accepted rules and repeatedly referred to astrology as the foolish stepdaughter of astronomy. However, casting horoscopes provided welcome supplementary income and later became a significant justification for his office as imperial mathematicus. Moreover, the profound feeling he developed for the harmony of the universe included a belief in a powerful accord between the cosmos and the individual. These views found their fullest development in the *Harmonicae mundi*, published towards the end of his life.

Meanwhile, just over a year after his arrival in Graz, Kepler's fertile imagination hit upon what he believed to be the secret key to the universe – the number, dimensions and motions of the planets. This theory, published in his decisively pro-Copernican treatise *Mysterium cosmographicum* of 1596, was based on the idea that the five regular solids space out the six known planets; each planetary orbit is circumscribed by a regular solid and has inscribed in it the solid of the next planet below. Although the principal idea was erroneous, Kepler established himself as the first (and, until Descartes, the only) scientist to demand physical explanations for celestial phenomena.

Kepler had submitted his manuscript to the scrutiny of the Tübingen senate because his publisher would not proceed without its approval. Although they raised no objection to the publication, he was requested to explain his discovery in a clearer and more popular style. When it appeared, the reasons for abandoning the Ptolemaic in favour of the Copernican system were set forth with remarkable lucidity. Kepler sent copies to various scholars, including Galileo and the great Danish astronomer Tycho Brahe. Its faults notwithstanding, *Mysterium cosmographicum* thrust Kepler into the front rank of astronomers. Seldom has so wrong a book been so seminal in directing the future course of science.

Meanwhile Kepler's friends arranged his marriage to Barbara Müehleck, the eldest daughter of a wealthy mill-owner. She was two years younger than Kepler and had been widowed twice. Early in 1596 Kepler sought her hand, but her family thought him beneath her and the negotiations were difficult and protracted. They insisted that the modest fortune she brought to their marriage be reserved for their children. The wedding took place the next spring, under ominous constellations, as Kepler noted in his diary. He soon realized that his wife would never understand anything of his work – 'simple of mind and fat of body' was Kepler's later description of her. Of their five children, one boy and one girl survived to adulthood.

The numerous Protestants in Graz remained unmolested by their Catholic rulers until 1598. Then, on a day in late September, all the teachers, including Kepler, were abruptly ordered to leave town before sunset. Although Kepler was allowed to return, unlike his colleagues, conditions remained tense. In the second half of the sixteenth century the Czech kingdom of Bohemia experienced great prosperity under the Habsburg emperor Rudolph II, who made Prague his capital and attracted to it a galaxy of artists, scholars, alchemists and magicians. In August 1599 Kepler learned that the wealthy, aristocratic Brahe had been appointed imperial mathematicus by Rudolph, with an exceptionally generous salary. Early in 1600, Kepler made an exploratory visit to the observatory Brahe had established at Benatky Castle, near Prague. It was equipped with scientific instruments of the highest quality, although telescopes had not yet come into use. Kepler respected the outstanding precision of Brahe's observational data and expected that he would be given access to them. However, Brahe treated him as a novice, rather than an independent investigator, and refused to share his results. The two astronomers soon quarrelled, but before Kepler returned to Graz they had achieved some degree of reconciliation.

In Graz, by this time, the counter-Reformation was taking effect, and in August 1600 Kepler and other Protestants were expelled from the predominantly Catholic city. Already deeply depressed by the death of his first two children, he decided to go back to Prague, with his family. When they arrived, Kepler found that Brahe's chief assistant, Longomontanus, had just died. Kepler was appointed in his place, but Brahe still refused to share his observational data, and there was further friction when payment of Kepler's salary was delayed. He returned to Graz in April 1601 on an extended visit occasioned by his father-in-law's death and the need to safeguard his wife's interests.

Eventually the differences between the two astronomers were patched up; then in the autumn Brahe was suddenly taken ill and towards the end of October he died. Almost at once Kepler was appointed to succeed him as imperial mathematicus, although five months passed before he received his first instalment of salary. One of his various duties was to complete Brahe's work on what became known as the *Tabulae Rudolphinae*, giving the positions of a great many stars and perpetual tables for calculating the positions of the planets on any date in the past or future. This task, involving enormous quantities of laborious calculations, was by no means congenial to Kepler, and, even when it had been completed, publication was delayed, as we shall see.

Kepler's main interest remained more in theoretical astronomy. He began to speculate that the solar system might be held together by magnetic attraction. Although this was not right, it represented an imaginative leap in the direction of universal gravitation. He also began to consider the possibility that the planetary orbits might be elliptic, with the sun at one focus, and here of course he was right. These ideas appeared in his next important book, the *Astronomia nova* of 1609. Unfortunately publication was held up, partly by the lack of imperial financial support but also by opposition from the heirs of Brahe. They took away Brahe's scientific instruments and allowed them to decay unused. They also tried to remove the vital records of his observations, but Kepler managed to prevent this.

Despite poor eyesight, Kepler was one of the pioneers of research into optics. He found a good approximation to the law of refraction; Descartes, the discoverer of the precise law, said that Kepler was his true teacher in optics, who knew more about this subject than did any of those that preceded him. This research was published in his *Dioptrice* of 1611, which also contains an account of a new astronomical telescope with two convex lenses. Towards the end of his life he wrote a small work on the gauging of wine casks, which is regarded as one of the significant works in the prehistory of the integral calculus. In lighter vein he also wrote a paper discussing why snowflakes are hexagonal.

Unlike Kepler himself, Catharina did not like living in Prague. She never felt comfortable in court circles. Moreover, she was often homesick and became upset when they ran short of money. He began to search for suitable employment in a place she would find more congenial. The need became pressing in May 1611 when Rudolph was deposed and Prague became a scene of bloodshed in the struggle for the kingdom. Moreover, Kepler's wife became seriously ill and their three children were stricken with smallpox, from which his favourite son died. Throughout his life Kepler kept trying to obtain a position in Protestant Württemberg, but without success, and now any remaining hopes of this were finally dashed when the theologians of Württemberg raised objections to his Calvinistic sympathies. He declined the offer of a professorship at the University of Bologna. Instead he decided to move to Linz, the chief city of Upper Austria, where he had been offered the specially created post of provincial mathematicus. However, before the move could take place his wife died from the typhus brought to Prague by the troops. It was not until January 1612 that he was able to leave for Linz; by then his appointment as imperial mathematicus had been renewed, so he was able to hold this as well as his provincial post, which was virtually a sinecure.

Soon after he arrived in Linz Kepler began to look for a new wife. In a letter he listed in detail eleven possibilities, and explained how God led him to choose the fifth, a woman who had evidently been considered beneath him by his family and friends. She was Susanna Reuttinger, a twenty-four-year-old orphan; the marriage was far happier than the first, but, of their seven children, five died in infancy or childhood, as had three of the five children of his first marriage. Then his aged but meddlesome mother was accused of and tried for witchcraft, and Kepler had to travel to Württemberg to arrange for her defence, which occupied much of his time and energy over the following three years. She was imprisoned and threatened with torture but in the end set free; she died shortly afterwards.

Kepler, a peaceful and deeply religious man, suffered greatly for the sake of his conscience throughout his life, particularly in Linz. His long stay there had started badly, for the local Lutheran pastor, who knew the opinion of the Württemberg theologians, excluded him from holy communion because of his Calvinistic tendencies. Kepler did not accept the exclusion willingly and made repeated appeals to the Württemberg consistory, but always in vain. While his co-religionists considered him a renegade, the Catholics tried to win him to their side.

All these troubles notwithstanding, Kepler published two major works during his fourteen years in Linz. The more important was his *Harmonicae mundi*, a work that had occupied him on and off for many years; this was published in 1618, with a dedication to King James the First of England. This has been described as a great cosmic vision woven out of science, poetry, philosophy, theology and mysticism. Kepler believed that the archetypal principles of the universe were based on geometry rather than on number, and it is in this work that the regular polyhedra known as the stellated dodecahedra make their debut. His other major work of this period is his *Epitome astronomiae Copernicanae*, a textbook of the Keplerian system. In the dedication he wrote 'I like to be on the side of the majority', but in his Copernicanism and in his deep-felt religious convictions he rather learned the role of being a member of a staunch, lonely minority. However, it was Galileo, a far bolder polemicist, who became the persuasive purveyor of the new cosmology.

When the counter-Reformation swept into Linz in 1625 an exception was made so that he was not banished, but his library was temporarily sealed and his children forced to attend Catholic services. By the summer of 1626 Linz was blockaded and Kepler's house, alongside the city wall, was burnt down. As soon as the long siege had been lifted, Kepler petitioned the emperor for permission to move to Ulm, where he knew that there were printers who could undertake the composition of the *Tabulae Rudolphinae*. Although he had worked in Linz longer than he had in any other place, Kepler was not sorry to leave. He packed up his household effects, books, manuscripts and printing equipment and travelled by boat up the Danube to Regensburg. After settling his wife and children he continued by road to Ulm to see the *Tabulae Rudolphinae* through the press. Even before that task had been finished, Kepler began to search for a new base. England was one possibility; in 1620 the English Ambassador Sir Henry Wootton had called on him in Linz and invited him to England, but nothing came of this. In fact Kepler never moved out of the region consisting of southern Germany, Bohemia and adjacent Austria.

In the end, reluctant to lose the financial security provided by his salaries as provincial and imperial mathematicus, Kepler went back to Prague to apply to Rudolph's successor for these appointments to be continued. The newly crowned king received him graciously, promising a reward for the dedication of the *Tabulae Rudolphinae*, but making it clear that the astronomer needed to become a Catholic if he wanted to remain in the imperial service. The imperial commander-in-chief, Albraecht von Wallenstein, was more accommodating. Wallenstein, then at the height of his power, had just been granted the duchy of Sagan in Silesia as a personal fief. Anxious to raise its status, as well as to have close access to an astrologer, he appointed Kepler as his personal mathematicus. Kepler objected that he was unwilling to 'let himself be used as an entertainer' and would not compromise his own scientific convictions to satisfy his astrologically minded patron. However, Wallenstein, who had no real interest in science, compromised by employing Kepler to calculate the precise positions of the planets and then obtaining the predictions from less-inhibited astrologers.

Kepler collected his family at Regensburg, settled his affairs in Linz and finally reached Sagan in July 1628. He found the inhabitants unfriendly and the local dialect almost incomprehensible. Before long religious strife broke out when, for political reasons, Wallenstein started to press Catholicism on his subjects. Although Kepler was not personally affected, the persecutions made it difficult to attract printers to work on the *Tabulae Rudolphinae*. He secured an assistant by the name of Jacob Bartsch, a young scholar who had studied astronomy and medicine at Strasbourg, who later became his son-in-law. Kepler wrote another book, the *Somnium*, which described an imaginary journey to the moon and used this to present an ingenious polemic on behalf of the Copernican system. The idea of universal gravitation, which 'vexed and haunted his mind', seems implicit in his description of the journey.

In Sagan Kepler waited in vain for the payment of his claims for arrears of salary, the responsibility for which had been transferred to Wallenstein. When the latter lost his position as commander-in-chief Kepler returned to Regensburg, presumably intending to consult the emperor and friends at the imperial court about his future and to collect at least some of his arrears of salary. However, a few days after arriving there Kepler became sick with an acute fever; his condition steadily worsened, he became delirious and, on November 15, 1630, he died. The symptoms are those of typhus, which was prevalent during the Thirty Years War. He was buried in the Protestant cemetery, soon to be completely destroyed in the conflict. His wife and children were left almost destitute, but Jacob Bartsch helped them collect the money owed to Kepler's estate by the state treasury. A wealth of papers left by the great astronomer passed through various hands; much has been lost but the remainder is to be found in libraries in Austria, Germany and elsewhere. The thousands of manuscript sheets left at his death went to his son Ludwig, who promised publication but lacked both the time and the knowledge for such an undertaking. A monumental *Gesammelte Werke* in nineteen volumes has been published, as well as a great deal of secondary literature.

Kepler's scientific thought was characterized by his profound sense of order and harmony, which was intimately linked with his theological view of God the creator. He saw in the visible universe the symbolic image of the Trinity. Repeatedly he stated that geometry and quantity are co-eternal with God and that mankind shared in them because man is created in the image of God. From these principles flowed his ideas on the cosmic links between man's soul and the geometrical configurations of the planets. Today, when physicists are said to be searching for a 'theory of everything' that would allow them to 'read the mind of God', we may be reminded of Kepler's indefatigable search for the mathematical harmonies of the universe. Yet contrasting with this mysticism was his insistence on physical causes.

Kepler never rid himself of a feeling of dependence; neither could he exhibit the imperious self-assurance of a Brahe or a Galileo. Nevertheless, his ready wit, modest demeanour and scrupulous honesty, as well as his wealth of knowledge, won him many friends. Although Newton seemed reluctant to acknowledge his influence in the *Principia*, that great work was presented to the Royal Society of London as 'a mathematical demonstration of the Copernican hypothesis as proposed by Kepler', and Halley, in reviewing the *Principia*, wrote that Newton's 'first eleven propositions were found to agree with the phenomena of the celestial motions as discovered by the great sagacity and diligence of Kepler'. In one of Galileo's letters to Kepler he states 'I thank you because you were the first one, and practically the only one, to have complete faith in my assertions.'

Although Kepler today is remembered chiefly for his three laws of planetary motion, these were but the elements in his much broader search for cosmic harmonies and a celestial physics. With the exception of Rheticus, he became the first enthusiastic Copernican after Copernicus himself. Kepler has been described as an astronomer's astronomer; he found an astronomy whose clumsy geocentric or heliostatic planetary mechanisms typically erred by several degrees and he left it with a unified and physically motivated heliocentric system nearly a hundred times more accurate. The writer Coleridge, in his *Table Talk*, gave it as his opinion that Galileo was a great genius and so was Newton, but it would take two or three Galileos and Newtons to make one Kepler. Few would agree with this sweeping statement but nevertheless Kepler's enquiring mind helped to break the mould of mediaeval cosmology.

CHRISTIAAN HUYGENS (1629–1695)

Unlike his English counterpart Isaac Newton, Christiaan Huygens came of a distinguished family. His paternal grandfather had been secretary to William the Silent during the eventful years after 1578, when he had accomplished

his mission of establishing a free commonwealth in defiance of the most powerful empire then existing. The last quarter of the sixteenth century saw the independence of the seven northern provinces of the Netherlands completed after an eighty-year struggle with Spain. The father of the scientist, Constantin Huygens, showed ability in mathematics but his education was directed towards a career as a courtier and diplomat. As secretary to the Prince of Orange he did much to guide his country through difficult times. A man of outstanding ability and brilliance, he became a close friend of René Descartes; after their first meeting Descartes wrote of him 'I could not believe that a single mind could occupy itself with so many things and acquit itself so well with all of them.' Constantin Huygens was a poet, student of natural philosophy and classical scholar, as well as courtier and diplomat.

Constantin Huygens married his cousin, Susanna van Baerke, daughter of a wealthy merchant of Amsterdam and by all accounts an intelligent and cultivated woman. Christiaan, their second child, was born on April 14, 1629, only a few months before the death of Kepler. There were four other children, of whom Constantin the younger, born the previous year, is the only one that concerns us here. When their mother died in 1637, after only ten years of married life, another cousin took care of the family, which moved to a house near The Hague. The two eldest sons, who early showed brilliance, were taught at home by a private tutor until 1645. Their education included singing, playing the lute and the composition of Latin verse. Like Newton as a boy Christiaan loved drawing and the making of mechanical models, on which he spent much labour and ingenuity. From the beginning, however, he showed special promise of ability in geometry, whereas his brother Constantin excelled in literary compositions. Christiaan was rather delicate and by nature gentle, and his sensitivity seemed almost feminine to his father. Descartes was much impressed by some early exercises of Christiaan and saw that great things might be expected from this rather serious boy with the pale face and the large dark eyes.

In 1645, when Christiaan was sixteen, both brothers entered the University of Leyden, where they studied jurisprudence and mathematics. The mathematics professor, a protégé of Descartes, regarded Christiaan as his best pupil. In 1644 Descartes had published his *Principia philosophiae*, an attempt to reduce all the changes of nature to mechanical processes. Later Christiaan recalled that 'it seemed to me when I first read this book that everything in the world became clearer and I was sure that when I found some difficulty it was my fault that I did not understand this thought. I was then only fifteen or sixteen years old.'

In 1647, after almost two years at Leyden, Christiaan Huygens joined his brother in studying law at the College of Orange at Breda. This college, of which his father was curator and in which Descartes seems to have taken a personal interest, achieved a temporary fame but did not survive into the next century. After his studies there were over, Christiaan Huygens began to visit some of the neighbouring countries, notably Denmark. He made plans for a visit to Paris in the company of his father but France, following the death of Louis XIII, was in a state of disorder and it was not until 1655 that he first went to the French capital. Before that he was able to establish contact with some of the French scientists through correspondence with Père Mersenne, another friend of his father's. Although Mersenne died in 1648, his influence on the young Huygens was important.

Any ambition Huygens might have retained for a diplomatic career was abandoned when William II died in 1650. Instead he found his metier in scientific research. In the next sixteen years he proved himself to be as good at it as anyone else at this time, with the possible exception of Newton. He studied telescopes and microscopes and introduced improvements in their design. His studies in mechanics touched on statics, hydrostatics, elastic collisions, projectile motion, pendulum theory, gravitational theory and an implicit concept of force, including centrifugal force. He pictured light as a train of wave fronts, transmitted through a medium consisting of elastic particles. Fundamental research in pure and applied mathematics, optical studies and the discovery of the large satellite Titan of Uranus all belong to this period. Thanks to his success in designing a more powerful telescope than anyone else had managed, he was able in 1655 to detect the rings of Saturn for the first time.

From 1655, when he settled in Paris, Huygens moved in an elegant and leisured society, occasionally visiting the salons. Thanks, no doubt, to his father's influence, he became a protégé of the powerful Jean-Baptiste Colbert. He toured the chateaux of the Ile de France and of the Loire Valley and accompanied his father on a brief visit to London. Constantin was well known in England; he had studied at Oxford, played the lute at the court of James the First and received an English knighthood. When he returned to England later, Christiaan was able to build on these contacts. His long stay in Paris was interrupted in 1664, when he went back to The Hague for two years.

On his return to Paris in 1666, Huygens was elected to membership of the Paris Academy, which had just been established officially, and for the next seventeen years he made the French capital his home. At the same time, however, he was developing his contacts in London, where the informal society of men of science which had been meeting in Gresham College had become recognized as the Royal Society. On his second visit to London Huygens was very impressed by this lively new body and thought that what it was doing surpassed anything happening in Paris. He arranged to be kept informed about scientific work in England, especially the discoveries of Newton, whose investigations in many respects ran parallel to his own.

Having dealt with the Fronde and established himself in power the young Louis XIV declared war on the new Dutch Republic. Rather surprisingly, Huygens remained in Paris while his homeland was in danger. Accepted as the most distinguished of the academicians, he presided over the Paris Academy until 1675, using his diplomatic skills to see the new institution through its formative years. In research he became interested in the problem of determining longitude at sea, so important for navigation. He invented the pendulum clock, intended for use on board ship, but this was not a success. He then developed the use of springs as regulators in clocks. His research in this area was published in his celebrated *Horologium oscillatorium* of 1673.

I digress at this point to say a few words about the mathematical work of the polymath Gottfried Leibniz, since it was Huygens who was his mentor during the period when Leibniz invented the differential and integral calculus. Leibniz started work on this about 1673, some years after Newton but independently. Both worked out complete algorithms that, except in their foundation, are substantially those in use today. Although he is not usually regarded as a physicist, Leibniz made some notable contributions to natural philosophy, but to go into these here would be too much of a digression.

Huygens never enjoyed good health. From early youth he suffered from some kind of disability, perhaps migraine, accompanied by severe headaches. A serious illness in 1670 brought about complete prostration and he clearly believed himself to be close to death. Whatever it was, it lasted in acute form for several weeks and it was three months before he was able to return to work. Early in 1676 there was a recurrence, and this time he showed greater caution in meeting the danger. Life in Paris, he decided, seemed to be bad for his health, so he returned to The Hague for treatment. To his brother he confessed his doubts about whether he would ever return to the French capital and, even when he had recovered a year later, he procrastinated under the pretext of uncertain health, while continuing his scientific work in The Hague. It was not until the middle of 1678 that he returned to Paris; before long he was taken ill again.

Recurrent ill-health no doubt accounts for the reduction in his mathematical and scientific work after 1680. Early in 1681 he was taken ill again but not until September was he able to return to The Hague, where he slowly recovered. He had hopes of returning to Paris but, after his patron Colbert died in 1683, Catholic intolerance in France was undoing much that Colbert had been at pains to build. Huygens' position at the academy was undermined; his nationality and religion told against him. In 1687 his father died. His brother Constantin accompanied William of Orange to England the next year, leaving Christiaan feeling alone. He made a short visit to England himself in 1689, when he went to Cambridge to see Newton and to lend his support to Newton's bid to become Provost of King's College, but too little is known about this. The illness which had dogged him throughout his life again recurred in a severe form and, in March 1695, Huygens felt it necessary to summon his lawyer and make final corrections to his will. The following month he became worse and, from then until July, pain and sleeplessness spared him hardly at all; his days were filled with deep despair. He thought he was being poisoned, kept hearing voices and lived in fear of losing his reason. Weakened by suffering, he died on July 9.

The professional and serious interests of Huygens are the ones which are foremost in his correspondence. Nevertheless, it would be a mistake to consider him as always having been nothing but a patient researcher. He was a man of wide culture and acquaintance throughout Europe. Neither was he averse to feminine society. Marianne Petit, daughter of one of Louis XIV's engineers, seems to have had an attraction for Huygens; their separation was due to her withdrawal from society when she entered a religious order. There were also some distant cousins he visited in Paris and there is no doubt he felt considerable attraction for the eldest of these. The parallel between Newton and Huygens in natural philosophy is striking. No other natural philosopher of the seventeenth century even approached their level. In matters relating to physics, their intellectual menus are strikingly similar. Working within the same tradition, they dealt with the same problems in many cases and pursued them to similar conclusions. Beyond mechanics, there were also parallel investigations in optics. At nearly the same time and stimulated by the same book, Robert Hooke's *Micrographia*, they thought of identical methods for measuring the thicknesses of thin coloured films. In his own world of abstract thought he was incomparable, as Leibniz said, his loss inestimable. Yet Huygens' influence beyond his own century was slight, whereas Newton's was enormous. One of his limitations was that he worked alone, with few disciples. Also, like Newton, he often hesitated to publish, and, when the work finally saw print, others had covered the same ground. More important, however, was his philosophical bias. He followed Descartes in the belief that natural phenomena must have mechanistic explanations. He dismissed Newton's theory of universal gravitation as absurd, because it was no more than mathematics and proposed no mechanisms.

Isaac Newton (1642–1726)

Isaac Newton was born on Christmas Day 1642 in Woolsthorpe Manor, a farmhouse near the Lincolnshire village of Colsterworth, sixty miles northwest of Cambridge. The baby was premature and at first was not expected to survive. His yeoman father, who had only recently married, had died three months before the happy event, leaving his mother Hanna to run the family farm. In 1645 when the boy was three his mother married the elderly Reverend Barnabas Smith, with whom she went to live at his rectory in nearby North Witham, leaving her son in the charge of his maternal grandparents. As a boy Isaac Newton appears to have had little affection for

his stepfather, his grandparents and their children. He grew up lonely and loveless.

At the age of twelve, after early education at local schools, Newton was sent to King's School at Grantham. Since it was too far away for him to live at home during school terms, he lodged with an apothecary named Clark, who seems to have been very kind to him and, in particular, encouraged him to make things with his hands. On the death of her second husband in 1656, his mother returned to Woolsthorpe with the three children from her second marriage. Two years later she took the fourteen-year-old Isaac away from school to help her manage the farm. He proved a somewhat incompetent farmer, his mind too much on other things. On the advice of his mother's brother he was sent back to King's School to prepare for entry to the University of Cambridge.

Again he lodged with the Clarks; their stepdaughter Catherine Storey became an intimate friend. According to the antiquarian William Stukely, who had a long conversation with her in old age, Isaac was 'always a sober, silent thinking lad, and was never known scarce to play with the boys abroad, but would rather choose to be at home, even among the girls.' While he was preparing for Cambridge his childhood affection for Catherine deepened and it seems they became engaged to be married. Stukeley continues 'Sir Isaac and she being thus brought up together, 'tis said that he entertained a passion for her, nor does she deny it; but her portion being not considerable, and he being a fellow of a college, it was incompatible with his fortunes to marry; perhaps his studies too. 'Tis certain he always had a great kindness for her. He visited her whenever in the country, in both her husbands' days, and gave at a time when it was useful to her, a sum of money.'

Newton was admitted to Trinity College in 1661 at the age of nineteen. He began as a subsizar and then sizar, which meant that he had to perform menial duties for his seniors in return for free board and tuition, although his mother, who had been left comfortably off as the heir of her second husband, could have paid for her son's expenses as a commoner. Although he was studying hard, he was not following the syllabus. As a result, when he tried for a scholarship in his second year, he failed the examination in geometry.

Owing to an outbreak of the bubonic plague the Cambridge colleges were suspended for the years 1665/6 and so the young man went home. It was at Woolsthorpe that he conceived the theories which were to revolutionize science. When he returned to the university in 1667 Newton's abilities were starting to be recognized. Trinity elected him to a minor fellowship, against strong competition, and, after this had been converted into a major fellowship the following year, he was entitled to reside in the college indefinitely. He acquired a patron in Isaac Barrow, the Lucasian professor of mathematics, who was later to become Master of Trinity, a Crown appointment. To become known in Court circles, Barrow secured the position of Royal Chaplain and vacated the Lucasian chair in Newton's favour.

Newton's first lecture as Lucasian professor took place at Trinity College in January 1670. It was about his research on optics, material which would find its way into his book *Opticks* of 1704, a much more accessible work than the *Principia* of 1686. The audience was small, no-one came to the second lecture, and he continued talking to an empty room throughout almost every lecture he gave for the next seventeen years. After that he gave up all pretence of teaching, which he never enjoyed. Only three students ever came to him for tuition; they were intellectually undistinguished and nothing is known about how they found him.

Newton's early enthusiasm for making mechanical models had developed into a passion for making scientific instruments, especially optical instruments. The celebrated reflecting telescope was a notable example of his extraordinary skill. It was this that first brought him to the attention of the Royal Society. When they heard about the instrument the fellows of the Royal Society asked to see it. Newton sent them an improved model, which was received with acclamation. When he followed it up with a paper about his optical research he was elected to the society forthwith. His fame quickly spread throughout Europe, but unfortunately the next few years were marred by troublesome controversies over his optical discoveries, particularly with Robert Hooke over priority of discovery and with Christiaan Huygens over theory.

As a young man Newton was an orthodox member of the Anglican church but during this period he had come to doubt the doctrine of the Holy Trinity. For the rest of his life he firmly adhered to the Arian heresy, in which Christ occupies an intermediate position between God and man and, unlike the Father, has no foreknowledge of the future. Newton became convinced that Trinitarianism was utterly false and spent much time and effort studying the Scriptures and the history of the early Christian church. Like many others at this time, he interpreted the Book of Revelation in a way that identified the Whore of Babylon with the Church of Rome but as an Arian he also regarded the Anglican church as heretical. This created a problem over his tenure of the Lucasian chair. It was a statutory condition that those elected to college fellowships should present themselves for ordination after a certain number of years. This applied to Newton, but as an Arian he no longer felt able to reaffirm his belief in the articles of the Anglican church. He consulted Barrow, who was still Royal Chaplain, and took his advice to seek a dispensation from the requirement directly from the King. This was forthcoming, not just for Newton but for the Lucasian professors in perpetuity.

It was in 1670, or thereabouts, that Newton first became interested in alchemy, which began as an offshoot of chemical experiments but soon developed into an obsession. When he retreated into his academic sanctuary at the end of 1676 it was in order to pursue his alchemical studies. He was in contact with a number of other enthusiasts and studied the literature exhaustively. The practice of alchemy was by its very nature mysterious and the mass of writings Newton left on the subject has left scholars perplexed.

Newton had not seen much of his mother since he left home. She died at Stamford of a fever in June 1679 during which her son 'sate up whole nights with her, gave her all the physick himself, dressed all her blisters with his own hands & made use of that manual dexterity for which he was so remarkable to lessen the pain which always attends the dressing the torturing remedy usually applied in that distemper with as much readiness as he ever had employed it in the most delightful experiments.' Newton was her executor and heir; as a result he became a man of substance.

In 1684 the astronomer Edmond Halley came to visit him at Cambridge, and this marked a turning point in Newton's scientific career. Spurred on by Halley, he composed the *Principia* between the autumn of 1684 and the spring of 1686. It was presented to the Royal Society in June 1686 and published the following year under its full title *Philosophiae naturalis principia mathematicae*. It is said he deliberately made it abstruse, using mathematical arguments to put off the uninitiated. The treatise brought Newton great fame in Britain and abroad. The great French scientist Pierre-Simon Laplace wrote 'the Principia is pre-eminent above any other production of the human genius.' The demand for his masterpiece grew steadily as its importance became more widely understood.

The 1680s were a difficult time for Trinity College financially; the magnificent new library, designed by Christopher Wren, was costing the foundation more than it could easily afford, and fellows no longer received their full dividend; moreover, the depredations of the Civil War had reduced the endowment income of colleges generally. Newton was of a seniority to have influence in the college and to play an active part in university affairs. He was one of eight college fellows who accompanied the vice-chancellor to London in 1687 to present the university's case in connection with the illegal encroachments by James II. There was an attempt to have him appointed Provost (i.e. head) of King's College, Cambridge. He was elected as a member of parliament for the university; he seems to have made no use of the position before giving it up at the end of the year, but it left him dissatisfied with academic life. At Cambridge he became friendly with the Whig politician Charles Montague, later Lord Halifax, and through him Newton made repeated attempts to obtain a position in the capital.

It was about this time that the young Swiss mathematician Fatio de Duillier entered his life. Fatio seems to have won his way into Newton's affections like no-one else. Fatio was twenty-two years younger than Newton and was the son of a wealthy Swiss landowner who wanted him to study divinity, but his more intellectual mother insisted that instead, or as well, he should study science in Paris. Although he was not without talent, Fatio showed an early flair for self-promotion and, soon after arriving in London in 1687, was elected to the Royal Society. He soon acquired the nickname 'the ape of Newton'. He was acquainted with many of the continental scientists personally, notably Huygens. Although much of the correspondence between Fatio and Newton has been lost, what survives seems much more intimate than Newton's other correspondence.

The relationship came to a sudden end in June 1693, when Newton experienced something in the nature of a nervous breakdown, with depression and paranoia. The episode lasted three or four months; after it was over he never resumed creative scientific work. The cause is unknown; one suggestion is that it was due to metallic poisoning, particularly poisoning by the mercury often used in alchemical experiments, but there are objections to this. When Newton resumed his career it was in an entirely new role, which involved a complete change in his mode of life. In 1696, through Montague's good offices, he was appointed a Warden of the Royal Mint in the Tower of London. This office had previously been treated as little more than a sinecure but Newton plunged himself into the work. Because the coinage had been debased by counterfeiting and clipping, the government decided to call in the old coins in exchange for new ones, which would be more difficult to counterfeit and impossible to clip.

Under Newton the Mint operated with greater efficiency than ever before and it was to be a long time before it did so again. It was one of his duties to ensure that counterfeiters and other miscreants were prosecuted. To obtain a conviction required witnesses and Newton put his energy into the task of producing them. He had himself commissioned as justice of the peace in the home counties and operated a network of paid informers. All this kept him busy for the next few years, at the end of which he was promoted to the lucrative office of Master of the Mint. He resigned his chair and fellowship at Cambridge; for some years his professorial duties had been performed by a deputy.

When Newton moved to London, after thirty-five years in Cambridge, he settled in a modest Jacobean house in Jermyn Street, near Piccadilly. He persuaded a seventeen-year-old niece of his named Catherine Barton to come and preside over his household. A witty and charming girl, she soon became very popular in London society, the toast of the Kit-Kat Club. Although her relationship with Newton seems to have been platonic, it may have been otherwise with his patron Lord Halifax, who in his will left her well provided for.

For some years, despite living nearby, Newton had not attended meetings of the Royal Society. This was partly because he would encounter the curator Robert Hooke, one of his principal adversaries. In 1703, after Hooke's death, Newton was elected president, an office to which he was re-elected annually for the rest of his life. The status of the society, which had been declining, soon began to revive under his presidency. He was knighted in 1705, during a royal visit to Cambridge, for his services to science. However, even when president, Newton did not find it easy to express fundamental convictions in public. He preferred silence to the risk of criticism in which he might find himself made an object of ridicule. Even so, he could not escape controversy and his bitter quarrels with Leibniz and others loom large in scientific history. Notoriously they, or rather their respective supporters, quarrelled over priority over the invention of the infinitesimal calculus. Just when Newton developed his direct and inverse theory of fluxions it is hard to say, but Leibniz started work on the differential and integral calculus about 1673, some years after Newton but independently. The ideas involved were foreshadowed in the work of various earlier mathematicians but Newton and Leibniz can nevertheless be regarded as the founders of the subject in a way that would not be reasonable with their precursors. Both worked out complete algorithms, which, except in their foundations, are substantially those used today. Although the notation used by Leibniz is somewhat illogical, it has displaced that used by Newton almost entirely.

By this time Newton was a wealthy man, and if he lived parsimoniously it was entirely his own wish, for he was a generous contributor to deserving persons and other good causes. He was scrupulously exact and regular in business matters and spent very little on himself, except that he entertained as befitted his official standing. Towards the end his memory was much decayed, although earlier it had been exceptional. His hearing was good and some myopia in early manhood had corrected itself later, as it so often does. Apart from the nervous breakdown, he enjoyed good health until his eightieth year, when he began to suffer from gout and other troubles and was often in great pain due to stones in the bladder. In 1709 he moved to Chelsea for a year, then to a house near Leicester Square until 1725. Then, at the age of eighty-three, he moved again to Kensington, at that time in the countryside, because he was having some trouble breathing; he was thinking of retiring to Grantham, near his birthplace. In February 1727 he presided for the last time at a meeting of the Royal Society, after which he became very ill and died on March 20. Following a state funeral, he was buried in Westminster Abbey, where his baroque monument lies in a prominent position in the nave while clustered around his feet lie memorials to other illustrious British scientists, such as Charles Darwin and James Clerk Maxwell.

Since Newton died intestate, his valuable property was divided up among surviving relatives according to the law. The bulk of his fortune went to 'an idle fellow who soon spent it in cocking, horse racing, drinking and folly'. After passing through various hands, most of his surviving papers ended up in Cambridge. Books entitled *The Chronicles of the Ancient Kingdoms Amended* and *Observations on the Prophecies of Daniel and the*

Apocalypse of St John, which appeared after his death, indicate the questions on which he had spent so much of his time; no-one took these very seriously, but, together with his alchemical investigations, they have earned him the title of 'the last of the magicians'.

Biographers of Newton are handicapped by the shortage of useful information about his life before he moved to London, the most interesting period from the scientific point of view. Of his contemporaries Humphrey Newton of Grantham (no relation), who served as his assistant and amanuensis for five years, recalled:

> He always kept close to his studies, very rarely went a-visiting & had as few visiters . . . I never knew him take any recreation or pastime, either in riding out to take the air, a-walking, bowling or any other exercise whatever, thinking all hours lost that were not spent in his studies, to which he kept so close, that he seldom left his chamber, unless at term time, when he read in the schools, as being Lucasian professor . . . he very rarely went to dine in the hall unless upon some public days, then, if he had not been minded, he would go very carelessly, the shoes down at heels, stockings untied, surplice on, and his head scarcely combed.

'Newton would with great acuteness answer a question', he added, 'but would very seldom start one'. During five years, he said he saw Newton laugh only once. 'He had loaned an acquaintance a copy of Euclid. The acquaintance asked what use its study would be to him. Upon which Sir Isaac was very merry.' He also described how excited Newton became when writing the *Principia*:

> so intent, so serious upon his studies that he sat very sparingly, nay oftentimes he forgot to eat at all, so that going into his chamber I have found his mess untouched of which I have reminded him, he would reply 'Have I?' and then making to the table would eat a bit or two standing. At some seldom times when he designed to dine in the hall, he would turn to the left hand and go out into the street, where making a stop, when he found his mistake, would hastily turn back, and then sometimes instead of going into the hall would return to his chamber again . . . when he had sometimes taken a turn or two [in the garden] he would make a sudden stand, turn himself about, run up the stairs, like another Archimedes with an eureka!, fall to write on his desk standing, without giving himself the leisure to draw a chair to sit down in.

These recollections were echoed by Stukeley:

> As well as when he had been in the hall at dinner, he has quite
> neglected to help himself, and the cloth has been taken away before he
> has eaten anything. Sometimes on surplice days, he would go forward
> to St Mary's church, instead of college chapel, or perhaps go in his
> surplice to dinner in hall. When he had friends to entertain at his
> chamber, if he stept into his study for a bottle of wine, and a thought
> came into his head, he would sit down to paper and forget his friends.

Newton was stocky in his youth, stout later on. In dress he was usually untidy and slovenly. Among his other peculiarities was a compulsion
to make draft after draft of many of his papers, for example as many as
eighteen, differing only slightly from each other, for the first chapter of his
Chronology, and he even felt the need to copy routine documents relating to
the business of the Mint. There are a number of portraits, also a death mask.
Descriptions of his appearance by those who knew him at various periods
are not entirely in agreement. John Conduitt, who eventually succeeded
Newton at the Mint and married Catherine Barton, tells us that

> [Newton] had a lively and piercing eye, a comely and gracious aspect,
> with a fine head of hair as white as silver, without any baldness, and
> when his peruke was off he was a venerable sight.

However, others said:

> In the whole of his face and make, there was nothing of that
> penetrating sagacity which appears in his compositions; he has
> something rather languid in his look and manner, which did not raise
> any great expectation in those who did not know him.

and

> Sir Isaac was a man of no very promising aspect. He was a short
> well-set man. He was full of thought and spoke very little in company,
> so that his conversation was not agreeable.

Against this, Stukely said that

> although he was of a very serious and composd frame of mind, yet I
> have often seen him laugh, and that upon moderate occasions. He had
> in his disposition a natural pleasantness of temper, very distant from
> moroseness, attended neither with gayety nor levity. He usd a good
> many sayings, bordering on joke and wit. In company he behavd very

agreeably; courteous, affable, he was easily made to smile, if not to laugh . . . he could be very agreeable in company, and even sometime talkative.

The farm tenant of Woolsthorpe described Newton as a man of very few words:

> he would sometimes be silent and thoughtful for above a quarter of an hour together, and all the while almost as if he was saying his prayers; but that when he did speak, it was always very much to the purpose.

To his successor in the Lucasian chair he was 'of the most fearful, cautious and suspicious temperament that I ever knew'.

Early attempts to collect more material about Newton's life yielded disappointing results. For example his first assistant John Wickins, who shared rooms with him for sixteen years, added little of substance to what was known from other sources; and it is the same with his niece Catherine Barton, except that some information given by her husband John Conduitt may have come from her. As a result biographers, while repeating the well-known anecdotes, have not succeeded in providing a satisfactory account of his private life. The early biographies have a tendency to hagiography and do not give a true picture. Scholars of today have concluded that Newton's side of the argument in the controversies in which he became involved was not always as convincing as his supporters made out.

The mathematician Augustus De Morgan was one of the first to try to penetrate more deeply:

> [Newton] had not within himself the resource from whence to inculcate high and true motives of action upon others. The fear of man was before his eyes. All his errors are to be traced to a disposition which seems to have been born with him.

More recently, Louis Trenchard More wrote

> He was singularly unable to form intimate friendships. Morbidly suspicious and secretive, he was subject to peevish outbreaks of ill-temper, even towards those who were his best friends. On such occasions he stooped to regrettable acts which involved him in a succession of painful controversies that plagued his life, robbed him of the just fruits of his work, and disheartened his sincere admirers . . . The Gods had showered on him at birth extraordinary

gifts such as have been given to almost no other man, but some evil fate cursed him with a suspicious and jealous temperament which marred his life. This taint in his blood did not show itself in the form of ordinary vanity but in an inordinate sensitiveness to any personal criticism or to a reflection on his personal honour. In spite of his love of meditation and of peace free of all distractions it involved him in constant quarrels and altercations; and during a long and illustrious life it raised an impenetrable barrier between him and other men. To his friends he was never more than lukewarm and he kept them constantly uneasy lest they had offended him; to his rivals he was, at times, disingenuous, unjust and cruel.

DANIEL BERNOULLI (1700–1782)

The remarkable Bernoulli family, originally from Antwerp, left the Spanish Netherlands in the late sixteenth century to escape the persecution of Protestants and settled in Basel, where they married into the merchant class. Generation after generation they produced remarkable mathematicians, beginning with the brothers Jakob and Johann. It is Johann's irascible son, the polymath Daniel, who is profiled here, although we could wish to know much more about him. He was arguably the ablest Bernoulli of them all; he contributed ideas and insights that not only shaped eighteenth-century science but also presaged future discoveries. Like Newton, he was

even more famous as a physicist than as a mathematician, but he flourished at a time when the two subjects were closely inter-related.

Johann Bernoulli was professor of mathematics at the University of Groningen, in the Netherlands; his wife Dorothea was the daughter of the patrician Daniel Faulkner. Their second son Daniel, the subject of this profile, was born on February 8, 1700. Five years later Johann was appointed to replace his elder brother Jakob as professor of mathematics at the University of Basel and so, with his family, he returned to the main base of the Bernoulli clan. The brothers Jakob and Johann were involved in several quarrels. The most notable of these started with misunderstandings on both sides, but quickly exploded into a public dispute, in which each opponent challenged the other, found mistakes in the work of the other and expressed his low opinion of the work of the other, not only in private letters but also in print. In 1699 both brothers were elected to the Paris Academy on condition that they would cease their disputes.

The young Daniel was a precocious student who studied logic and philosophy at the university, earning a master's degree at the age of sixteen. At the same time his father Johann and elder brother Nikolaus helped him to learn some mathematics. However, as a career for his son Johann first tried to interest him in commerce and then, when that was a failure, grudgingly allowed him to study medicine, first in Basel, then in Heidelberg and then in Strasbourg, before returning to Basel. Daniel Bernoulli graduated from the University of Basel in 1721 by writing a dissertation on the mechanics of respiration. Having tried without success to obtain a teaching position in Switzerland (the final decision among qualified candidates was decided by lot, and Daniel was unlucky), he followed in his brother's footsteps by moving to Venice. There he gained some experience of practical medicine and was going on to continue his medical studies in Padua when he was taken seriously ill. At the same time he was pursuing research in mathematics and the result of this was his important *Exercitationes quaedam mathematicae* (*Mathematical Exercises*), published in Venice in 1724. In this he discussed a wide variety of scientific subjects, including probability and fluid dynamics. This attracted a great deal of attention and on the strength of it he was offered a teaching position at the St Petersburg Academy.

Prize competitions were an important feature of scientific life at least until the end of the nineteenth century. Originally they were a way of seeking solutions to specific problems. They usually emanated from the royal academies, notably those in Berlin and Paris, and, although they provided an opportunity for an unknown young researcher, it was quite normal for the well-established to enter. In the case of the Paris Academy, for example, prizes were awarded for memoirs addressing specific problems in the mathematical and physical sciences. Among the rules of procedure, each entry had to be under a pseudonym or motto, accompanied by a sealed envelope similarly inscribed containing the name of the author, although this could often be guessed by the judges. The Bernoullis were often successful in these competitions.

In 1725 Daniel, who had returned to Basel, won the prize (later he was to win nine more) and then took up his appointment in St Petersburg, accompanied by his brother Nikolaus. Although he was officially a professor of mathematics, he worked in many different fields. For example, his medical publications include important papers on muscular contraction and the optic nerve, and his writings on physics include a paper on oscillation. In mathematics he was particularly interested in probability and statistics. He corresponded with d'Alembert about the correct way to assess the value of risky medical procedures for patients of various ages. He demonstrated the importance of probability for economics, and its relevance to gambling.

Unfortunately the harsh climate of the Russian capital proved too much for his brother, who died in 1726 of a 'hectic fever'. However, the next year Daniel's young friend and compatriot Leonhard Euler came out to join him in St Petersburg. Daniel Bernoulli provided Euler with accommodation, and they regularly took meals together. The two men often worked together during the next six years, Daniel's most creative period. Daniel's research interests at this stage lay mainly in mechanics, physics and particularly hydrodynamics. Although his work in St Petersburg was highly successful, he began to look for an opportunity to return to Basel. The only vacancy at the university was a chair in anatomy and botany. Although these were not subjects he was much interested in, he took the position. Accompanied by his shy younger brother, named Johann after their father, Daniel went to Basel via Danzig and Hamburg, then across the Netherlands to Paris, where he was given a particularly warm reception.

Daniel continued to win the prize competitions of the Paris Academy; he won ten altogether, on subjects as various as astronomy, gravity, tides, magnetism, ocean currents and the behaviour of ships at sea. In 1734, when the subject was planetary orbits, father and son had separately entered their work and when they were successful they were told that they could share the prize, which was larger than usual because there had been no award the previous time one was offered. As a result a bitter dispute arose between them. In this case Johann had behaved badly, but he was often involved in priority disputes in which he had a valid case, particularly with his brother Jakob. Several of the discoveries to which the name Bernoulli is attached were Johann's but it was Jakob who got the credit.

Daniel Bernoulli's masterpiece, the seminal *Hydrodynamica, sive de viribus et motibus fluidorum commentarii* (*Hydrodynamics, or Commentaries on the Forces and Motions of Fluids*) was published in 1738, although it was largely completed in St Petersburg. This confirmed his reputation as one of the leading scientists of his time. Although he missed finding the basic partial differential equations of hydrodynamics, the *Hydrodynamica* contained other important advances. One is the principle, to which the name of Bernoulli is attached, that the pressure of a fluid diminishes as its velocity increases. This is fundamental to aerodynamics and much other modern industrial design. Even more important is his explanation of the mechanics of gases, which he regarded as composed of fast and randomly moving particles. He established the theory underlying Boyle's law, which had been deduced experimentally, and realized that the pressure of a gas is in direct proportion to its temperature. Altogether he may be said to have laid the foundations of the modern kinetic theory of gases.

Unfortunately publication of the *Hydrodynamica* was delayed, leaving his claim to priority open to attack, and the one to take advantage of this was his own father, who rather resented his son's success. They were already on bad terms because father and son took different sides in the priority dispute between the followers of Leibniz and Newton regarding which of them invented the differential and integral calculus. Johann was an ardent disciple of Leibniz while Daniel, who was more of a physicist, adhered to the side of Newton. Johann attempted a blatant priority theft by publishing a book on hydrodynamics in 1743 and dating it 1732. Daniel was understandably upset and wrote to Euler 'Of my entire *Hydrodynamica* not one iota of which do I in fact owe to my father, I am all at once robbed completely and lose thus in one moment the fruits of the work of ten years. All propositions are taken from my *Hydrodynamica*, and then my father calls his writings "Hydraulics, now for the first time disclosed, 1732", since my *Hydrodynamica* was printed only in 1738.' The situation was not quite as clear cut as Daniel claimed, but at any rate Johann Bernoulli's deception backfired. His reputation was so tarnished by the episode that he did not even receive credit for the parts of the work which were original.

In 1737 Daniel Bernoulli had delivered a historic lecture about the work done by the action of the heart. Six years later he was appointed professor of physiology at Basel, a field he much preferred to botany. Finally, in

1760 he obtained the chair of natural philosophy, or mathematical physics, which best matched his scientific interests. He applied himself to solving, with the aid of the new analysis, difficult mechanical problems that had defeated the more geometrical methods of Newton and Huygens. Although it would be going too far to say that he enunciated for the first time the principle of the conservation of energy, perhaps the most useful idea in all of science, he came closer to this than did anyone else until the 1840s. It was in the field of physics that he most brilliantly applied his mathematical genius. He is regarded as one of the main pioneers of that branch of science later known as mathematical physics.

Daniel Bernoulli also applied his mathematical insights to natural phenomena. While studying the nature of sound, for example, he discovered distinct mathematical regularities in the shape of sound waves and so was able to calculate the natural frequencies of a variety of musical instruments. His seminal work in the field of acoustics directly presaged such great discoveries of nineteenth-century mathematics as the harmonic analysis of Fourier. What he believed intuitively, namely that a sound can be represented as a trigonometric function, Fourier was able to demonstrate mathematically.

Daniel Bernoulli, who was an immensely popular lecturer, especially on experimental physics, continued as professor of natural philosophy until his retirement in 1776. Five years later he died in Basel on March 17, 1782, at the age of eighty-two, and was buried in the Peterskirche. He regularly corresponded with other leading scientists of the period. This was the age of the scientific society and academy; Daniel Bernoulli was a member of many of the most important, including those of Bologna (1724), St Petersburg (1730), Berlin (1747), Paris (1748), London (1750), Bern (1762), Turin (1764), Zürich (1764) and Mannheim (1767). In contrast to the other physicists profiled in this chapter, we really know very little about his personal life; we do not even know whether he was married. Of the sons of Johann, Daniel's younger brother was also an able mathematician, who was awarded four prizes by the Paris Academy, but, although he lived to old age, a frail constitution limited his scientific output. Historians of science are preparing an edition of the Bernoulli papers and, when that is complete, we shall know much more about the relationships of members of the family with each other and with other scientists of the period.

2 From Franklin to Laplace

Our next five remarkable physicists were born in the thirty-one years from 1706 to 1736. Two came from France and one each from America, Dalmatia and England.

BENJAMIN FRANKLIN (1706–1790)

It has been said of Benjamin Franklin that he found electricity a curiosity and left it a science. He was undoubtedly the first important American scientist and also the first international statesman of note whose reputation was partly gained through his scientific work. Of course it is quite impossible to do justice to the life-story of such an interesting person in the few pages available here, but fortunately there are plenty of full-scale biographies.

The Franklins came from Northamptonshire and Oxfordshire, in the Midlands of England. About 1685 Benjamin's father Josiah, with his first wife Anne, emigrated from Banbury to escape from religious persecution. She died in America after giving birth to her seventh child. Josiah then married again, this time marrying Abiah Folger of Nantucket; her father Peter Folger was a weaver, schoolmaster and writer of verses. Of their ten children, the youngest son Benjamin, born on January 17, 1706, was the eighth. Josiah had been a dyer in Banbury and became a tallow chandler and soap boiler in Boston. He had a mechanical aptitude and sound judgement, qualities that Benjamin both recognized and inherited. On both sides of the family the boy had forebears skilled in the use of their hands and with literary or intellectual gifts.

Franklin was writing an autobiography towards the end of his life. Although it is not considered entirely trustworthy, this evergreen work in homespun language is nevertheless a useful source of first-hand information. Franklin writes that he 'was put to the grammar school at eight years of age' but remained 'not quite one year'. His father then sent him 'to a school for writing and arithmetic'; he failed arithmetic. Apprenticed at the age of ten to his father's business, which he did not care for, Franklin dreamed of the sea; his father, fearful of losing him, then apprenticed the boy for five years to his half-brother James, who was a printer in Boston. Later, working in Philadelphia, he was noticed by the Governor of Pennsylvania,

William Keith, who promised him some financial backing and persuaded him in 1724 to go to London to complete his training as a printer and to collect materials for setting up a press in Philadelphia. Once at sea Franklin discovered that the Governor had sent him off without any letters of introduction and without funds for purchasing printing equipment, merely 'playing pitiful tricks on a poor ignorant boy'.

Franklin found himself in the London of Newton, Swift, Defoe, Fielding and Samuel Richardson. As a young printer he aroused the admiration of his fellow journeymen with his physical strength. He was a powerful swimmer and in America sometimes drifted for hours floating on lakes, towed by a string of kites. He began to question the validity of what he was asked to print and was therefore moved to write a book of his own on liberty and religion. The hundred copies that he then printed brought him to the notice of the coffee-house clientele, and he thereby almost realized an ambition to meet Newton. He was beginning to thrust himself forwards and, by offering some North American curiosities to Sir Hans Sloane, the Secretary of the Royal Society, he won an invitation to see the latter's collection at his house in Bloomsbury Square. After nineteen months in London, Franklin returned to Philadelphia in 1726. Almost at once he founded a self-improvement club, called the *junto*.

Franklin was a little under six feet tall and strongly built, with a large head and square deft hands. His hair was blonde or light brown, his mouth wide and humorous. He told later of 'that hard-to-be-governed passion of youth [which] hurried me frequently into intrigues with low women that fell in my way, which were attended with some expense and great inconvenience, besides a continual risk to my health by distemper which of all things I dreaded, though by great good luck I escaped it'. One such woman bore him a son, named William Franklin, of whom more later. On September 1, 1730 he married a woman named Deborah Read, whose previous husband, a bigamist, was presumed dead. She was a sturdy, handsome, highly coloured woman, untaught and sometimes turbulent, little interested in her new husband's studies or speculations but economical, sensible and devoted to him. 'She proved a good and faithful helpmate, assisted me much by attending the shop; we throve together, and have ever mutually endeavoured to make each other happy.' Together they raised his natural son William, although she was not the mother. The formal gentry of Philadelphia, disapproving of informal marriages, never accepted his wife socially. He settled down and over the years cultivated a reputation for diligent respectability. 'In order to secure my credit and character . . . I took care not only to be in reality industrious and frugal but to avoid all appearance to the contrary. I dressed plainly; I was seen at no place of idle diversion. I never went out a-fishing or shooting . . .'

Within two years of returning to the land of his birth Franklin set up a press that over the next decade became the most flourishing in the colonies. With his lively outlook and a press at his disposal, he began to write his own material and publish his own political studies, such as *A Modest Enquiry into the Nature and Necessity of a Paper Currency*. By 1732 he had produced the first edition of his *Poor Richard's Almanack*, illustrated with maxims on the virtues of thrift and industry. This annual publication ran for 25 years and made his fortune. The indefatigable Franklin became Public Printer to the province of Pennsylvania, Deputy Postmaster of Philadelphia and Clerk to the Assembly, publisher of *The General Magazine* and organizer of the Union Fire Company. He helped to establish an institution of higher education, called the Academy of Philadelphia, where the central themes were to be the English language and practical studies. Later he was annoyed to find that it degenerated into an old-fashioned classical school through local paedagogic snobbery, but later still it evolved into the present University of Pennsylvania. He also established a subscription library, later known as the Library Company, where members of the general public were welcome to read the books, although only members could borrow them; the idea was widely imitated in the American colonies.

About 1732 the Franklins had a son, christened Francis Folger Franklin, who died of smallpox at the age of four; long afterwards Franklin wrote to his sister Jane that a grandson of his 'brings often to my mind the idea of my son Franky, though now dead thirty-six years, whom I have seldom seen equalled in everything, and whom to this day I cannot think of without a sigh'. A daughter, Sarah Franklin, known as Sally, was born at the end of August 1743.

Franklin was largely self-taught in science – as he was in other subjects – but this does not mean that he was uneducated. He had rigorously studied the science of the day in the writings of the best masters available. In 1741 he had organized the American Philosophical Society, the first of many established at this period. It was based in Philadelphia, where there were to be seven members – a physician, a botanist, a mathematician, a chemist, a mechanician, a geographer and a general natural philosopher – besides the president, treasurer and secretary. The Philadelphia members were to meet at least once a month to transact their philosophical business, consider such reports and queries as might have been sent in by correspondents and arrange to keep all the members informed of what they were doing. The society got off to a slow start and was moribund for a time but eventually became one of the institutions of Philadelphia, as it still is.

In 1746 Franklin met a certain Dr Archibald Spencer, an itinerant lecturer on popular science, who had just arrived from Scotland, and who demonstrated some electrical experiments to him. Fascinated, Franklin purchased Spencer's apparatus and sent for more from London. The same year an account of various experiments made with the Leyden jar, together with one of the jars, was sent in 1746 to the Library Company by a friend of Franklin who lived in London and was interested in science. Franklin, aided by a few friends, at once began to perform experiments and sent his London friend an account of these and the view he had reached regarding the nature of electricity. In his view, electricity is to be regarded as a fluid whose particles repel each other. Matter when not electrified contains a definite quantity of this fluid; if it contains more than this quantity it is electrified negatively if the fluid is regarded as consisting of negative electricity, positively if it consists of positive electricity, whereas if it contains less than this quantity it is positively charged in the first case, negatively in the second. The electrification of a body is due to the passage of this fluid into or out of it.

Two bodies, each of which has more, or less, than the normal amount of the fluid, repel each other, whereas two bodies, one of which has less and the other more than the normal amount, attract each other.

A similar view of the nature of electricity had been put forward in the *Philosophical Transactions* of the Royal Society by Sir William Watson, but Franklin was unaware of this. However, the clarity of Franklin's exposition and his demonstration that it gave simple explanations of many of the phenomena associated with the Leyden jar led to the discovery of many new effects and established his position as a physicist. The service the one-fluid theory rendered to the science of electricity can hardly be overestimated.

Franklin was also enthusing about the 'wonderful effects of pointed bodies, both in drawing off and throwing off electrical fire' which were later to suggest lightning conductors. 'The doctrine of points is very curious, and the effects of them truly wonderful; and from what I have observed on experiments I am of the opinion that houses, ships, and even towers and churches may be effectively secured from the strokes of light.' He proposed an experiment:

> To determine the question, whether the clouds that contain lightning are electrified or not, I would propose an experiment where it may be tried out conveniently. On the top of some high tower or steeple, place a kind of sentry box big enough to contain a man and an electrical stand. From the middle of the stand let an iron rod rise and pass bending out of the door, and then pass upright 20 or 30 feet, pointed very sharp at the end. If the electrical stand be kept clean and dry, a man standing on it when such clouds are passing low, might be electrified and afford sparks, the rod drawing fire to him from a cloud. If any danger to the man should be apprehended (though I think there would be none) let him stand on the floor of his box, and now and then bring near to the rod the loop of a wire that has one end fastened to the leads, he holding it by a wax handle; so the sparks, if the rod is electrified, will strike from the rod to the wire and not affect him.

In 1752 the experiment was first carried out successfully in France, then repeated in England. A few months later Franklin, in America, was urging that another, more dramatic, experiment be carried out, the celebrated experiment of generating sparks and shocks by flying a kite carrying a pointed rod during a thunderstorm:

The doctor, having published his method of verifying his hypothesis concerning the sameness of electricity with the matter of lightning, was waiting for the erection of a spire in Philadelphia to carry his views into execution, not imagining that a pointed rod of a moderate height could answer the purpose, when it occurred to him that by means of a common kite he could have better access to the regions of thunder than by any spire whatever. Preparing, therefore, a large silk handkerchief and two cross-stocks of a proper length on which to extend it, he took the opportunity of the first approaching thunderstorm to take a walk in the fields, in which there was a shed convenient for his purpose. But, dreading the ridicule which so commonly attends unsuccessful attempts in science, he communicated his intended experiments to nobody but his son who assisted him in raising the kite. The kite being raised, a considerable length of time elapsed before there was any appearance of its being electrified. One very promising cloud passed over it without any effect, when, at length, just as he was beginning to despair of his contrivance, he observed some loose threads of the hempen string to stand erect and to avoid one another, just as if they had been suspended on a common conductor. Struck with this promising appearance, he immediately presented his knuckle to the key, and (let the reader judge of the exquisite pleasure he must have felt at that moment) the discovery was complete. He perceived a very evident electric spark. Others succeeded, even before the string was wet, so as to put the matter past all dispute, and when the rain had wet the string he collected electric fire copiously. This happened on June 3, 1752, a month after the electricians in France had verified the same theory, but before he heard of anything they had done.

It was characteristic of Franklin that he should at once apply this discovery to useful purposes by inventing lightning conductors, long pointed rods reaching beyond the highest point of a building and in metallic communication with the ground. He had not only established that lightning was in the nature of an electrical discharge but also demonstrated that buildings could be protected from its effects by the installation of a lightning conductor. Before long this became standard practice for churches and other major buildings, so that, when Franklin returned for his second visit, Europe was well prepared to receive the electrician from the American colonies.

Experiments like these made his reputation as a man of science. He published articles in the *Philosophical Transactions* of the Royal Society and wrote a little book about electricity that went through three editions in England and two in France. In 1709 Sir Godfrey Copley had bequeathed to the Royal Society a fund of one hundred pounds to be spent on experiments or otherwise. In 1736 the Society had voted to 'strike a gold medal in the amount of five per cent of the fund as an honorary favour for the best experiment carried out during the year'. Franklin was awarded the gold medal 'on account of his curious experiments and observations on electricity' and was elected to the Society. There was similar enthusiasm in Paris for what he had achieved. In America, Harvard, William and Mary and Yale Colleges awarded him honorary degrees.

Franklin was becoming more deeply involved with civic affairs and public life. He was elected a member of the Assembly of Pennsylvania in 1751 and two years later was appointed deputy Postmaster-General for the British colonies in North America, a royal office he held for over twenty years. As a member of the Assembly, he represented the colony in various ways. In 1751 came the first of his warnings to the British government, concerning the transportation of convicts. 'If felons were the scourge of Britain', he wrote, 'rattlesnakes were the scourge of America' and he proposed to have some sent over to be liberated in St James' Park and other such places throughout Britain.

Since his marriage to Deborah in 1730 Franklin had not sought intimate friendships with other women. However, at the beginning of 1755, when he was approaching fifty, he met one Catherine Ray, the first of a series of younger women who fell under his spell and found him adorable. After the initial meeting they did not meet again for over eight years and after that they met only twice more. However, they continued to exchange letters for almost forty-five years, letters that are a delight to read. At eighty-three he wrote to her for the last time 'among the felicities of my life I reckon your friendship, which I shall remember as long as life lasts'.

Much has been written about Franklin, mainly about his political activities, especially his prominent role in the events leading up to the Declaration of Independence. The general history of this period is so well known that just an outline of Franklin's part in these events will be sufficient. It is necessary to begin by recalling that William Penn, the True and Absolute Proprietary of the province, had been granted various special privileges, which passed to his heirs when he died. The governor of the province was a nominee of the Proprietaries, as they were called; although

Franklin was initially one of their supporters, by this time he had turned against them. Earlier in the century there had been major conflicts with the French, who, with some Indian support, were threatening the British colonies from the west. For many years British subjects in the northern colonies had also felt threatened by the constant encroachments the French were making on Nova Scotia and along their northern frontiers. Organized fighting began in 1755 and was not formally ended until eight years later. These immensely expensive French and Indian wars, as they were called, were ruinous to the British economy. Since Pennsylvania was threatened, Franklin took the initiative of raising a militia from the colonists to support the British troops involved. The Governor objected to this, especially when Franklin was elected colonel of the militia, and from then on the Penns and their supporters took every opportunity to discredit him.

One of the grievances of the people of Pennsylvania was that the Proprietaries, although wealthy, refused to contribute to the cost of the provincial government, particularly in its dealings with the Indians. In 1757 Franklin was sent by the Assembly to London to see whether something could be done about this. The mission was not an easy one; it lasted for five years, and his efforts did not achieve much, as we shall see.

When Franklin sailed for England for the second time he took his son William with him. In London they lodged at the house of a widow named Margaret Stevenson (this has recently been converted into a small Franklin museum). He enjoyed her company and enjoyed even more that of her daughter Polly, of whom more later. His first political action was to see the Lord President of the Council in Whitehall, who gave him a lecture about the constitutional position, to the effect that the Assembly had no right to tax the Proprietaries. Then he went to see William Penn's son Thomas, who represented the interests of the Proprietaries. He too was most unhelpful. After that Franklin succumbed to a serious illness, which lasted eight weeks.

Franklin's scientific reputation had preceded him, and, once he had recovered, he spent some time on experimental work. In London he demonstrated an electrical machine he had built, which threw a spark nine inches long. He went to Cambridge with his son and performed some more experiments. They went on to Northamptonshire, to call on some of the English Franklins, and to Birmingham, to look up some of Deborah's relations. They made a tour of Scotland, meeting men of distinction, in the course of which he received an honorary degree from the University of St Andrews.

On his return to London Franklin tried to see William Pitt the Elder, the prime minister, but found him too preoccupied with other problems. Unable to interest the government in the colonists' cause, he turned to the press for support, and as a result public opinion began to turn against the Penns. Eventually a compromise was reached, but it was not at all what the Assembly had hoped for. After all his efforts, Franklin's mission essentially ended in failure. In no hurry to return home, he remained in London, increasingly the unofficial spokesman for the American colonists. Meanwhile his son William, who had been studying law at the Middle Temple, had made such a good impression that, after he qualified as a barrister, he was appointed Governor of the State of New Jersey. Franklin went to visit the Low Countries, meeting the men of science there, and was back in London in time for the coronation of King George III. Just before he was due to return to America he received the honorary degree of Doctor of Civil Law from the University of Oxford.

On his return to Philadelphia Franklin became Speaker of the Assembly. He had been away five years. He had many friends but also powerful enemies who went to great lengths to discredit him. The Assembly decided to petition the King to govern the province directly, rather than through the governor, and Franklin was sent back to England in 1764 to deal with this. However, when he arrived in London he found that matters had taken a turn for the worse. The British government had decided that the colonies should contribute to the cost of their own defence. Stamp duty, on newspapers and publications, was the method chosen. Although the same tax was imposed at home it was deeply resented in America; there was widespread disorder, but in Britain this was not understood. Franklin was called to the Bar of the House of Commons to explain the reasons for American opposition to the Stamp Act and succeeded in having it repealed, but only to see the principle of taxation without representation reaffirmed by the imposition of customs duties on imports to the colonies.

Up to this point Franklin had been strongly pro-British: 'I have long been of the opinion', he said, 'that the foundations of the future grandeur and stability of the British Empire lie in America; and though, like other foundations, they lie low and are little seen, they are nevertheless broad and strong enough to support the greatest political structure human wisdom has ever yet erected.' However, America and Britain were rapidly drifting apart, and there was little that someone in Franklin's position could do about it. 'In England I am regarded as too much an American', he declared, 'and in America as being too much an Englishman.' The assemblies of

Georgia, New Jersey and Massachusetts asked him to represent them, as well as Pennsylvania. Edmund Burke, who had been briefed by Franklin, spoke persuasively in Parliament in favour of conciliation with the colonists, but all in vain.

While this was happening he managed to find a little time for science. In 1766 he had met Joseph Priestley, a Unitarian minister with an interest in science, who repeated some of Franklin's electrical experiments and performed others, which led him to formulate the inverse square law of electrostatics. Priestley and Franklin became close friends. Franklin spent his last day in London before returning home discussing scientific matters with Priestley. (It is to Priestley that we owe the account of the kite experiment quoted above.) Earlier Franklin had been to France and met some of the scientists in Paris, and had also visited Germany.

Once the War of Independence finally broke out, Franklin's life became bound up with the course of the revolution. As is well known, he was one of the founding fathers of the American Republic. When Thomas Jefferson's draft of the Declaration of Independence was submitted to Congress Franklin was much involved in the resulting deliberations. He was a member of the Second Continental Congress and drew up a plan of union for the colonies. He was then sent to France, with two other commissioners, to try to enlist French support for the revolutionaries. Although the French were naturally pleased that the British had this problem on their hands, they had no desire to support what might prove to be a losing cause. Franklin was lionized in Paris, thanks to his scientific reputation but not only that: to many Frenchmen his simplicity of dress, his native wit and wisdom, his gentle manners free of affectation seemed to exemplify the virtues of the natural man, the personification of many ideas cherished in the Age of Enlightenment. He had the distinction of being elected one of the eight foreign associates of the Paris Academy (not to be confused with the corresponding members, who were much more numerous).

Franklin settled down in Palsy, on the western outskirts of Paris, where he was joined by Polly Stevenson. His flirtatious reputation seems mainly to refer to this period, when he lived an active social life despite being plagued by gout and kidney stones. Franklin enjoyed contact with the many scientists in Paris and made the acquaintance of Alessandro Volta, a strong supporter of Franklin's 'one-fluid' theory. Volta initiated the next stage of electrical science with his invention of the battery, which made possible the production of a continuous electric current. Franklin was also a leading member of the commission appointed to investigate the claims of Franz Anton Mesmer; its report gave the death blow to Mesmerism, the belief that magnets could be used to cure sickness.

The mission to Paris was successful, bringing to the colonists badly needed financial support in their struggle. He organized pro-American propaganda and was not above a little intelligence work. In 1778 Congress appointed him Minister Plenipotentiary to the Court of France; two years later John Adams was also sent to London to negotiate a peace settlement with Great Britain. Although Adams and Franklin had previously been on good terms, they now began to clash; the diplomatically inept but socially superior Adams did not wish for Franklin's help in the negotiations. However, in 1781 Adams and Franklin were appointed as joint commissioners to negotiate the final peace.

By this time Franklin was already seventy-seven and increasingly concerned about his health. He asked Congress for permission to retire, but, although his duties were lightened somewhat, it was to be two more years before he was able to return to America. Just as the war was ending Franklin was able to observe the first manned balloon flights take place in France and was impressed by their military possibilities. Never a man to miss an opportunity for a little scientific work, he made some observations of the Gulf Stream on the way back to America; he appears to have been the first man of science to study the great current. He arrived back in the land of his birth to a hero's welcome. The last five years of his life were spent as President of the Province of Pennsylvania; he also had a seat on the Constitutional Convention. He disinherited his son William as a pro-British renegade. He found time to write some more of his autobiography, which was first published in 1868.

Benjamin Franklin died of pleurisy in Philadelphia on April 17, 1790, having just received the news of the outbreak of the French Revolution, which took such a very different course from the American. More than twenty thousand people attended his funeral in Philadelphia. The French National Assembly went into mourning; in the United States Congress the House of Representatives did so, the Senate did not. In his youth he wrote an epitaph for himself:

> The Body of
> B Franklin Printer
> (Like the Cover of an old Book
> Its Contents torn out
> And stript of its Lettering & Gilding)

Lies here, Food for Worms.
But the Work shall not be lost:
For it will, (as he believ'd) appear once more,
In a new and more elegant Edition
Revised and corrected,
By the Author.

ROGER JOSEPH BOSCOVICH (1711–1787)

For one who participated in the actual work of science the interests of Boscovich were extraordinarily varied even by the standards of the eighteenth century. Despite the limitations imposed upon his thought by religious obedience, he is increasingly recognized as a natural philosopher of world stature, one of the last of the polymaths. In science his interests ranged through mathematics, physics, astronomy and geodesy, but there was so much else that he put his mind to.

Roger Joseph was the son of Nikola Boscovich, a Croatian merchant of the Dalmatian port of Dubrovnik, then known as Ragusa, and Paula Bettera, the daughter of a merchant whose family came from the city of Bergamo, east of Milan. He was born on May 18, 1711; his father died when he was ten; his mother, a robust and active woman with a happy personality, lived until she was 103. His eldest sister was the only sibling who married; another sister became a nun, his eldest brother became a soldier, a second brother a Jesuit priest, a third brother a monk. The family was noted for its seriousness, piety and literary interests.

Roger Boscovich was drawn to a religious career and began his education in the Jesuit college of Dubrovnik, following the standard curriculum, and then offered himself for more protracted and severe religious and

intellectual training by the Jesuits in Rome. So in September 1725 he crossed the Adriatic to Ancona and then went on by land to Rome, entering Sant'Andrea delle Fratte on the Quirinal as a novice at the age of fourteen. After four years he was ready to enter the Gregorian University, the apex of the Jesuit system of education.

Boscovich was extraordinarily sharp of mind, comprehensive in intelligence and tireless in application – in short an outstanding student. He learned science in a way characteristic of his later career, through independent study of mathematics, physics, astronomy and geodesy. In 1735 he began studying Newton's *Opticks* and *Principia* at the Gregorian University, where he made himself an enthusiastic propagator of the new natural philosophy. He composed a lengthy scientific poem on eclipses, in Latin hexameters, which he went on revising for many years, until it was eventually published in 1779. The exact sciences were what always appealed to him – in the first instance mathematics. He was directed to spend some time teaching novices at Sant'Andrea, and then, although he had not yet completed his theological studies, he was appointed professor of mathematics at the Gregorian University itself. This post, which he held for nineteen years from 1740, largely determined the course of his future career. Teaching interested him as method as well as for content. In this respect, as in others, his spirit was progressive. He published a textbook of his teaching in 1754 – *Elementa universae matheseos* – of which the third and final volume contains an original theory of conic sections.

The year 1740 was when Benedict XIV succeeded to the papacy. His reign recalled the best years of the Renaissance. Benedict XIV was a profound scholar and a vivid personality, magnanimous and virtuous in his private life, who surrounded himself with the leading intellectuals of Rome. The Papal Secretary of State, the able and versatile Cardinal Valenti Gonzaga, shared Dubrovnik connections with Boscovich, who naturally came to his attention, and Valenti's residence soon became like a second home to him. During this period of his life he was entrusted with several practical and diplomatic commissions for lay and ecclesiastical authorities, as was not unusual among qualified clergymen of his time.

In 1735 there had been a resurgence of rumours that the long-standing cracks in the dome of St Peter's basilica – which had been completed in 1590 – presaged an imminent collapse. The new Pope consulted various experts, one of whom recommended demolishing the huge structure and rebuilding it. Others reported that no danger threatened the dome. However, for further reassurance a commission consisting of Boscovich and two

French scientists was appointed to investigate the causes and make recommendations. Boscovich drafted the report, which, by analysing the problem in theoretical terms, achieved – despite certain errors – the reputation of a minor classic in architectural statics. He undertook further such work for the Pope later.

In 1744 Boscovich, now thirty-three, finally completed his theological studies and, after nineteen years of training, was ordained priest and so became a full member of the Society of Jesus. He could now devote himself to professorial duties, especially research. He was a scholar with a growing reputation among the brilliant intellects of Rome, but his work was also becoming known elsewhere. The next fourteen years, from 1744 to 1758, were his most prolific period of mature scholarship, but there was much else. For example, he was interested in archaeology. In 1743 a Roman villa was discovered and excavated above the town of Frascati, the ancient Tusculum, on the western slopes of the Alban hills. Boscovich published a description of what was found there, particularly a sundial. In 1750 he published a critical study of a red granite obelisk with hieroglyphic writing that had just been discovered in the Campus Martius. He composed a series of memoirs on the practice of hydraulic engineering and on the regulation of the flow of the river Tiber and other watercourses. He made a survey and map of the Papal States and made a plan for the harbours at Rimini and Savona. Parma, Genoa, Lucca, the Venetian Republic – all had occasion to seek his advice on questions of practical hydraulics.

Boscovich was particularly interested in astronomy and used to spend much of his time observing the heavens. The solar eclipse of 1748 provided him with a fine opportunity to demonstrate the spectacle at the Gregorian University before a brilliant assembly of notables, including cardinals, princes and numerous prelates. In 1735 the Paris Academy had decided to test Newton's theory that the earth was flattened at the poles by comparing careful geodetic measurements made in the tropics and the arctic. This necessitated international cooperation in geodesy, in which Boscovich was a leading participant. He collaborated with an English colleague, then rector of the English Jesuit college in Rome, on the arduous task of surveying two and a half degrees of the meridian between Rome and Rimini. On his initiative, meridians were also measured in Austria, Piedmont and Pennsylvania.

It was during this fruitful period that Boscovich was active in the Accademia degli Arcadi. This exotic society had been founded by Queen Christina of Sweden when she was in Rome. Only poets were admitted as members; the men were called shepherds, the ladies nymphs. Each member took a classical name suggestive of life in Arcadia; Boscovich was known as Numenius Anigreus. Scientific interests were strong in the academy but poetry predominated. Boscovich took the opportunity to regale the members with passages from his long poem about eclipses. The meetings of the society enabled him to get to know a wide circle of influential people in a stylized Arcadian setting, making links that provided valuable introductions when he travelled abroad. His career as professor of mathematics at the Gregorian University was now coming to an end and he was preparing for a new role as scholar-diplomat. Experience of social life in high ecclesiastical, academic and diplomatic circles in Rome later proved to be an important aspect of his training for the wider field of activity that his Jesuit superiors were planning for him or approved as occasion arose.

In 1757 Boscovich undertook the first of several diplomatic missions when he represented the interests of the Republic of Lucca before the Imperial court in Vienna in a dispute with Tuscany over water rights. Vienna was a Jesuit stronghold and Boscovich had excellent contacts there. With the Seven Years War about to begin, it was a difficult time to settle such a dispute, but he won the case. The Empress Maria Theresa consulted him on the structure of the thirty-year-old Imperial Library, which was already showing defects.

However, his principal occupation in Vienna was the completion of his great work in the field of natural philosophy, *Theoria philosophiae naturalis, redacta ad unicam legem virium in natura existensium* (*A Theory of Natural Philosophy, Reduced to a Single Law of the Forces Existing in Nature*), which appeared in the autumn of 1758. This daringly original work, the mature expression of ideas that Boscovich had put forward in a series of papers from 1745 onwards, was well known and influential for 150 years thereafter. Faraday, Clerk Maxwell and Kelvin were all interested in his ideas, as were many of the leading continental scientists of that period. That it should be so neglected today, at least in the western world, is ironic since Boscovich's ideas are in several respects in tune with modern thought. The Boscovich theory is developed mainly by geometrical methods. It is concerned with observable phenomena, not their causes. No quantitative predictions are made, the aim being to demonstrate that a single interaction law of the type proposed can in principle account for a wide range of physical properties. Thus he presented, in the form of a logical and mathematical scheme, a programme for point atomism, in which all primary particles are identical, a single oscillatory law determines their interactions, only relational quantities enter and the distinction between empty and occupied space disappears. The essentials of the theory were summarized by the English physicist Henry Cavendish as follows:

> matter does not consist of solid impenetrable particles as commonly supposed, but only of certain degrees of attraction and repulsion directed towards central points. They also suppose the action of two of these central points on each other alternately varies from repulsion to attraction numerous times as the distance increases. There is the utmost reason to think that both these phenomena are true, and they serve to account for many phenomena of nature which would otherwise be inexplicable. But even if it is otherwise, and if it must be admitted that there are solid impenetrable particles, still there seems sufficient reason to think that those particles do not touch each other, but are kept from ever coming into contact by their repulsive force.

Although Boscovich was dedicated to the Society of Jesus, he was out of sympathy with certain policies of his ecclesiastical superiors. He was disappointed by the negative attitude that a number of Jesuit philosophers adopted towards his own system of natural philosophy; some Jesuit fathers even regarded the theories of Newton as heretical. He resented their rejection of proposals he had advanced in arguing for the modernization of education both in method and in subject matter. While he was in Vienna his patron Cardinal Valenti had died. Benedict XIV had also died; although the next Pope was also well disposed towards him personally, it seemed that he could be more useful to his Order away from Rome, so his superiors decided to send him to Paris. In 1759 Boscovich set off on his travels, seeing various notables on the way. In Marseilles he met some of the Jesuits recently deported from Portugal; the Vatican's hesitant reaction to the persecution of his Order there was an indication of what was to come. As soon as he arrived in Paris he found the Order under attack in France as well. *Philosophes* such as d'Alembert were attacking the power of the Catholic Church generally, but the Jesuits were a particular target. Nevertheless, Boscovich was personally well received in aristocratic, scientific and literary circles. The Paris Academy had long before elected him a corresponding member following the publication of a discourse of his on the *aurora borealis*.

Boscovich had come to Paris during one of the most disastrous years in the history of France. The country was in financial crisis, and the whole population was longing for peace, but the war was dragging on. France had been defeated on land and sea. All Canada, except Montréal, had been lost. Boscovich's reputation and influential contacts opened doors for him and his reports back to Rome indicate that he was discharging his mission successfully, whatever that might have been (the Jesuits were accused of being spies for the Vatican). During the six months he stayed in Paris he took the opportunity to attend the meetings of the Academy. The academician Alexis Clairaut wrote to a colleague 'he is one of the most amiable men I have ever known and I cannot compare him to anyone but yourself for the combination of knowledge and social qualities. We saw one another very often and I introduced him to all my friends who thought the same.' A diplomatic intervention on behalf of his native city of Dubrovnik took him to the court at Versailles; the Jesuits were well in with the royal family.

While Boscovich was in Paris he was elected to honorary membership of the St Petersburg Academy and made plans to visit Russia at an early date. Although he was happy enough in the French capital, and was to return there later, in 1760 he crossed over to London, where again his reputation had preceded him among literary and scientific cognoscenti. He stayed in England (then still at war with France) for seven months, during which he had discussions with representatives of the Anglican Church; met Benjamin Franklin, who showed him some electrical experiments at the house of the painter Richard Wilson; and visited the universities of Oxford and Cambridge. When Boscovich was in London, Cavendish invited him to the Royal Society Dining Club, where he met John Michell, a near contemporary of Cavendish from Cambridge, who made important observations and discoveries at his home at Thornhill, near Wakefield in Yorkshire. It seems possible that one or both of Cavendish and Michell had arrived at the same theory as Boscovich independently. He met many other leading personalities of the time: for example he dined with Samuel Johnson on several occasions, conversing in Latin, for, although he could read English, he was unable to speak it. He sat for a portrait, now only known through a copy. At the start of 1761 the Royal Society elected him a fellow, and, in recognition of the honour, he dedicated to it the poem on the eclipses of the sun and moon, on which he had been working for so many years. He then lent his weight to efforts to persuade the society to organize an expedition for the purpose of observing the transit of the planet Venus in June 1761.

Boscovich left England suddenly on December 15, 1760. It had been arranged that he would join the new Venetian ambassador to the Sublime Porte in Venice and that they would then travel to Istanbul together. On the way to Venice he was received as a favoured guest by Duke Stanislas of Lorraine and Bar, who was still styled King of Poland. The duke was one of the most cultured rulers of the time; he surrounded himself with a constellation of artists and men of letters, instituted coveted prizes for the encouragement of the arts and sciences and founded a royal society of letters, which became known as the Academy of Stanislas. Boscovich was elected a foreign associate during the four days he spent at Nancy, the capital of the Duchy.

Boscovich had planned to arrive in time to make observations of the transit of Venus in Istanbul. However, he was still in Venice when the transit occurred, but, since the local weather was overcast, he was unable to observe it. Fortunately the British ambassador to the Porte, an amateur scientist himself, was able to observe the transit in Istanbul and report on this important event to the Royal Society. When Boscovich and Correr, the Venetian ambassador, reached Istanbul in November 1761, Boscovich fell seriously ill with a leg infection and had to remain there for several months of recuperation. When he had partially recovered he set off again, this time in the company of the British ambassador, and travelled through Turkey and Bulgaria to Moldavia, but, after an accident that further injured his leg, he abandoned the original plan of going on to Russia. Instead he went to Poland, where he spent several months. In Warsaw he was received in ecclesiastical and diplomatic circles, and afterwards wrote a careful assessment of the political situation in that country. His diary of the tours he made through Bulgaria and Moldavia amounts to a systematic description of the region; it was published in Italian in 1784, having already been translated into French and German.

From Poland Boscovich finally returned to Rome – by way of Silesia, Austria and Venice – arriving back there in November 1763 after an absence of over five years. The Vatican consulted him on the age-old problem of draining the Pontine marshes. The report he produced for the papal government served as the basis for all subsequent work on the drainage problem.' I am afraid that the drying up of your marshes is taking up a great deal of your time', wrote his friend Alexis Clairaut, 'You take on all sorts of duties. Take great care I beg of you not to overtax yourself and don't go and ruin your health which is precious to mathematicians.'

Boscovich was now fifty-two years old and had reached another turning-point in his active life. He received an invitation to become professor of mathematics at the University of Pavia, which, after a period of stagnation, was being revived under Austrian administration. At Pavia he organized his department realistically and lectured himself, with an emphasis on applied mathematics. In research he concentrated his efforts mainly in the field of optics and the improvement of lenses for telescopes and played a leading role in the organization of the observatory at the Jesuit College in Brera near Milan in 1764. The observatory when completed was a remarkable achievement, appraised by all the experts as the finest yet constructed. Its design benefited from his knowledge both of architecture and of astronomy; he spent quite a lot of his own money on the building and its equipment. However, he was irked when he found that his work on the observatory was not sufficiently appreciated by the authorities at the Jesuit college. His expert advice was also sought in connection with a structural problem concerning the dome of Milan Cathedral.

Recalling his interests in the transits of Venus, the Council of the Royal Society invited him to lead an expedition to California for the purpose of observing the second of the famous pair of transits, that of 1769. Unfortunately political conditions prevented him doing so; the Jesuits had been expelled from Spain and Naples in 1767 and from Parma and Malta the following year, and their total suppression was being demanded; in those circumstances it was not thought wise for him to leave Lombardy.

In his old age Boscovich was becoming increasingly irritable; he often took offence over real or imaginary slights. Partly this may have been due to his ulcerated leg, which was getting worse. He went to Paris for treatment, but it was not the doctors of Paris but an unqualified barber-surgeon in Brussels who finally managed to cure him. In 1770 there was a reorganization, which resulted in Boscovich moving to the department of optics and astronomy at the Scuola Palatina in Milan, where he would be near the observatory he had built and considered he should control. Unfortunately the transfer provoked opposition among his colleagues at the observatory and in the university at large, resulting in petty annoyances that he was unable to rise above. In 1772 the court in Vienna yielded to the demands of the majority and relieved Boscovich of his 'concern' for the observatory. In despair he resigned his professorship as well. All his world was dissolving; the final blow came the next year, when the Pope ordered the suppression of the Society of Jesus.

By this time Boscovich was in his sixty-third year and had spent most of his savings on the observatory. In the hope of receiving a pension, he applied to some of the institutions he had served at various times in his life, but without success. He thought of retiring to Dubrovnik, his birthplace, where his mother was still alive. However, influential friends persuaded him to return to Paris instead, and this time he stayed in the French capital for nine years. With the approval of Louis XV a well-paid post was arranged for him as director of optics for the French navy; consequently he became a subject of the French crown. An opportunity for him to return to the Brera observatory arose, but by that time he felt committed to France. In Paris, during this, the last productive period of his life, he mainly worked on problems of optics and astronomy, for example the theory of the achromatic telescope. He renewed his acquaintance with the leading French scientists of the day and with visitors such as Franklin and Priestley. In search of health and tranquillity, Boscovich spent the greater part of each year in the countryside residing at the estates of one or other of his friends among the rich and powerful.

Priority disputes were all too common in the eighteenth century. One of Boscovich's more tiresome disputes was with the young Laplace over a method that Boscovich had devised for determining the paths of comets; another was on priority over the invention of a device that became important in the design of geodetic telemeters. Laplace wrote an account of the history of the Brera observatory without even mentioning Boscovich. However, these were relatively minor distractions in his last years of constructive work. Despite failing health, he continued to lead an active social life among the good and the great. A new edition of the poem on eclipses, with Latin original and French translation on opposite pages, was published in Paris, with a flattering dedication to Louis XVI.

Boscovich published over a hundred dissertations, papers and books. In 1782 he received leave from the king to return to Italy in order to prepare a collection of his works on optics and astronomy for the press. He settled in Bassano, north of Rome, where he had assistance in the task, but the strain of reading proofs told on his health. Once again he set off on his travels, without leaving Italy, to visit old friends and make peace with his enemies. He found a cordial welcome in Milan, where former opponents were inclined to let bygones be bygones, and settled down to work in the Brera observatory; but sadly his mental powers were leaving him, forgetfulness, anxiety and fear for his scientific reputation grew on him.

When it became clear that his mind was failing, Boscovich was moved to the Jesuit college in Monza. His condition rapidly worsened and was accompanied by other problems. He died of a lung ailment on February 13, 1787 and was buried in the church of Santa Maria Bordone in Milan. No trace of his tomb can be seen nowadays. Today the citizens of Dubrovnik claim him as their most illustrious son. The Serbs claim him as a Serb, on his father's side; the Italians as an Italian, on his mother's side, and describe him as largely Italian by culture and career; while the French point to his adoption of French nationality.

HENRY CAVENDISH (1731–1810)

As a fellow-scientist wrote, Henry Cavendish possessed a clarity of comprehension and an acuteness of reasoning that have been the lot of very few of his predecessors since the days of Newton. At home and abroad he was regarded as the most distinguished British man of science of his day. Among his many achievements are the demonstration of the existence of hydrogen as a distinct substance, the demonstration that water is a compound and the determination of the density of the earth. He was also one of the pioneers of electrical research, presaging much of the work of Coulomb, Faraday and Ohm. Clerk Maxwell, who edited some of his papers, was fascinated by his character: 'Cavendish cared more for investigation than publication. He would undertake the most laborious researches in order to clear up a difficulty which no-one but himself could appreciate, or was even aware of. And we cannot doubt that the result of his enquiries, when successful, gave him a certain degree of satisfaction. But it did not excite in him that desire to communicate the discovery to others which, in the case of ordinary men of science, generally ensures the publication of their results.'

Lord Charles Cavendish, the third son of the second Duke of Devonshire, married Lady Anne Grey, the fourth daughter of Henry, Duke of Kent. She was living in Nice, owing to frail health, when her first child

Henry was born on October 10, 1731. A second child, Frederick, was born in England two years later, but their mother died shortly afterwards. Little is known of the early years of the two boys, except that they attended the Hackney Academy, a London school well thought of in its day for the education of children of the upper classes in sound classical learning. Each of the brothers went up to the University of Cambridge, matriculated as a nobleman and resided there for four years, but left without taking a degree. The college to which they belonged was St Peter's, commonly known as Peterhouse. Shortly after the younger brother had left Cambridge they made the customary tour on the continent; apart from Paris, it is not known where they went. Henry may well have studied mathematics and physics when he was in Paris. The brothers did not have much to do with each other later in life, although they remained on good terms.

After returning to England, Henry Cavendish went to live with his father at a house in Great Marlborough Street, in the Soho district of London, and apparently continued to do so until his father died. It was during this period of almost thirty years that he carried out the fundamental electrical research which so impressed Clerk Maxwell. He began his research career by assisting his father, a gifted experimental physicist, who was a prominent fellow of the Royal Society. Lord Charles made some valuable investigations into heat, electricity and terrestrial magnetism. Franklin remarked that 'It were to be wished that this noble philosopher would communicate more of his experiments to the world, as he makes many, and with great accuracy.'

Lord Charles was not a wealthy man but the financial allowance he made to his eldest son was so small as to be described as niggardly by contemporaries. It is not known just where the money came from, but, in 1783, when his father died, or even before, Henry Cavendish became extremely wealthy, apparently through a succession of legacies from relatives. However, by this time he had become accustomed to living parsimoniously. His large library of scientific works, housed in Bedford Square, was open to any serious scholars. At one time it was in a somewhat neglected state, so, having been told of a German scholar in straitened circumstances who was capable of classifying the books in a satisfactory manner, Cavendish arranged for him to act as his librarian; in return Cavendish gave him the princely sum of £10 000, with which to purchase an annuity. He could be remarkably generous when he felt so inclined.

Cavendish's principal residence was a large villa at Clapham, then just a village south of London. Most of its rooms were equipped with scientific apparatus. It was at Clapham that he made his discovery of the composition of water and measured by means of a torsion balance the density of the earth. There was a ladder up a large tree in the garden, from the top of which he made astronomical and meteorological observations. He was very much a man of habit, invariably dining off leg of mutton and taking exactly the same walk every day on his own. His pathologically shy and nervous disposition, on which anyone who had any contact with Henry Cavendish was apt to remark, has been attributed to his comparative poverty during the first forty years of his life.

In appearance Cavendish was tall and thin, his face intelligent and mild. His voice was hesitant and somewhat shrill. He retained the dress of his youth – faded violet suit with high collar, frilled shirt-wrists and a knocker-tailed periwig. Each year on a fixed day his tailor provided him with a new suit that was a replica of the old one. When out-of-doors he was to be seen wearing a three-cornered hat. He would often be accompanied by Sir Charles Blagden, who for seven years acted as his assistant. As secretary of the Royal Society, Blagden made frequent visits to the continent, usually to Paris, where he was a friend of Berthollet and Laplace, amongst others, and courted the lively widow of Antoine Lavoisier. When eventually Cavendish parted with Blagden's services he provided him with an annuity of £500 and left him a legacy of £15 000 in his will.

Cavendish's interests extended over a wide field of natural philosophy, and every subject of investigation was subjected to a rigorous quantitative examination. The results he obtained with simple methods and apparatus were amazing. He was not only a highly skilled experimentalist but also a capable mathematician. In common with others in England during this period, he employed the methods of Newton, for example the fluxional notation for differentiation. In chemistry he adhered to the old caloric theory of heat, although the experiments he performed were helping to undermine it. Like Newton, he had a deep dislike of controversy. As a result he published remarkably little; for example only two research papers on electricity, although, when Clerk Maxwell was editing Cavendish's electrical researches for publication, after his death, he found twenty packages full of manuscripts on mathematical and experimental electricity.

The vast bulk of the Cavendish papers must have given Maxwell pause, but, once he had begun, he found them fascinating. Cavendish had quietly presaged many of the important results of the following century. He had performed some extraordinarily accurate experiments with the crudest of equipment, using a pair of pithballs, on strings, which repelled each other to measure charge and his own body to measure resistance. In going through the papers Maxwell found many of the experiments mentioned so original that they seemed worth repeating, checking or improving. Cavendish had done his experiments by making himself part of the electric circuit and noticing how intense the electric shocks he felt under different circumstances were. To check his conclusions he would summon his servant Richard to replace him and then observe his servant's reactions. Visitors to his laboratory were often pressed into taking part instead of Richard; Cavendish offended a visiting American physicist, who refused to act as a guinea pig and went off saying 'when an English man of science comes to the United States we do not treat him like that'.

Although Franklin's work had been published twenty years earlier, Cavendish's paper 'An attempt to explain some of the principal phenomena of electricity by means of an elastic fluid' involves basically the same idea, but gives it a mathematical treatment, quantitative rather than qualitative. Both Cavendish and Franklin served on a committee of the Royal Society to report on the best way of protecting buildings from lightning; they recommended the installation of pointed conductors. However, others were in favour of blunt ends, and George III agreed. A notorious controversy erupted, with political overtones, since pointed ends were thought somehow to be unpatriotic.

Although Cavendish mainly lived as a recluse owing to a morbid dislike of society, he nevertheless participated in the intellectual life of London. He was a member of the Royal Society of Arts, a trustee of the British Museum, a fellow of the Society of Antiquaries, a manager of the Royal Institution and a foreign associate of the Paris Academy. Like his father, he was prominent in the Royal Society, to which he was elected in 1760, served on the Council and some of its committees; and regularly attended the Dining Club, to which he often brought guests. They were advised that it was useless to try to engage him in conversation on any non-scientific topic. The only known portrait of him, now in the British Museum, was drawn surreptitiously at one of the club dinners.

Henry, later Lord, Brougham recalled seeing him at a Royal Society Conversazione and hearing 'the shrill cry he uttered as he shuffled quickly from room to room, seeming to be annoyed if looked at, but sometimes approaching to hear what was passing among others. His walk was quick and uneasy. He probably uttered fewer words in the course of his life than any man who lived to fourscore years, not at all excepting the monks of La Trappe.' Of the many stories told about his idiosyncrasies, one concerns a distinguished foreign scientist who said that he wished to meet 'one of the

greatest intellectual ornaments of this country, and one of the most profound philosophers of all time'. Cavendish was so embarrassed that he was
reduced to total silence and escaped in his carriage at the first opportunity.

Cavendish made a number of journeys by carriage within Britain,
always in the summer, when conditions of travel were least difficult, and
generally accompanied by Blagden. Although usually their main purpose
was to visit other men of science, generally some scientific work was done
en route; for example they studied the variation of barometric pressure with
altitude, or collected specimens of minerals to be examined at leisure on
their return. They inspected many of the places where science was being
applied in industry, as the industrial revolution began to gather momentum.
Often the people he met were later guests of his at the Royal Society Dining
Club.

Cavendish died on February 24, 1810, at the age of seventy-eight,
and was buried in All Saints Church, Derby, now designated the cathedral,
where his famous ancestor Bess of Hardwick had built an elaborate tomb
for herself. Owing to his frugal life-style, he had accumulated a fortune of
over a million pounds, a huge sum in those days; he was one of the richest
men in England. When he died none of this wealth went directly to support scientific research; he believed that it should return to the family from
which it came. However, many years later the University of Cambridge benefited from the generosity of the Cavendish family through the endowment
of the Cavendish Professorship of Experimental Physics and the Cavendish
Laboratory.

CHARLES AUGUSTIN COULOMB (1736–1806)

The end of the Thirty Years War left France the most powerful nation in
Europe. Although the golden age of French science was yet to come, some
remarkable physicists were already distinguishing themselves before the
end of the *ancien régime*. One of the first was Coulomb, the subject of our
next profile. He has been described as the complete physicist, rivalled in
the eighteenth century only by Henry Cavendish, combining experimental skill, accuracy of measurement and great originality with mathematical
powers adequate to all his demands. He invented the torsion balance and
used it to show that the force between electrically charged particles is proportional to the product of their charges and inversely proportional to the
distance between them. This fundamental result is known as Coulomb's
law; the unit of electrical charge is also named after him.

Charles Augustin Coulomb was born in Angoulême on June 14, 1736. Little is known of his mother, Catherine Bajet, except that her mother was from the wealthy de Senac family. Charles' father, Henry Coulomb, after a period of military service, had been appointed to a minor government administrative post, that of 'Inspecteur des Domaines du Roi'. Charles was born away from his ancestral province, for the Coulombs came from Languedoc, and the family had lived at least several generations in Montpellier. They had traditionally been lawyers; Charles' older cousin Louis was head of a branch of the family that was active in politics and finance throughout the eighteenth century.

As an inspector of the king's lands, Henry Coulomb was liable to be transferred from place to place in the course of royal business; thus, early in Charles' childhood, the family moved to Paris, where his father became involved in tax-farming. His mother, who wished her son to become a physician, arranged for him to attend the Collège des Quatre Nations, probably between the ages of ten and fifteen. The college had been founded by the will of Cardinal Mazarin upon his death in 1661 and had a good reputation for mathematics. Charles also attended lectures on the subject at the Collège Royale de France and decided that he wanted to become a mathematician.

At some point Charles' father lost all his money through unwise investments and returned to Montpellier, leaving the rest of the family behind in Paris. Since his mother still wanted him to study medicine, and he did not, Charles went to join his father and other relatives, including a cousin who introduced him to the Société Royale des Sciences de Montpellier, which was modelled on the Paris Academy but on a smaller scale and without funding of its own. He was not yet old enough to become a full member, but in 1757 was admitted as an adjunct. Charles read his first paper, on geometry, at one of its meetings and later two more on mathematics and three on astronomy.

For a career Charles considered engineering, either civil (Ecole de Ponts et Chaussées) or military (Ecole du Génie). Having decided on the military option, he returned to Paris for further study to pass the entrance examination and then entered the school at Mézières. He graduated after a year with the rank of *lieutenant en premier* and then, after a short period of leave, was posted to Brest and assigned to minor duties. Suddenly, however, his career took an unexpected turn. Three years after the start of the Seven Years War, the British navy had appeared before Port Royal, the chief town of the French colony of Martinique, with seventeen ships and 8000 men. After destroying the town and its fort they withdrew, but later they returned and controlled Martinique until 1763. When the Treaty of Paris restored the island to France, the French Minister of War decided that the islands in the French Antilles must be put in such a position that they could be defended in future engagements with the British and called for major work on fortifications to be carried out in Martinique.

It was not originally intended that Coulomb would play a large role in the fortification of this island in the Caribbean; he was conscripted at the last moment to replace another newly qualified engineer who had fallen ill. 'I was responsible for eight years', Coulomb wrote later, 'for the construction of Fort Bourbon and for a work-gang of 1200 men, where I was often in the situation of discovering how much all the theories, founded on hypotheses or on experiments carried out in miniature in *cabinets de physique*, were insufficient guide in practice. I devoted myself to every form of research that could be applied to the enterprises that engineering officers undertake.' Coulomb was inexperienced but his work was exemplary and he was promoted to captain. By 1770, with the fort almost complete, he became seriously ill and asked to be allowed to return to France. Unfortunately his superior officer wished to do the same, with the result that Coulomb had to remain another year, to the long-term detriment of his health.

On his return from Martinique in 1772 Coulomb was posted to the inland French town of Bouchain. Since no engineering works were in progress there, he had time to write a memoir concerning the mechanics of civil engineering, an *Essay on an Application of the Rules of Maxima and Minima to some Problems in Statics, Related to Architecture*, which he presented to the Paris Academy in the spring of 1773. It dealt with the strength of materials, the design of arches and similar matters. The report on the essay said that 'under this modest title Coulomb encompassed all the statics of architecture', and this led to him being placed on the first step of the ladder which was to lead to full membership of the academy eight years later. Coulomb's next posting was to the Channel port of Cherbourg, where again his duties allowed time for scientific work, and this was when he turned his attention to physics.

In 1773 the Paris Academy had announced the subject of one of its regular prize contests to be the best means of constructing magnetic compasses. After no award had been made in 1775, the academy doubled the prize and reset the contest for 1777. Coulomb submitted a memoir entitled *Investigations of the Best Method of Making Magnetic Needles*, one of his most important works. What he wrote contained the elements of all his major physical studies: the quantitative study of magnetic phenomena, torsion and the torsion balance, friction and fluid resistance and the germ of his theory of electricity and magnetism. He shared the prize with another candidate. Another memoir he had written, *On the Service of Officers of the Corps du Génie*, addressed various problems within the corps and in its relationship with the rest of the French army and played a significant role in the reorganization which took place in 1776. The following year he was transferred from Cherbourg to Besançon, where he returned to writing about civil engineering, notably his method for executing under water all types of hydraulic works without the necessity of drainage, which was also presented to the academy.

While their son was in Martinique Coulomb's father Henry had died, and his mother died in 1779. He went to Paris to monitor the administration of his mother's substantial estate, of which he was one of the beneficiaries, and while this was in progress was ordered to proceed to the Atlantic port of Rochefort, where he would be involved in constructing a new type of fort on an island in the estuary of the River Charente. Unfortunately the overall commander was the marquis de Montalembert, and this was a pet project of his. Montalembert had little regard for the Corps Royal du Génie and was engaged in a long-standing quarrel with the chief military engineer, into which Coulomb was dragged. Montalembert's controversial design for the fort was like a land-based man-of-war, made entirely of timber, expensive to construct and full of defects; moreover, he kept changing his mind about the details.

Coulomb won another academy prize in 1781, this time for the solution of problems of friction of sliding and rolling surfaces, and the resistance to bending in ropes, and the application of these solutions to simple machines used in the French navy. He developed a generalized theory of friction and a series of empirical formulae that soon became standard. When on leave in Paris to sort out further problems related to his mother's estate he read a memoir to the academy, *On the Limits of Man's Force and on the Greatest Action that One Can Exert for some Seconds, from which is Concluded the Impossibility of Flying in the Air like Birds.*

Like his contemporary the mathematician Gaspard Monge, Coulomb always kept himself busy. When his next posting turned out to be Lille, he wrote a memoir on the theory and design of windmills. His ambition, never fulfilled, was to incorporate his various investigations into a comprehensive new course of engineering – not so much a course of mathematics or engineering-drawing for the student but more a handbook for the practising engineer. He persuaded the corps to post him to Paris, essential if he was to become a full academican. Coulomb was finally elected in 1781, after being front-runner in 1779. In the corps he was promoted to *capitaine en premier de la première classe.*

Coulomb was called upon to advise on engineering projects outside Paris, notably the construction of canals, which had become strategically important at a time when the British navy, with its command of the seas, prevented coastal shipping from operating normally. In 1784 he was named *intendant des eaux et fontaines du roi*, particularly concerned with waterworks in the Paris region. This was no sinecure; there were not only problems in engineering but also litigation arising from private concessions, and by 1790 he had given the office up. Coulomb was also active in the field of public health, in hospital reform, in the construction of relief maps of places of strategic importance and on the commission established to reform the chaotic systems of weights and measures. However, before the commission reached any conclusion the Revolution began and Coulomb lost his place in the resulting purge. He resigned from the Corps du Génie in 1791, after 31 years of service, partly in protest at changes made by the National Assembly. To escape the Terror he went to his house north of Paris, close to the château Chaumontel near Luzarches, and then went on further to another property he owned near Blois in the Loire valley. This was just a farmhouse where he often went on holiday and liked to entertain friends from Paris, often fellow-scientists. He loved the countryside and took a particular interest in plant physiology.

For some years Coulomb had been living with a young woman from Doué, thirty years younger than he was, named Louise Françoise LeProust Desormeaux; it is not known when he married her, but their first son, Charles Augustin II, had been born in 1790 and their second, Henry Louis, in 1797. In Paris they lived on the rue du Chantre, close to the cathedral of Notre Dame. In 1801 he was elected to the largely honorary presidency of the Institut de France and the following year he was appointed by Napoleon to a commission to reconstitute the French educational system, the last public service of his career. In his four years as *Inspecteur-Général de l'Instruction Publique* he participated in the foundation of the new semi-militarized system of French education, reflecting the ideas of Napoleon. He moved to an apartment on the left bank of the Seine, where in June 1806 he con-tracted the fever which led to his death on August 23, 1806.

PIERRE-SIMON LAPLACE (1749–1827)

The early nineteenth century was a golden age for science in France. The mathematical physicist Laplace was one of the leading figures. He sought to reduce physical phenomena to mechanical theories involving particles of matter and the forces acting between them, on the model of physical

astronomy. Laplace and his followers, such as Biot and Poisson, achieved great successes with the corpuscular theory of light and with fluid theories of electricity and magnetism, in which the postulated electrical and magnetic fluids consist of particles acting on one another through short-range forces. Most of all, however, he was renowned for his success in applying Newtonian gravitational theory to problems in celestial mechanics.

Beaumont-en-Auge is a small town in the Calvados district of Normandy, not far from Pont l'Evêque. Pierre Laplace (or La Place) was engaged in farming and was also an official of the local parish. The family of his wife Marie-Anne (née Sochon), who came from nearby Tourgeville, were rather more prosperous farmers. Their second child Pierre-Simon was born on March 28, 1749. Between the ages of seven and sixteen he attended the local Benedictine school, where his paternal uncle Louis Laplace was a teacher. Apart from this uncle, who was interested in mathematics, there is no sign of intellectual distinction on either side of the family.

Laplace's father wanted him to make a career in the Church, and in 1766 the young man entered the University of Caen for theological training. His mathematical interests were already apparent, however, and he was encouraged in this by his teachers. In 1768, at the age of nineteen, he went to Paris with a letter of recommendation to d'Alembert. The story of how d'Alembert gave Laplace difficult mathematical problems to solve as a test of his ability has often been told: the young man was able to solve them overnight. Much impressed, d'Alembert used his influence to secure Laplace a position at the Ecole Militaire in Paris, teaching the cadets elementary mathematics. This lasted for the next seven years.

The immediate aim of Laplace was to become a member of the Paris Academy. So he began to submit research papers to the permanent secretary, who wrote afterwards that he had never before received in so short a time so many important papers on such varied and difficult topics from such a young man. Laplace was first proposed for election in 1771. After two unsuccessful attempts, he achieved adjoint membership in March 1773, at the early age of twenty-four.

Laplace's first contributions to mathematics involved using integral calculus to solve difference equations, but his main interest lay elsewhere. He was a great admirer of the work of Newton and in 1774, having come of age intellectually, he set himself the task of perfecting the Newtonian world-picture. The fundamental problem, the long-term stability of the solar system, is an unresolved question even today. Laplace investigated the outstanding problems of celestial mechanics and, in a remarkable series of memoirs written between 1783 and 1786, he resolved many of them. For example, it had been observed that the orbit of the planet Jupiter was shrinking and the orbit of Saturn expanding; he demonstrated that the orbital eccentricities are self-correcting and that the mean motions of the planets are invariable. He also explained the acceleration of the moon around the earth, the perturbations produced in the motion of the planets by their satellites and the orbits of comets.

During these fruitful years Laplace collaborated with Lavoisier on important physical and chemical experimental work. He made a study of the vital statistics of Paris and used probabilistic techniques to estimate the population of France. As well as teaching at the Ecole Militaire he was appointed examiner of the artillery cadets; one of his first examinees was the future Emperor. Significantly Monge, just three years his senior, was the examiner of naval cadets at this time, and the experience they each gained in their respective capacities stood them in good stead when, later on, they cooperated in the planning of the Ecole Polytechnique.

Little is known of Laplace's private life at this time, but there are clear signs that d'Alembert was beginning to resent the way his protégé was superseding his own work in rational mechanics. Although Laplace had extensive knowledge of other sciences, in the Academy he wanted to pronounce on everything and already he was developing a reputation for arrogance, which made him unpopular with his fellow academicians. A visitor to Paris wrote 'I have seen much of M. de la Place, he is amiable and a great geometer, but he is tremendously punctilious and hasty, he hardly listens to anyone but himself.'

In 1788 Laplace married Marie-Charlotte de Courty de Romanges, whose family came from Besançon. She was twenty years younger than he was. They had two children, a son, Charles-Emile, who pursued a military career, ending up with the rank of general, and a daughter, Sophie-Suzanne, who married into the nobility but died in childbirth at the age of twenty-five.

The rationalization of the chaotic system of weights and measures in use in different parts of France had been under discussion for years. In May 1790 the Revolutionary Government charged the Academy with the task of making recommendations for reform and Laplace, together with Lagrange and Monge, served on the commission appointed to deal with the problem. They made recommendations for units of length, area, volume and mass, with decimal subdivisions and multiples. They also proposed decimal systems for money, angles and the calendar. It was at Laplace's suggestion that the basic unit of length was named the metre. The Revolutionary calendar lasted only a short while, even in France, and the decimalization of angle measurements was not generally accepted, but the other metric units were gradually adopted around the world.

In 1793 Laplace and others were expelled from the Académie des Sciences on political grounds, just before all the Academies were suppressed. To avoid the Terror, he took the precaution of moving his family from Paris to the nearby town of Melun. By 1795 the danger appeared to be over. The Thermidorean regime had placed the introduction of the metric system and all matters pertaining to navigation and official astronomy under the administration of the new Bureau des Longitudes. Membership of the well-funded Bureau was regarded as a full-time occupation and rewarded liberally. Laplace served regularly as a member, often used its meetings as the forum in which to present appropriate papers and published frequently in its journal, the *Con- naissance des temps*, which originally was just a nautical almanac.

In 1796, Laplace published his popular scientific classic, the *Exposition du système du monde* (*Explanation of the Solar System*), which eschews the use of mathematical formulae. In particular he discussed the nebular hypothesis (that the solar system condensed from a cloud of rotating gas) which had been proposed by the philosopher Immanuel Kant, albeit without mathematical underpinning. It is not known whether Laplace knew this; he seldom gave acknowledgements. Moreover, Laplace is even thought by some to have foreseen the concept of the black hole, whose characteristics were deduced much later in Einstein's general theory of relativity. The *Exposition* contains this statement of Laplace's general point of view:

The algebraic analysis soon makes us forget the main object [*of our researches*] by focusing our attention on abstract combinations and it is only at the end that we return to the original objective. But in abandoning oneself to the operations of analysis, one is led to the generality of this method and the inestimable advantage of transforming the reasoning by mechanical procedures to results often inaccessible by geometry. Such is the fecundity of the analysis that it suffices to translate into this universal language particular truths in order to see emerge from their very expression a multitude of new and unexpected truths. No other language has the capacity for the elegance that arises from a long sequence of expressions linked one to the other and all stemming from one fundamental idea. Therefore the geometers [*mathematicians*] of this century convinced of its superiority have applied themselves primarily to extending its domain and pushing back its bounds.

The mathematical arguments supporting Laplace's theories were kept for his masterpiece, the monumental five-volume *Traité de mécanique céleste* (*Treatise on Celestial Mechanics*), which was published between 1799 and 1825. In it he completed Newton's work in this field. Whereas Newton believed that divine intervention would be necessary in order to 'reset' the solar system periodically, Laplace argued that the law of universal gravitation implied its long-term stability. When Napoleon observed that Laplace's voluminous treatise did not mention God as creator of the universe, Laplace is said to have replied 'Sir, I do not have need of that hypothesis', and when Napoleon repeated this to Lagrange, the latter remarked 'Ah, but that is a fine hypothesis. It explains so many things.' When Laplace was finishing the *Mécanique céleste* he wrote to his correspondent Mary Somerville in London that the task had forced him to reread 'with particular attention the incomparable *Principia* of Newton, which contains the germ of all his investigations. The more I study this work the more I admire it, it takes me back above all to the time when it was published. But at the same time as I felt the elegance of the synthetic method by which Newton has presented his discoveries, I have recognized the unavoidable necessity of analysis to fathom questions which can only be dealt with superficially by his synthesis. I see with great pleasure that your mathematicians now devote themselves to analysis, and I no longer doubt that in following this method with the peculiar wisdom of your nation, they will be led to make important discoveries.'

Although the main field of Laplace's research was celestial mechanics, he also made important contributions to the theory of probability and statistical inference. In his *Théorie analytique des probabilités* (*Analytical Theory of Probability*) of 1812 he summarized, in a masterly introduction, all that was then known in the area of probability and its applications. This work introduced the technique known later as the Laplace transform, a simple and elegant method of solving integral equations. Some of Laplace's contributions to the theory of probability were derived from questions in astronomy, for example the central-limit theorem as applied to the inclination of the orbits of comets. Laplace believed that, through probability, mathematics could be brought to bear on the social sciences and suggested

various applications. With his fellow academician Lavoisier he also worked
on various problems in physics, including thermal conductivity and capillary action. He is best known for his concept of potential, as an analytical
device, which proved to be invaluable in such a wide range of subjects, as
gravitation, electromagnetism, acoustics and hydrodynamics.

He seemed to have had the whole literature of the exact sciences at his
fingertips. In his own work, however, he frequently neglected to acknowledge the sources of his results and left the impression that they were his
own when they were not.

Laplace considered himself the best mathematician in France; his colleagues thought that, although this might well be true, a little modesty
would not have come amiss. There was great amusement when it emerged
that, due to administrative error, the proud Laplace had been enrolled as a
student rather than a professor at the short-lived Ecole Normale. If anything
his arrogance increased as he grew older. After he had become successful
he ignored his parents and elder sister. That Laplace was vain and selfish
can hardly be denied; his behaviour towards the benefactors of his youth and
his political friends was mean and ungrateful, while his appropriation of the
results of those who were comparatively unknown appears well-established.

After his death an anonymous critic wrote that 'The genius of Laplace
was a perfect sledgehammer in bursting purely mathematical obstacles, but
like that useful instrument, it gave neither finish nor beauty to the results . . .
nevertheless, Laplace never attempted the investigation of a subject without leaving upon it the marks of difficulties conquered: sometimes clumsily,
sometimes indirectly, but still his end is obtained and the difficulty is conquered.'

From the late eighteenth century until well into the nineteenth
Laplace dominated the Paris Academy, imposing his scientific preferences
and deterministic ideology on younger colleagues. When the suppressed
academies were revived, as part of the Institut de France, Laplace was elected
vice-president of the new Académie des Sciences at the organizational meeting in December 1795 and five months later became president. That office
was more than honorary; it ensured that he was a member, frequently chairman, of the numerous committees where policies were formulated and
decisions pre-empted. Laplace threw himself into this political work with
ability and enthusiasm. He now presided over the Bureau des Longitudes. He
taught at the Ecole Normale when it was opened briefly in 1795 and in 1800
was instrumental in forming the constitution of the Ecole Polytechnique,
where he served as a graduation examiner. Laplace was known for the
'rapidity' of his teaching and in his writing was notorious for his frequent use

of the phrase 'it is easy to see', by which he skipped steps in his exposition, confounding some of his later readers.

This was a period when prominent scientists were being increasingly called upon to undertake various forms of public service. Laplace was one of the most prominent. In 1799 Napoleon, by this time First Consul, appointed him Minister of the Interior, but he was not a success in this office and after six weeks Napoleon replaced him by his brother Lucien Bonaparte. As the former Emperor remarked after he had been exiled to St Helena, 'a mathematician of the first rank, Laplace quickly revealed himself as only a mediocre administrator . . . Laplace could never get a grasp on any question of its true significance, he sought everywhere for subtleties, had only problematic ideas, and in short carried the spirit of the infinitesimally small into administration'. Even so, under the Empire, Laplace became a senator and held the office of Chancellor, as the result of which he became wealthy. He was made a count of the Empire and decorated with France's highest honours, the Grand Cross of the Legion of Honour and the Order of the Reunion. There is a portrait bust of him by Houdon in the Institut de France.

From 1806 Laplace spent much of his time on his country estate at the village of Arcueil, about five miles south of the city of Paris. The great chemist Berthollet owned an adjoining estate. The two wealthy academicians organized a small, informal society for scientific research at Arcueil, and much good work was done by this Société d'Arcueil in its early years. Promising young scientists from the Polytechnique and elsewhere were encouraged to participate. Naturally the patronage of someone as influential as Laplace was a distinct advantage in career terms.

In the Senate, Laplace voted against the continuation of Napoleon's rule in 1814, supporting Louis XVIII instead. After the restoration of the Bourbon monarchy the following year, he was rewarded with the title of marquis and was appointed president of the committee to oversee the reorganization of the Ecole Polytechnique. In this capacity he reduced the length of the course from three years to two and made it a prerequisite for entry into the more specialist schools known as Ecoles d'Applications, such as the Ecole de Ponts et Chaussées, which had been revived. The emphasis on military engineering was reduced and new courses in history, morality and 'social arithmetic' were introduced. However, mathematics remained central to the course, and throughout most of the nineteenth century the Polytechnique, with professors of the calibre of Ampère, Cauchy, Fourier, Lagrange, Laplace, Legendre, Monge, Poisson and Poncelet, was the best place to study the subject in France or anywhere else.

Laplace's political opportunism allowed him to prosper and continue his scientific work. Yet his power was waning, as new discoveries started to undermine his beliefs and his theories began to be superseded. Laplace found himself increasingly isolated in the scientific community of the Restoration; he remained loyal to the Bourbons for the rest of his life. When the literary Académie Française, of which he was a member, issued a declaration in support of free speech, he refused to sign it.

During his declining years Laplace lived mainly on his estate in Arcueil. In the elegantly furnished mansion, set in a beautiful park, he welcomed many of the scientists who visited Paris and entertained them in style. One of his guests was Mary Somerville, who described him as 'not tall, but thin, upright and rather formal. He was distinguished in his manners and I thought there was a little of the courtier in them, perhaps from having been so much at the court of the first Napoleon.' Humphrey Davy also visited Laplace at Arcueil and described him as 'rather formal and grand in manner, with an air of protection rather than courtesy.' A more detailed picture of life at Arcueil was given on the occasion of a visit by John Dalton in July 1822 by one of his companions:

> At four in the afternoon, by coach with Dalton to Arcueil, La Place's country seat, to dine. Engaged the carriage to wait for our return at nine. On alighting we were conducted through a suite of rooms where in succession, dinner, dessert, and coffee tables were set out; and onwards through a large hall, upon a terrace, commanding an extent of gardens and pleasure grounds . . . as yet we had seen no-one, when part of the company came into view at a distance; a gentleman of advanced years and two young men . . . we approached this group, when the elderly gentleman took off his hat and advanced to give his hand to Dalton. It was Berthollet. The two younger were La Place's son and the astronomer royal, Arago. Climbing some steps upon a long avenue we saw at a distance La Place walking uncovered with Madame Biot on his arm; and Biot, Fourier and Courtois, father of the Marchioness of La Place. At the front of the house this lady and her granddaughter met us.
>
>   At dinner Dalton was on the right hand of Madame La Place and Berthollet on her left, etc. conversation on the zodiac of Denderah and Egypt, Berthollet and Fourier having been there with Napoleon . . . after dinner abroad in the beautiful grounds . . . Dalton walking with La Place on one side, and Berthollet on the other.

Laplace possessed a remarkably good memory, which he retained to an advanced age. As Fourier said, 'he has not cultivated the fine arts but he appreciated them. He was fond of Italian music and of the poetry of Racine, and he took pleasure in quoting from memory various passages of this great poet. Paintings attributed to Raphael adorned his apartments and they were found besides portraits of Descartes, François Vieta, Newton, Galileo and Euler.' After a generally healthy and vigorous life, he died after a short illness on March 5, 1827 at Arcueil, just before his seventy-seventh birthday. The date, it was noted, was almost exactly a century after the death of Newton, the scientist to whom he was so often compared. Berthollet, his partner in the Société, had died five years previously. Initially Laplace was buried in Paris, at the cemetery of Père Lachaise. In 1878 the monument erected to him was moved from there to his birthplace of Beaumont-en-Auge, and ten years later his remains were transferred from Père Lachaise to the family estate of St Julien de Mailloc, where they were re-interred with the remains of his wife and children. The marquise, who lived until 1862, endowed a fund to allow the highest-ranking student in each year of the Ecole Polytechnique to be given a complete set of the works of her late husband. Unfortunately, many of his unpublished papers were destroyed by a fire at the Château de Mailloc in 1925, which was then owned by his great-great-grandson, the Count of Colbert-Laplace.

3 From Rumford to Oersted

Our third five remarkable physicists were born in the twenty-five years from 1753 to 1777. Two came from France and one from each of America, England and Denmark.

Sir Benjamin Thompson (Count Rumford) (1753–1814)

We return to America for our next profile. It is impossible to give a full account of all the twists and turns of the picaresque life of Benjamin Thompson in just a few pages. A convenient starting-point, for our purposes, might be his arrival in London in the summer of 1776, when he was already twenty-three; except that the early years of his life are significant for what came later, so at least an outline is essential. Like Benjamin Franklin he was born in New England, of humble parents, and spent part of his career in America, part in Europe. However, Thompson and Franklin had very different characters and whereas Franklin became a leading American patriot, Thompson did not.

The future Sir Benjamin Thompson, Count Rumford, was born Benjamin Thompson at Woburn, Massachusetts, on March 26, 1753; his parents Benjamin and Ruth (née Simonds) were farmers in a small way. Owing to a boundary dispute, the same place was also known as Rumford (now Concord), New Hampshire, whence the title he took later on in life. In boyhood he early showed an inclination towards science. He was naturally skilful in drawing, lettering and mechanical techniques, as well as ingenious in the arrangement of the meagre apparatus at his disposal, and he devised experiments in electrostatics, chemistry and mechanics. His interest in the laws of nature extended to astronomy and, at the age of fourteen, he was able to predict the time of occurrence of an eclipse. From a local doctor with whom he boarded for a while he learned the rudiments of medicine and surgery. Through the acquaintance of a mutual friend, Loammi Baldwin, late of the colonial forces in the French and Indian War, young Thompson attended lectures at Harvard College given by the celebrated professor John Winthrop. Thompson and Baldwin learned from each other, in a self-improvement society of two. They tried to repeat Franklin's experiment of

flying a kite during a thunderstorm but merely 'felt a general weakness in their joints and limbs and a kind of lifeless feeling'. Baldwin went on to become a prominent civil engineer, but his friend's career developed quite differently.

First Thompson made an advantageous marriage with a wealthy widow, Sarah Walker Rolfe, who was considerably older than himself. They had a daughter, also called Sarah, of whom more later. His wife introduced him into her social circle, including the Royal Governor of New Hampshire, on whom he created such an impression that he was immediately offered a commission as major in the 2nd New Hampshire Regiment. It is thought that this must have been on the understanding that he would provide intelligence for the British forces in the War of Independence. However Thompson spent the next year or so, he thought it prudent to leave the Boston area when the city was taken by George Washington. Before long he was on his way to England, having left his wife and daughter behind; he made no enquiries after his wife for the next thirty years and none after his daughter for the next twenty. He contrived to keep his whereabouts secret from them so that the only way they could get in touch with him was through Baldwin.

When he arrived in London in the summer of 1776 the outcome of the War of Independence was still in doubt and many American supporters of the British side were in London expecting that a grateful nation would provide for them until they could safely return home. They were generally disappointed in this, but not Thompson. He had a way of furthering his fortunes by associating with those in power and before long he had become a protégé of Lord George Germain. Germain had been out of favour with the King and had allied himself with the Prince of Wales. When the Prince succeeded to the throne he rewarded Germain by making him Secretary of State for the Colonies, with responsibility for ensuring that the colonial rebellions were suppressed. Naturally Thompson, as his number two, was himself in a position to acquire power and influence. He secured the impor-tant, and lucrative, task of equipping and victualling large numbers of British troops.

Meanwhile he was developing a reputation in other ways. His inter-ests extended to military affairs, social reforms, animal husbandry and hor-ticulture; to each of these fields he applied the rigour of the experimental scientist. From the very first months of his residence in England he used his powers of observation to devise ways and means for the promotion of human comfort and safety; in the economical use of fuel, the better designing of chimneys and the more efficient use of firearms. This was technology rather than science, but he was applying his scientific knowledge, as he was later when he conducted experiments on ballistics. It was these activities that led to his election to the Royal Society in 1780, at the age of twenty-seven.

After Germain had fallen from power, Thompson, never backward in looking after himself, decided to move on. Sensing that the British cause was lost in America, he returned to New York, where he was temporarily given command of a Republican regiment. After returning to London Thompson decided, on the strength of some limited military experience acquired in America, to purchase a colonelcy; at this time commissions in the British army were available for purchase, although influence was also useful. He then set out for the continent in search of employment as a soldier of fortune. Karl Theodor, the Elector of Bavaria, was an old-style ruler who was looking for a scientist to ornament his court, and, when Colonel Thompson presented letters of introduction and offered his services, the Elector decided that he might be valuable both as a scientist and in a mil-itary capacity. Permission from London for him to enter the service of the Elector being forthcoming, together with a quite unexpected knighthood, Sir Benjamin, as he was now entitled to call himself, set out to make himself useful.

Among the services he provided for Karl Theodor was the clearance of beggars from the streets of the capital. Bavaria was a poor country and poverty was widespread. Rumford's solution was to use the army to force the beggars into purpose-built workhouses, where they were confined and set to work making military uniforms in return for their subsistence. This was characteristic of Rumford's attitude to social problems. Another service was the construction of the 'English Garden', a popular feature of Munich to this day. This was an attractive park laid out along the Neckar by the army for the Elector but open to the public. The grateful Karl Theodor heaped appointments on him; by 1790 he was privy councillor, major general of the cavalry and adjutant general; two years later he was also lieutenant general of the artillery, chief of staff and chief of the general staff. In 1791 the Elector conferred on him the title of Count, amongst other honours, and from now on we may refer to him as Count Rumford.

In Munich Thompson had two mistresses, who were sisters. The older, Countess Baumgarten, was plump and buxom, a celebrated beauty whose concurrent intimacy with Karl Theodor proved most useful to Thompson. She bore him a daughter, Sophia, who was raised as a Baumgarten child. The younger sister was Countess Nogarola. She was a slim, athletic-looking woman with a strong intelligent face, who was not particularly beautiful but whose keen mind and intellectual interests were quite a match for Thompson's own. They were intimate friends for many years. She helped him with his correspondence in French and German and translated some of his writings into Italian.

Meanwhile Rumford still found time for scientific work. The theory of heat and its applications were his speciality; and he helped to undermine the old caloric theory. His most important discovery was that a limitless amount of heat can be produced by friction, which was conclusively demonstrated in the cannon foundry of Munich, where the boring of gun barrels produced large quantities of heat. He was a careful and methodical experimenter, although the far-reaching implications of the results he had obtained were not fully appreciated at the time.

In March 1793 London heard that 'General Count Rumford' had obtained leave from the Elector to go to Italy for the sake of his health. Just before leaving he heard from his American daughter Sarah that her mother had died and Baldwin had revealed her father's whereabouts to her. A young American then arrived in Munich to ask for Sarah's hand in marriage; he was promptly sent away. Rumford then went off to Italy, first to Milan and Pisa, then to Pavia, where he had some scientific contact with Volta, and finally east to Verona, where his beloved Countess Nogarola lived with her children. After some months in that area he went down to Florence, where he had an affair with Lady Palmerston, who was then on the Grand Tour with her husband, and met her again in the south of Italy. After a year away he began the return journey, stopping for several months with Countess Nogarola en route.

After a year back in Munich Rumford found that life there was becoming increasingly unpleasant. As his political power grew, so did persistent rumours of mismanagement and self-aggrandizement; he reacted in a most vindictive manner to those who opposed him. He persuaded the Elector to grant him six months' leave to go back to London to oversee the publication of his new scientific work. He set out in October 1795, but, when he arrived in London, a large trunk, containing 'almost all the papers of any consequence I possess in the world', was removed from his carriage by thieves. Although he lost all his scientific papers, probably what the thieves were after was his private papers, which might have been useful to his enemies. Rumford soon discovered that he was a political and social outcast in London. He settled down to rebuilding his reputation in the capital and publishing his scientific work. He wanted someone to look after him and thought of his long-neglected daughter Sarah, with whom he had become reconciled. He invited her to come over and join him. How he thought a daughter he had abandoned years before in a New England village could suddenly be transported to the fashionable salons of London and be an ornament to society is hard to understand, but, if this is what he expected, Sarah was a disappointment to him, as he was to Sarah. After a month he put her into a fashionable girls' school to learn the manners and etiquette of London society.

Meanwhile Karl Theodor's troubles had escalated. To go back some years, when the Elector Maximilian II of Bavaria died in 1777, the junior line of the Wittelsbachs, which had ruled that country for four and a half centuries, became extinct. As Elector Palatine of the Rhine, Karl Theodor, head of the family's senior branch, was expected by virtue of arrangements of long standing to inherit the Bavarian electorate and thereby reunite the two Wittelsbach fiefs. This was acceptable to Austria only on conditions that were unacceptable to Frederick the Great in Berlin, but in the end Karl Theodor succeeded to the Munich throne and Bavaria lost some territory to Austria. By 1792, however, Karl Theodor found himself threatened not only by Austria, which remained dissatisfied, but also by the French revolutionary armies. The latter overran the Palatinate and in 1795 invaded Bavaria itself. The Elector, trapped between the forces of France and Austria, appealed to Rumford to return from London and prepared to take refuge in Saxony.

Rumford arrived, accompanied by his daughter Sarah, just before Karl Theodor fled to Dresden, and began by rallying the Bavarian troops. Thanks to his knowledge of large-scale cooking and to his improvements in the design of boilers and stoves, as well as his skill in conserving fuel, Rumford was able to provide cheaply and comfortably for the large numbers of soldiers, at first scattered and disorganized, who had converged on Munich to be lodged and fed. When the Austrians tried to use Munich as a base from which to resist the French, Rumford, with his customary firmness, tact and presence of mind, managed to persuade both the invading armies to refrain from molesting that city. This was no small achievement.

Rumford had served Bavaria in many capacities. On his return to Munich, Karl Theodor offered him one more appointment, as head of the General Police, but Rumford preferred to go back to Britain. The Elector sought to honour him by designating him as his ambassador to the Court of St James, and, under this supposed assignment, Rumford returned to London in 1798. However, the count, as Sir Benjamin Thompson, was already a British subject and could not therefore serve as a representative of a foreign power, be it ever so friendly.

Rumford was by this time a wealthy man. After a long absence he was able to mingle once again with his many friends and admirers in the scientific world of the city, drawing their attention to the measures of public and domestic economy which had made him famous in Germany. He was welcomed on all sides for his ingenuity and philanthropy. Both the Royal Academy and the Society for the Encouragement of Arts and Manufactures (later the Royal Society of Arts) elected him to honorary membership. In 1792 he had been awarded the Copley medal by the Royal Society for 'various papers on the properties and method of communication of heat'. Possibly through a desire to emulate Sir Godfrey Copley and hence perpetuate his own name in the British scientific world, Rumford wrote to the Royal Society in July 1796 offering 'a fund of one thousand pounds, the interest on which would be spent every second year as a premium to the author of the most important discovery or improvement in the subjects of heat and light, especially those applications which would contribute most to the good of mankind'. The offer was accepted and the dies for a Rumford medal ordered and made. Rumford himself received the first such medal in 1802.

Although the impossibility of serving as the Bavarian ambassador in London was at first a severe blow to Rumford, he did not let himself remain idle for long. His active mind turned towards a project that would afford at once activity and, he hoped, advancement. His interests now centred themselves on a philanthropic organization that was at that time attracting public notice, the Society for Bettering the Condition and Increasing the Comforts of the Poor. One of its leading members was William Wilberforce, anti-slavery leader and prominent evangelical layman in the Church of England. What this praiseworthy society was trying to do for the lower orders of population by lessening their misery in one way should, in the opinion of Rumford, be augmented by efforts towards increasing the efficiency of their labours, for example by improving their mechanical skills.

Acting with the support of the society, Rumford drew up plans 'for forming by subscription, in the metropolis of the British Empire, a public institution for diffusing the knowledge and facilitating the general introduction of useful mechanical inventions and improvements, and for teaching by courses of philosophical lectures and experiments the application of science to the common purposes of life'. A committee chaired by Sir Joseph Banks was formed, and the outcome was the Royal Institution of Great Britain, which was incorporated on January 13, 1800 and granted a Royal Charter by George III. Since the Royal Society had no laboratory facilities itself, it became customary for its fellows to turn to the Royal Institution if there were experiments they wished to see performed. Later it was often described as 'the workshop of the Royal Society'.

With characteristic enthusiasm and passion for details, Rumford threw himself into the organization of the new institution. A building in Albemarle Street was purchased and remodelled according to his plans – complete with lecture room, repository for models, workshops and kitchens. In order that possible patrons might not feel left out, Rumford engaged professors and lecturers to provide scientific instruction for those of the upper classes who might care to keep abreast of the times. Committees were set up to suggest and supervise research in science and the useful arts. The programme included a school for mechanics and a print shop for the publication of research articles. It seemed as though every one of the count's gamut of interests, save the military, was wrapped up in this new socio-scientific-philanthropic institution

For the educational programme the services of Thomas Young were secured; his profile is given later. Unfortunately he was too erudite to make a good lecturer for popular audiences, but his colleague Humphrey Davy made up for this. Davy, who was already renowned for his discovery of the benefits of nitrous oxide as an anaesthetic, was appointed assistant lecturer in chemistry, director of the laboratory and assistant editor of the publications of the Institution. He was allowed to occupy a room in the house, which was furnished with coals and candles, and paid one hundred guineas per annum. His lectures drew large crowds, for he had a natural gift for making science interesting to the layman and could hold his audiences by the infection of his own enthusiasm. What he started soon became the chief characteristic of the Royal Institution – a well-balanced combination of scientific research and the exposition of its results in the lecture theatre.

Unfortunately the entire scheme was altogether too ambitious. The institution could not be financially self-supporting, at least not in its early days. Rumford, with his care for details and rigorous attitude towards the fulfilment of his plans, could not cope with the situation he had created. Once he was out of the country, some of his pet schemes were abandoned on the grounds of economy. The school for mechanics, the culinary demonstrations and the workshops for model-makers were soon discarded. On reflection, education of the masses seemed less important than the popularization of science among the upper classes. Of course all strata of the population were welcome to visit the rooms and workshops, but the times were not ripe for the establishment of trade schools.

When Rumford returned to Munich, he was pleased to see his natural daughter Sophia Baumgarten had grown into a charming young lady. However, the pleasures of his stay in Bavaria were eclipsed by his flattering reception in Paris, his next port of call. He was presented to Napoleon as a Bavarian general; the First Consul took an immediate liking to him, showering him with special attention. He stayed altogether two months in the French capital, sending reports to London and Munich on politics, to Lady Palmerston on the social life and to Sir Joseph Banks on science and scientists. As for Napoleon: 'Bonaparte is a person endowed with very uncommon abilities. And the more I see of him, and the more I consider the wonderful things he has done, and is doing, the more I am disposed to admire his genius, and to give credit to the wisdom of his plans, and the purity of his intentions.' Napoleon made great use of natural philosophers, and Rumford as a famous visiting scientist found that he had an immediate entrée to the highest political circles. He met Lafayette and Talleyrand, dined with Berthollet, was introduced to Laplace, attended a session of the Paris Academy and was elected to membership in two different classes, namely the one dealing with mathematics and physics and the one dealing with political economy. Most of all he was fêted by the Parisian ladies.

When he returned to London Rumford found that the king ignored him, while Lady Palmerston, offended by his glowing accounts of the ladies of Paris, would not receive him. He tried to sort out the problems at the

Royal Institution, which was already in financial difficulties. Unfortunately differences kept arising between Rumford and the management of the institution, which became so acute that he left London in May 1802, vowing never to return. He again went first to Munich and then on to France. By this time Napoleon's campaigns were redrawing the map of Europe, and England was targeted. Napoleon no longer singled him out as a person of importance, never once spoke to him privately.

It was quite natural that the brilliant salons which Madame Lavoisier, the widow of the great chemist, held regularly at her home in Paris would attract a scientist of Rumford's stature. A veritable galaxy of scholarship was represented at these gatherings, not only French scientists such as Arago, Berthollet, Biot, Lagrange and Laplace, but also scientists of other nationalities, such as Charles Blagden, who courted her unsuccessfully, and the famous Baron von Humboldt. Rumford's international reputation, together with his courteous manner and interests so akin to those of her late husband, could not fail to impress the vivacious and yet serious-minded lady. She was no doubt impressed by her new guest's account of his work for the poor of Bavaria and his improvement of the living conditions of the indigent of London.

Marie Anne Pierrette Paulze had married Lavoisier when she was thirteen years old. For the twenty-three years they lived together she acted as his research assistant, librarian, collaborator and scientific confidante. She was a gracious hostess to their constant flow of visitors, both French and foreign, and, being the better linguist, translated papers for him. She knew enough chemistry to edit and publish his great *Mémoires de chimie*, which had been left in a very fragmented form at the time he was guillotined for being farmer-general in the oppressive taxation system. She continued to be famous for her social events: a formal dinner party every Monday, an 'at home' every Tuesday, a musical soirée every Friday. She was now 43 and had been a widow for nine years, Rumford was 50 and had been a widower for eleven. Before long their marriage was celebrated.

Sarah, who had returned to America in 1798, had not been told by her father of this impending development until he wrote to her early in 1804: 'I shall withhold this information from you no longer. I really do think of marrying, though I am not yet absolutely determined on matrimony. I made the acquaintance of this very amiable widow in Paris, who, I believe, would have no objection to having me for a husband, and whom in all respects would be a proper match for me. She is a widow, without children, never having had any; is about my own age, she enjoys good health, is very pleasant in society, has a handsome fortune at her own disposal, enjoys a most respectable reputation, keeps a good house, which is frequented by the first philosophers and men of eminence in the science and literature of the age, or rather of Paris. And what is more than all the rest, she is goodness itself . . . she has been very handsome in her day, and even now at 46 or 48 is not bad looking; of a middling size, but rather *en bon point* than thin. She has a great deal of vivacity and writes incomparably well.'

Rumford made another brief visit to Munich, where little had changed. Madame Lavoisier came out to join him and together they returned to Paris via Switzerland. She went on ahead of him to ensure that he would be allowed into France, because by this time England and France were at war again. They were married on October 24, 1805 and settled into a house in the rue d'Anjou close to the Tuileries Gardens and the Champs Elysées. It cost 3000 guineas and was set in a beautiful two-acre garden. She decided that she wished to be known as Countess Lavoisier de Rumford. Rumford seemed to be very happy and he wrote to Sarah saying that 'he had the best of pinned hopes of passing my days in peace and quiet in this paradise of a place'. However, before long she learnt that all was not well with the new marriage. Although they had known each other for almost five years, they soon discovered that they were incompatible and started arguing about almost everything. The countess liked entertainment and small-talk, but he preferred quiet contemplation and experiment. She liked good food and wine, but his stomach could not take it, so he generally sat at a separate small table when they had guests. He loved music but she did not care for it. Most of all she objected to the way he kept altering the house and its contents to suit his own tastes without any reference to hers.

Two months after the marriage he was writing to Sarah that, 'between you and myself, as a family secret, I am not at all sure that certain persons were not wholly mistaken in their marriage, as to each other's characters'. At the end of the first year: 'very likely she is as much disaffected towards me as I am towards her. Little it matters with me, but I call her the female dragon.' On the second anniversary: 'I am still here, and so far from things getting better they get worse every day. We are more violent and more open in our quarrels'. Things went from bad to worse. In April 1808 he wrote to Sarah, describing his wife as the most imperious, tyrannical and unfeeling woman that had ever existed, whose perseverance in pursuing an object was equal to her profound cunning and wickedness. He said that he could not call her a lady, that it was impossible to continue and spoke of separation.

Rumford had tired of social events and talking with people when he would rather have been working on his essays or in his garden. She was annoyed by his ceaseless alterations to their house, and there was friction between the German servants he brought from Bavaria and her own, who were French. Although they tried to make a success of the marriage, after nearly four years of endless quarrels, some violent, it was annulled by mutual consent. Countess Lavoisier de Rumford returned to her round of salons while the count retreated to a property at Auteuil near the Bois de Boulogne. Here he resided for the remainder of his days, devoting his energies to arranging, beautifying and improving the grounds. He paid a final brief visit to Munich, where he learned that Countess Nogarola had died and Sophia, his illegitimate child by her sister, was very ill. He summoned Sarah to Paris to keep him company. She arrived late in 1807, bringing the news that his old friend Baldwin had died. She was pleased to see her father so contented with his fine garden, string of lively horses and songbirds in the dining room. He had renewed his scientific interests and began to frequent the Paris Academy again. He was still on speaking terms with his ex-wife, whom Sarah found very charming: 'it was a fine match, could they but have agreed'.

Sarah was with her father at the last, acting as his hostess when he invited some of his old associates in the sciences to dine with him and discuss some of the problems at the Royal Institution. There was also another mistress named Victoire Lefèvre, who bore him a son, named Charles François Robert Lefèvre, and it was his son, Rumford's grandson, who succeeded to the title in due course. Rumford appeared to be in good health, but on August 21, 1814 he died suddenly of 'a nervous fever'. The funeral, which took place at the cemetery at Auteuil three days later, was a lonely affair with only a handful of people at the graveside. He left the residue of his estate to Harvard University to establish a Rumford Professorship, which still exists. Its purpose is 'to teach by regular courses of academical and public lectures, accompanied by proper experiments, the utility of the physical and mathematical sciences for the improvement of the useful arts, and for the extension of the industry, prosperity, happiness and well-being of society'.

In his declining years Rumford had lost most of his old friends and made hardly any new ones. The generally accepted view is that this was because he was an insufferable genius who treated people with whom he had to work with such disdain and contempt that he simply made more and more enemies as he grew older. One commentator wrote that 'he was utterly devoid of humour and humanism; hard, brittle and self-centred to the last'. Another wrote that he was 'unbelievably cold-blooded, inherently egotistic and a snob', and yet another that 'he was the most unpleasant personality in the whole of science since Isaac Newton'.

After her father's death Sarah stayed on at Auteuil until May 1815 but thereafter moved between London and Paris until she returned to Concord to end her days and died in 1852 in the very room where she had been born seventy-nine years before. When asked about her father she would say 'he was fond of having his own way, even, as I fancied, to despite me'. 'He could go one way or the other. And it was invariably the case, that when quiet and happy himself he was like others or, in other words, agreeable; but when perplexed with cares or business, or much occupied, there was no living with him.'

JEAN-BAPTISTE FOURIER (1768–1830)

Jean-Baptiste-Joseph Fourier, to give him his full name, is the most illustrious son of Auxerre, the principal city of western Burgundy, where he was born on March 21, 1768. Both his father Joseph, a master tailor originally from Lorraine, and his mother Edmie died before he was ten years old. Fortunately certain local citizens took an interest in the orphaned boy's education and secured him a place in the progressive Ecole Royale

Militaire, one of a number of such schools run by the Benedictine and other
monastic orders. Science and mathematics were taught there, among other
subjects, and, while the boy displayed all-round ability, he had a special gift
for mathematics. He went on from there to complete his studies in Paris at
the Collège Montagu. His aim was to join either the artillery or the engineers, the branches of the army supposedly open to all classes of society,
but when he applied he was turned down. Although he could have been
rejected on medical grounds, the reason given by the minister was that only
candidates of noble birth were acceptable.

After this setback Fourier embarked on a career in the church. He
became a novice at the famous Benedictine Abbey of St Benoît-sur-Loir,
where he was called on to teach elementary mathematics to other novices.
After taking monastic vows he became known as Abbé (Father or Reverend)
Fourier, but instead of pursuing a career in the church he returned to Auxerre
to teach at the Ecole Militaire. By this time he was twenty-one and had
already read a research paper at a meeting of the Paris Academy.

During the first tempestuous years of the Revolution, Fourier was
prominent in local affairs. His courageous defence of victims of the Terror
led to his arrest by order of the notorious Committee for Public Safety
in 1794. A personal appeal to Robespierre was unsuccessful, but he was
released after Robespierre himself had been guillotined. Fourier then went
as a student to the short-lived Ecole Normale Supèrieure. The innovative
teaching methods which had been introduced there made a strong impression on him and it gave him the opportunity to meet some of the foremost
scientists of the day, including Lagrange, Laplace and Monge. The next year,
when the Ecole Polytechnique opened its doors, under its original name of
the Ecole Centrale des Travaux Publiques, Fourier was appointed assistant
lecturer to back up the teaching of Lagrange and Monge. However, before
long he fell victim to the forces of reaction and was arrested again. He had
an anxious time in prison but his colleagues at the Ecole successfully sought
his release.

In 1798 Fourier was selected to join an expedition to an undisclosed
destination. This proved to be Napoleon's Egyptian adventure. Once the
military objectives had been secured, Berthollet and Monge established an
Egyptian Institute in Cairo, of which Fourier was made permanent secretary,
and the cultural arm of the expedition set to work studying the antiquities,
some of which were appropriated. On top of this activity Fourier was also
entrusted with some negotiations of a diplomatic nature and he even found
time to think about mathematics. He proposed that a report on the work

of the Egyptian Institute be published and, on his return to France, was consulted regarding its organization and deputed to write a historical preface describing the rediscovery of the wonders of the ancient civilization. When the *Description de l'Egypte* was published, Fourier's elegant preface, somewhat edited by Napoleon, appeared at the front of it.

Meanwhile Fourier had resumed his work at the Ecole Polytechnique. Before long Napoleon, who had been impressed by his capacity for administration, decided to appoint him prefect of the Département of Isère, based at Grenoble and extending to what was then the Italian border. The office of prefect was a demanding one, but it was during this period that Fourier wrote his classic monograph on diffusion of heat entitled *Théorie analytique de la chaleur* (*On the Propagation of Heat in Solid Bodies*) and presented it to the Paris Academy in 1807. It was refereed by Lagrange, Laplace, Lacroix and Monge. Lagrange was adamant in his rejection of several of its features (especially the central concept of trigonometric or, as we say, Fourier series), so its publication in full was blocked; only an inadequate five-page summary, written by Laplace's protégé Poisson, appeared. Later Fourier received a prize from the academy for the work, but it was not until 1822 that his theory of diffusion of heat was published. To quote from the preface to the *Théorie analytique de la chaleur*, this 'great mathematical poem' as Clerk Maxwell described it:

> First causes are not known to us, but they are subjected to simple and constant laws that can be studied by observation and whose study is the goal of Natural Philosophy . . . Heat penetrates, as does gravity, all the substances of the universe; its rays occupy all regions of space. The aim of our work is to expose the mathematical laws that this element follows . . . But whatever the extent of the mechanical theories, they do not apply at all to the effects of heat. They constitute a special order of phenomena that cannot be explained by principles of movement and of equilibrium . . . The differential equations for the propagation of heat express the most general conditions and reduce physical questions to problems in pure Analysis that is properly the object of the theory.

As prefect, Fourier's administrative achievements included securing the agreement of thirty-seven different communities to the drainage of a huge area of marshland to make valuable agricultural land and the planning of a spectacular highway between Grenoble and Turin, of which only the

French section was built. Napoleon conferred on him the title of baron, in recognition of his excellent work as prefect.

Fourier was still at Grenoble in 1814 when Napoleon fell from power. The city happened to be directly on the route of the party escorting the Emperor from Paris to the south and thence to Elba; to avoid an embarrassing encounter with his former chief, Fourier negotiated a detour in the route. No such detour was possible when Napoleon returned on his march to Paris in 1815, so Fourier compromised, fulfilling his duties as prefect by ordering the preparation of the defences – which he knew to be futile – and then leaving the town by one gate as Napoleon entered by another. His handling of this awkward situation did not adversely affect their relationship. In fact the Emperor promptly gave him the title of count and appointed him prefect of the neighbouring Département of the Rhône, based at Lyon. However, before the end of the Hundred Days Fourier had resigned his new title and appointment in protest against the severities of the regime and returned to Paris to concentrate on scientific work.

This was the low point in Fourier's life. For a short while he was without employment, subsisting on a small pension, and out of favour politically. However, a former student at the Ecole Polytechnique and companion in Egypt was now prefect of the Département of the Seine. He appointed Fourier director of the Statistical Bureau of the Seine, a post without arduous duties but with a stipend sufficient for his needs.

Fourier's last burst of creative activity came in 1817/8 when he achieved an effective insight into the relation between integral-transform solutions to differential equations and the operational calculus. There was at that time a three-cornered race in progress among Fourier, Poisson and Cauchy to develop such techniques. In a crushing response to a criticism by Poisson, Fourier exhibited integral-transform solutions of several equations that had long defied analysis, and paved the way for Cauchy to develop a systematic theory, en route to the calculus of residues.

In 1816 Fourier was elected to the reconstituted Académie des Sciences, but Louis XVIII could not forgive his acceptance of the Rhône prefecture from Napoleon and at first refused to approve the election. Diplomatic negotiation eventually resolved the situation and his renomination the next year was approved. He also had some trouble with the second edition of the *Description de l'Egypte* (for now the references to Napoleon needed revision), but in general his reputation was recovering rapidly. He was left in a position of strength after the decline of the Société d'Arcueil and gained the support of Laplace against the enmity of Poisson. In 1822 he was elected permanent secretary of the Académie des Sciences. In 1827, like d'Alembert and Laplace before him, he was elected to the literary Académie Française; and he also succeeded Laplace as president of the council of the Ecole Polytechnique. Outside France he was elected to the Royal Society of London.

Fourier's health was never robust and, towards the end of his life, he began to display peculiar symptoms that are thought to have been due to a disease of the thyroid gland called myxoedema, which he had possibly contracted in Egypt. As well as causing certain physical symptoms, this disorder can lead to a dulling of the memory, which may account for the mishandling of some of the memoirs he received as permanent secretary and for the rambling character of those he wrote himself towards the end of his life. Early in May 1830 he suffered a collapse and his condition deteriorated until he died on May 16, at the age of sixty-two. The funeral service took place at the church of St Jacques de Haut Pas, and he was buried in the cemetery of Père Lachaise, close to the grave of Monge.

Throughout his career, Fourier won the loyalty of younger friends by his unselfish support and encouragement; in his later years he helped many mathematicians and scientists, including Abel, Dirichlet, Oersted and Sturm. His scientific achievements lie mainly in the study of the diffusion of heat and in the mathematical techniques he introduced to further that study. His interest in the problem may have begun while he was in Egypt, but the main work on it was done at Grenoble. Profound study of nature, he used to say, is the most fertile source of mathematical discoveries, but the mathematical discoveries he made have found a wide variety of applications, as well as playing a highly important role in mathematical theory.

Fourier's viewpoint was that of rational mechanics. He had a superb mastery of analytical technique and this power, guided by physical intuition, brought him success. Previously the equations used in the leading problems of rational mechanics had usually been non-linear and they were solved by *ad hoc* methods. Fourier created a coherent method by which the various components of an equation and its solution in series were neatly identified with the various aspects of the physical solution being analysed. He also had a uniquely sure instinct for interpreting the asymptotic properties of the solutions of his equations for their physical meaning. So powerful was his approach that a full century passed before non-linear equations regained prominence in rational mechanics.

In mathematical physics Fourier's ideas have proved to be much more fruitful than those of Laplace. He also had a long-standing interest in the theory of equations and was trying to complete a book on the subject towards the end of his life. Some of his ideas on the subject evolved into Sturm's theorem, whereas others have recently found applications in the theory of linear programming. He was also interested in dynamics and in the theory of probability. As a mathematician, Fourier had as much concern for practical problems of rigour as anyone in his day except Cauchy and Abel, but he could not conceive of the theory of limiting processes as a meaningful exercise in its own right. Poisson and Biot, outclassed as rivals in the theory of diffusion of heat, tried for years to belittle Fourier's achievements.

The later history of thermodynamics is complex. After Fourier, the next important advances in understanding the nature of heat were due to the theoretician Sadi Carnot, set out in his great memoir of 1824 *Reflexions sur la puissance motrice du feu* (*Reflections on the Motive Power of Fire*). Although Fourier, with other academicians, heard him lecture on his theories, he did not seem to have appreciated their importance; after Carnot died at the age of thirty-six they were neglected for years.

Thomas Young (1773–1829)

We now return to England. Quakers, members of the Religious Society of Friends, were widely respected for their integrity as men of business; they adopted a distinctive mode of dress, refused to take oaths and contributed many distinguished men to science in eighteenth-century England. Thomas Young, mentioned earlier, was one of them. His limited influence on science was the result of his personality, his poor choice of methods of communication and his frequent changes of occupation. He published sporadically, often in obscure and inappropriate ways; his prose was awkward and his

mathematics inadequate. He was a gentleman scientist with many good, even brilliant, ideas who lived to see others receive the credit and fame for completing what he had begun. He made acute suggestions, but left them for others to develop. The German scientist Helmholtz, whose work was considerably influenced by Young's ideas, gave his opinion that

> he was one of the most clear-sighted men who have ever lived, but he had the misfortune to be too greatly superior in sagacity to his contemporaries. They gazed at him with astonishment, but could not always follow the bold flights of his intellect, and thus a multitude of his most important ideas lay buried and forgotten in the great tomes of the Royal Society of London, till a later generation in tardy advance made his discoveries and convinced itself of the accuracy and force of his inferences.

Thomas Young was born in the village of Milverton, near Taunton, Somerset, on June 13, 1773. His father, of the same name, was a cloth merchant, banker and landowner. He and his wife Sarah (née Davis) were Quakers. Thomas was their eldest son, in a family that grew to nine. Perhaps because of the size of the family, he was sent to live until the age of seven with his maternal grandfather, Robert Davis of Minehead, who was an enthusiastic classicist.

Thomas was a precocious child, with a phenomenal memory, especially for languages. He could read by the age of two. The first school he went to, a boarding school in Bristol, had nothing to teach him. After a year back in Milverton, where he read science books borrowed from a neighbour, he went to a school at Compton, Dorset, where he not only studied classics, mathematics and natural philosophy but was also taught practical skills. Young very early showed real mathematical ability. Prodigies tend to excel in languages, mathematics or music.

After leaving school, at the age of 13, he returned to Milverton and studied various Near-Eastern languages, including Syriac (western Aramaic) and Hebrew. He also began making optical instruments and is said before long to have mastered Newton's *Opticks* and *Principia*. In 1787, aged only fourteen, he went to join the grandson of the Quaker banker David Barclay. The boy, named Hudson Gurney, was being educated privately, and it was thought that he should have a companion in his studies, but to some extent Thomas Young also acted as tutor to the younger boy. The two became life-long friends; later Hudson Gurney became well known as an antiquary, writer and politician.

Although Barclay usually spent most of the year at his country house near Ware in Hertfordshire, he generally resided for four of the winter months in London. So during the five years Thomas Young spent in the Barclay household, he was partly in London, where there were lectures he could attend and libraries he could use. It was during one of the London visits that he came to the notice of his great-uncle Richard Brocklesby, who was prominent in the medical world. A fellow of the College of Physicians and of the Royal Society, he had written on the therapeutic value of music and had produced a standard treatise on military hygiene. Brocklesby was immensely impressed by the accomplishments of his brilliant young relative.

In 1789, in the middle of his time with the Barclays, Young was taken ill with suspected consumption. Brocklesby succeeded in restoring him to health and advised him to take more care of himself. At Brocklesby's London residence Young met most of the distinguished literary men of the day and impressed them with his classical scholarship. On the advice of his great-uncle, he took up medicine with a view to succeeding to his practice once he qualified. In 1793 he entered St Bartholomew's Hospital and in the following year was elected to the Royal Society for an original paper on the action of the ciliary muscles in the accommodation of the eye. The Duke of Richmond, a man of the world, wrote to Brocklesby, who was his medical adviser:

> But I must tell you how much pleased we all are with Mr Young.
> I really never saw a young man more pleasing and engaging. He seems to have already acquired much knowledge in most branches, and to be studious of obtaining more: it comes out without affectation on all subjects he talks upon. He is very cheerful and easy without assuming anything; and even on the peculiarity of his dress and Quakerism he talked so reasonably, that one cannot wish him to alter himself in any one particular.

The duke, who was Master-General of the Ordnance at the time, offered to appoint the young man as his private secretary, but, as Young wrote to his mother,

> I have very lately refused the pressing offer of a situation which would have been the most favourable and flattering introduction to political life that a young man in my circumstances could desire. I might have lived at a duke's table, with a salary of £200 a year as his secretary, and with hopes of a more lucrative appointment in a short time. I should

have been in an agreeable family, have had enough time to study, a library, a laboratory and philosophical apparatus at my service; and I was not ashamed to allege my regard for our Society as a principal reason for not accepting the proposal.

As Quakers are pacifists, it might have been the military connection which inspired his refusal.

Although body-snatching had reached scandalous proportions, Edinburgh at this time was the most highly regarded place in Europe for a medical education. Young spent the next year in the Scottish capital, studying modern languages as well as medicine. While he was there he ceased to display the outward characteristics of Quakerism, mixed in society and learnt to play the flute, to sing and to dance. Having decided to spend the next academic year in Göttingen he ended his stay in Scotland with an ambitious tour of the Highlands, still something of an adventure.

The University of Göttingen, known as the Georgia Augusta, had been founded in 1727 on the model of Oxford and Cambridge by King George II of Great Britain, who was also Elector of Hanover. The new foundation attracted students from all over Germany and from elsewhere in Europe as well. Although its endowments were more generous and it enjoyed greater autonomy, it was still in many respects a typical German university of that period. According to the writer Heinrich Heine, 'Göttingen is renowned for its university and its sausages; the inhabitants are divided into four classes: students, professors, Philistines and cattle.'

In Göttingen Young took full advantage of the excellent library and other facilities. He was struck by the way that professors might entertain students in their homes and yet completely ignore them if they passed them in the street. He commented that 'science here has one advantage that the doctrines of both countries are well known here, while the English attend very little to any opinion but those of their own country'. In addition to medical studies he learnt horsemanship and devoted considerable attention to music and other arts. At the end of the year he left with the degree of doctor of physic, surgery and midwifery from the Georgia Augusta, having submitted a thesis *De corporis humani viribus conservatricibus*. He travelled, mainly on foot, to Dresden, where he spent a month studying the art collections, and then returned to London, via Berlin and Hamburg, a remarkably accomplished and educated young man. 'His language was correct and his utterance rapid.' He was 'emphatically a man of truth' and 'could not bear the slightest degree of exaggeration'; he was 'accustomed to reciprocate visits with the best society' and was 'always ready to take part in a dance or a glee, or to join in any scheme of amusement calculated to give life or interest to a party'.

In 1797, on Brocklesby's recommendation, Young entered Emmanuel College, Cambridge, where he spent two years as a fellow-commoner, which gave him the right to dine with the fellows rather than with the other undergraduates. He had been caught by a change in the regulations of the College of Physicians; previously two years of university study had been required of a medical practitioner, now it had to be two years at the same university. The alternative was to return to Edinburgh for a second year, which would have been far better professionally and avoid the problem of the religious test which involved taking an oath. As a Quaker, in the first case of its kind, he was given leave not to take the oath. Being technically a student, Young found that the more distinguished senior members of the university were just as aloof as they had been at Göttingen, but he made some friends among the junior members, among whom he was known as 'Phenomenon Young'. He was seldom seen in the libraries, being mainly occupied with experimental work on sound and light, including the phenomenon of interference, the results of which he communicated to the Royal Society. He described what he was doing in a letter to one of his scientific friends:

> I am ashamed to find how much the foreign mathematicians for these forty years have surpassed the English in the higher branches of these sciences. Euler, Bernoulli and d'Alembert have given solutions to problems which have scarcely occurred to us in this country. I have had particular occasion to observe this in considering the figure of vibrating chords, the sound of musical pipes and some similar matters in which I fancied I had hit on some ideas entirely new, but I was glad to find them in part anticipated by Daniel Bernoulli in 1753 and 1762. There are still several particulars respecting the gyration of chords and formation of synchronous harmonics, the combination of sounds in the air, the phaenomenon of beats, on which I flatter myself that I shall be able to throw some new light.

Already he was beginning to follow Hooke and Huygens in regarding light as a wave motion, like sound.

Not long after he had gone up to Cambridge his great-uncle died, leaving Young a fortune of £10 000, a large London house, a library and works of art. In preparation for taking over his benefactor's medical practice, Young gained some further experience in the London hospitals and then established himself in practice in Welbeck Street. While he continued his contributions to literary scholarship and science, he often published anonymously to avoid any suspicion that he might be neglecting his patients.

In 1800 Young, as we know, was appointed professor of natural philosophy at the Royal Institution; he was also editor of its publications and superintendent of its premises. As a lecturer to a popular audience he was not a success; he displayed extraordinary erudition but his style was too didactic and condensed. Friends advised him that his professorial responsibilities were interfering too much with his medical work and so he resigned his chair after only two years, to devote himself largely to medicine. In 1802 he was appointed secretary of the Royal Society, became foreign secretary two years later and held that office for the rest of his life. In 1804, at the age of thirty, he married an 'extremely young' lady named Eliza Maxwell, who had Scottish aristocratic connections.

Young was awarded the degree of M.B. at Cambridge in 1803 and that of M.D. five years later. He was elected fellow of the College of Physicians in 1809, became censor in 1813 and 1823 and was Croonian Lecturer in 1822 and 1823. During the winters of 1809 and 1810 he gave courses of lectures at the Middlesex Hospital, but these again were not a success with the students. In 1811 he was appointed physician at the highly respected St George's Hospital, a position he held until his death. Yet as a medical practitioner success eluded him and in 1814 he retired from practice. This was not only an exciting time for medical research but also a period when medical practice was being modernized. Perhaps Young was ahead of his time.

Major changes of occupation were a feature of Young's later life. He published several papers dealing with life insurance and was appointed inspector of calculations and physician to the Palladium Insurance Company. He became secretary of the Commission on Weights and Measures and of the Board of Longitude. In 1818 he was appointed superintendent of the Nautical Almanac; his view that the almanac should, as in the past, supply only information of importance in navigation brought him into conflict with many astronomers of the day, who wished it to cater for their needs as well. To a large extent these various activities were related to his position as foreign secretary of the Royal Society, but he also returned to his early interest in philology.

All knowledge of the meaning of the hieroglyphic inscriptions found on Egyptian remains had been lost for 1300 years. Many unsuccessful attempts to interpret them had been made during the eighteenth century, but it was thought that at least some of the characters represented sounds and that those enclosed in an oval line represented proper names. When, in 1790, a tablet, somewhat damaged, was discovered at Rosetta, at the mouth of the Nile, with a decree of the priests inscribed on it in hieroglyphic (sacred), enchorial (cursive) and Greek characters, it was realized that the Greek text might provide a clue to the interpretation of the Egyptian inscriptions. The tablet was placed on display in the British Museum. Impressions of the inscription were circulated to scholars all over Europe, but twenty years went by without much progress in solving the problem of interpretation. Meanwhile more hieroglyphic inscriptions had been discovered following Napoleon's expedition to Egypt.

In 1813 Young started attempting to decipher the inscriptions on the Rosetta stone and by the following year he had translated the 'enchorial', or domestic, running script and had concluded that the enchorial was derived from the hieroglyphic. Pressure of other work prevented him from doing much more for some time, but he returned to the problem in the closing years of his life and made remarkable progress. The illustrious Champollion was a specialist in a field where Young was a gifted amateur, but at least he set Champollion on the right track. Young was still working on an *Enchorial Egyptian Dictionary* up to the end of his life; it was published in 1830, shortly after his death. This was the most notable of a number of works of scholarship dating from his last decade. Another that should be mentioned is his work as a major contributor to the fourth edition of the *Encyclopaedia Britannica*, published in 1819. His lengthy article *'Egypt'* became famous; he also contributed a large number of the biographical notices.

In the latter part of his life Young usually spent the period from November to June each year in London and the rest of the year in the fashionable south-coast resort of Worthing, where he tried unsuccessfully to build up a medical practice. After 1815, when peace had been restored on the continent, he made several visits to Paris. The first was in 1817, when he met Arago, Laplace and Baron von Humboldt amongst others. Four years later he returned to the continent for a more extended tour, including Italy. In 1827 he received international recognition when, in succession to Volta, he was elected to the select band of eight foreign associates of the Paris Academy. The following year when he was in Paris again it was noticed that his strength was declining. In London he moved from Welbeck Street to a house in Park Square, not far away, and it was there that he died of heart disease on May 10, 1829, at the age of fifty-six. He was buried in the vault of the family of his wife Eliza at Farnborough. A profile medallion was placed in Westminster Abbey with an inscription referring to his achievements.

In his essay *On the Cohesion of Fluids* Young gave in non-mathematical language the theory of capillary action soon after brought forwards by Laplace independently. He was also the first to use the term 'energy' for the product of the mass of a body and the square of its velocity. In the theory of elasticity he introduced the formula relating stress to strain which we know as Young's modulus, but characteristically his definition is hopelessly obscure. His theory of tides explained more tidal phenomena than had any previous one. Apart from his contributions to the theory of light and sound, he wrote many important papers on medical subjects.

Young has been called the founder of physiological optics, building on foundations laid by Kepler, Descartes, Huygens and others. He was the first to prove conclusively that the accommodation of the eye for vision at different distances was due to a change in curvature of the crystalline lens. His memoir on the mechanism of the eye contained the first description and measurement of astigmatism, a condition from which he suffered himself. He also gave an explanation of colour blindness. When Young began to write on physical optics the wave theory of light had made little headway against its rival the corpuscular theory, which was favoured by Newton. Young developed the wave theory, the vibrations being transverse to the ray, and obtained many experimental results supporting that view, but his work was little understood in his lifetime. Not long afterwards Fourier's disciple Augustin Fresnel contributed his mathematically sophisticated wave theory of light, which went far beyond Young's ideas.

ANDRÉ-MARIE AMPÈRE (1775–1836)

It is perhaps unfortunate that, because of the dates of birth, the profile of the French physicist André-Marie Ampère must precede that of the Danish physicist Hans Oersted, because it was Ampère who gave the first mathematical treatment of electromagnetism, the marvellous discovery of Oersted. The son André-Marie of Jean-Jacques Ampère and Jeanne-Antoinette (née Desutières-Sarcey) was born about January 20, 1775 in Lyon, the second city of France, where both sides of the family were dealers in silks. Just before marriage Jean-Jacques had acquired a country estate at Poleymieux-les-Monts-d'Or, a small village in the picturesque hills a short distance up the river Saône: this was used initially as a summer retreat from the city and later as the family's permanent home after he retired from business in 1782. Thus, for the first seven years of his life, their son and his

elder sister Antoinette were able to enjoy both the busy life of the city and
the peace of the country. Three years after this he acquired a younger sister
Josephine, of whom more later.

The interests of Jean-Jacques included Latin and French literature,
as well as several branches of science, and, being a follower of the social
philosopher Rousseau, he encouraged his son to educate himself by reading
the books in his extensive library (although an education on the lines laid
down in *Emile* might be all very well, according to Arago the discipline
of a public school might have had a most salutary influence on Ampère's
character). The boy, who displayed unusual powers of concentration and
a prodigious memory, read his way through Diderot's great encyclopaedia.
Having been forbidden the rigours of geometry because of his tender years,
he defied parental authority and worked out the material in the early books
of Euclid by himself. When he found that the classics of science tended to
be in Latin, he taught himself the language in order to be able to study the
works of Euler and his contemporaries. He read the *Mécanique analytique*
of Lagrange and worked through all the calculations it contains.

Later he described the three most influential events in his life as his
first communion, which established him in the faith of his fathers, the
Eulogy of Descartes, which instilled in him a belief in the nobility of a life
in science, and the fall of the Bastille, which decided his political sentiments

for his entire life. The first two years of the Revolution, ending in the attempted flight of the royal family, had some impact, even in Poleymieux, but at first Lyon was spared the excesses which culminated in the Terror. The city, a royalist stronghold, was in the hands of the relatively moderate Girondins. Jean-Jacques Ampère took on the responsibilities of magistrate and presiding legal functionary of the police tribunal, but this led him into trouble when the Jacobin government in Paris, supported by sympathisers in Lyon itself, took control of the city by force. Jean-Jacques was arrested and imprisoned, tried for having approved a warrant of arrest for one of the leading Jacobins, found guilty and promptly guillotined.

This devastating experience left Ampère averse to violence and militarism. His mother, a deeply religious person, now took charge of his education. Fortunately the family property had been transferred into her name, so that at least some of it was shielded from confiscation. A lengthy period of depression, during which he occupied himself mainly with botanical studies, was broken by a love affair with the daughter Cathérine-Antoinette Julie Carron of a local family in somewhat similar circumstances to his own. After he had courted her assiduously for three years, they were married in August 1799; their first and only child Jean-Jacques was born the next year.

Already Ampère had begun publishing his first memoirs in mathematics and these brought him to the notice of the highly centralized French scientific community. At the same time he made some close and enduring friendships among the intelligentsia of Lyon. He also undertook some private tutoring of mathematics students and some teaching at local schools. In 1802 he left the city to become professor of physics and chemistry at the Ecole Centrale du Département de l'Ain at Bourg-en-Bresse, which was to be the scene of much of his early scientific work. Before the Revolution it had been staffed by Jesuits, but, after a transitional stage, it was secularized and renamed the Lycée Lalande. The plan was for him to gain some teaching experience before returning to Lyon to work in a new Lycée that was being established by Napoleon. Meanwhile he was living in Bourg while his wife Julie was forty miles away in Lyon. In 1803 she died of what may have been a malignancy, leaving their three-year-old son in his care.

Following this second major tragedy in his life, Ampère was appointed to a post at the new Lycée in Lyon, but after less than a year he decided to move to the capital to make more of a name for himself in the world of science. He was appointed *répétiteur* for analysis at the Ecole Polytechnique. This meant that he was essentially a tutor to the students who were lectured

to by the professor of analysis, initially Augustin Cauchy. By 1809 he had
been promoted to professor of analytical mathematics and mechanics and
had become a member of the Legion of Honour. All too soon he became
involved in the quarrels of the scientific community in the French capital
and came to regret leaving Lyon.

The period between 1804, when Ampère arrived in Paris, and his death
in 1836 was a time of complex political and social change. He witnessed
the consequences of Napoleon's coronation as emperor, his military successes and subsequent defeat, the initial restoration of Louis XVIII in 1814,
Napoleon's return for the Hundred Days, the second restoration of 1815 and
the revolution of 1830. No supporter of Napoleon, Ampère favoured the idea
of a paternal and enlightened sovereign as head of state.

As well as his chair at the Ecole Polytechnique, Ampère, as was normal, collected a number of other posts. Notably, in 1808 he was appointed
inspector-general of the newly formed university system, a post that he held
almost continuously for the rest of his life. In 1814 he was elected to the
Paris Academy and ten years later he was appointed professor of physics
at the Collège de France. This was a most fruitful period for his scientific
work, especially after 1820 when, inspired by Oersted's discovery of electromagnetism, he made most of the discoveries for which his name is so
renowned and collected them together in his great 1827 *Mémoire sur la
théorie mathématique des phenomènes électrodynamiques, uniquement
déduite de l'expérience (Memoir on the Mathematical Theory of Electro-
dynamic Phenomena, Uniquely Deduced from Experiments)*. In this work,
which has been described as the *Principia* of electrodynamics, he introduced the important distinctions between electrostatics and electric currents and between current and voltage, demonstrating that current-carrying
wires exert a force on each other, and gave an explanation of magnetism in
terms of electric currents.

Unfortunately there was to be further misery in Ampère's private life,
to some extent self-imposed. Another Lyonnais living in the capital, Jean-
Baptiste Potot, had a 26-year-old daughter Jeanne-Françoise, who went by
the name of Jenny. The surviving descriptions of her family are uniformly
disparaging: 'it was a household as bourgeois as possible, in the pejorative
sense in which artists understand that word: narrow ideas, prejudices,
pretensions, living only for money and vanity, not having the least idea
of the sciences, exactly the opposite of what it should have been for a big
child as simple and modest as Ampère'. He courted Jenny with the same
passion that had animated his relationship with Julie. As soon as they were

married and had started living together she began to distance herself from him. Her interest in the marriage seemed to be mainly financial: after the birth of a daughter, Josephine-Albine, Ampère was granted a separation and obtained custody of the child. To take care of his household, he persuaded his mother and his sister Josephine to come to Paris, bringing his son Jean-Jacques with them. Ampère's friends in Lyon considered that he was being selfish, and in fact his mother did not survive the move for more than eighteen months. Meanwhile Ampère was having another love affair, but the object of his attentions married someone else. In addition to all this, his relations with his seven-year-old son by his first wife became strained.

Ampère found some solace in his mathematical work, writing to a friend 'I am going to take up mathematics again. I have some trouble at first, but when I have overcome the initial repugnance, I no longer want to leave the calculations. I still experience a great charm there when I can eliminate every other thought and occupy myself with it alone, absolutely alone.' He completed a significant memoir on partial differential equations, presented it to the Academy, made the customary social calls and, when the vote on membership was taken in November 1814, was successful, beating Cauchy. Ironically Cauchy was just at the beginning of his most creative period whereas Ampère hardly touched mathematics afterwards, concentrating on natural science instead. Unfortunately Ampère lacked perseverance; he was always flying off to something new.

The middle-aged Ampère returned to the religious practices of his youth. He sold off the family property at Poleymieux and bought a house in Paris, where he took scholarly boarders who constituted a ready audience when he needed someone to talk to, as he often did. He began taking more interest in the education and ambitions of his son; the boy was quite bright, learning to read at an early age, but in childhood was subject to violent tantrums. By this time he was well-trained in languages, literature and the sciences. Initially Ampère encouraged him to make a career in the chemical industry, but Jean-Jacques was more interested in becoming a writer.

In the last ten years of his life Ampère gradually lost interest in science; for example, he did not keep up with Faraday's discoveries. Financial problems became a daily concern. His sister ran up large debts maintaining his household, while his son used the inheritance from his grandmother to enjoy leisurely journeys abroad. Both father and son were temperamental, given to bursts of anger interspersed with long periods of silence; it was impossible for them to live under the same roof. In 1820 Jean-Jacques met the celebrated Madame de Récamier and fell under her spell. This stimulated him to complete his first play, *Rosamunde*, which his father tried to get produced. Madame de Récamier went off to Italy for a year, pursued by Jean-Jacques. On his return to Paris he lost his interest in writing historical dramas but gradually found a new one in literary history. He embarked on an academic career, with such success that by 1833 he was teaching foreign literature at the Collège de France. He never married but maintained a strained and unsatisfying relationship with Madame de Récamier until her death in 1850.

Meanwhile Ampère's daughter Albine became a source of concern. She had married a military man with a predilection for strong drink, violence and gambling. He also had a dangerous habit of trying to avoid payment of his gambling debts. After a tempestuous period in Paris he was sent off to Louisiana, where two of his brothers had settled. Ampère himself displayed remarkable patience and sympathy while all this was happening. After his death Albine's husband returned to Paris quite unreformed and before long he was institutionalized. Albine herself became increasingly deranged. She died in 1842; her father did not live to witness the final years of her unhappy life.

Ampère was one of those who could obtain no inspiration when seated; he preferred to stand up or walk around when thinking. Later in life his behaviour became distinctly odd; his pupils made fun of him. His study was open to all, but visitors found it difficult to leave it without playing a game of chess with him. The serious financial problems he experienced in later years were due partly to his expenditure on scientific instruments. His scientific work came under criticism; among the academicians only Fourier received his theories favourably. Outside France criticism was even more severe. After 1829 Ampère's years of scientific creativity had come to an end and his health began to deteriorate. Suffering from bronchitis, laryngitis, rheumatism and occasionally pneumonia, he wintered in the south of France, with his son for company. He was in Marseille when he died from pneumonia on June 10, 1836. Ampère's remains were transferred to the cemetery in Montmartre in 1869. The house at Poleymieux is now a national museum, dedicated to his life and work.

HANS CHRISTIAN OERSTED (1777–1851)

At the opening of the nineteenth century very little was yet known about electricity and magnetism. Hans Christian Oersted, who discovered, almost by chance, the fundamental relation between them, was born in the small Danish town of Rudjøbing on August 14, 1777. His father Søren

Christian Oersted was an impecunious pharmacist. His mother Karen (née Hermansen) gave birth to another son, Anders Sandoe, the next year and other children later. No formal schooling was available in Rudjøbing, which is on the Baltic island of Langeland, but the resourceful pharmacist overcame this problem. The two boys, while they were still young, were placed with a German wigmaker, Christian Oldenberg, and his Danish wife. Their father arranged for Oldenburg to teach them the German language and his wife to help them learn to read and write. As for arithmetic, Oldenburg's knowledge was limited to addition and subtraction, but happily a local youth imparted the arts of multiplication and division. From the mayor of Rudjøbing they learnt a smattering of French, from the baker a little drawing and from the local surveyor some geometry. Moreover, when Hans was twelve years old he assisted his father in the pharmacy and thus acquired a useful grounding in chemistry.

Various difficulties notwithstanding, Hans and Anders studied hard; their intelligence was above average, and their parents used every available means to encourage them. The two boys were so successful that they passed with honours the entrance examination for the University of Copenhagen, the only one in the country. There they received a little financial support from the state and made up the remainder by teaching. They were both interested in science, but the ambition of Anders was to become a lawyer, whereas Hans was more inclined towards literature and philosophy, subjects in which he won prizes for his work. For his doctoral thesis he wrote about the architectonics of natural metaphysics, for he was a passionate disciple of Kant. It must be remembered that in much of Europe a doctorate is a first degree.

The year 1800, memorable for Volta's invention of the battery as a source of electric current, saw Oersted again working in a pharmacy, this time in Copenhagen, while the owner was away on a European tour. The young man, aware of the discoveries which had just been made by Volta, welcomed the opportunity this gave him for scientific research. On the return of the owner of the pharmacy, who was professor of surgery at the university, Oersted organized a scientific tour for himself, with the help of a travel grant from the government. In Germany he found an abundance of theory, but it was experimental work that interested him more. In Paris he rashly gave a lecture, which was greeted by ridicule; it taught him to be more careful in future. On the return leg of the tour he visited Brussels, Leiden, Haarlem and Amsterdam. When he returned to Copenhagen he found that the university building had been damaged by fire, and the physical laboratory destroyed. While waiting for a suitable opening he gave some public lectures, which showed that he had a gift for teaching. Within two years Oersted was assistant professor of physics at the university and also held a position at the military school. Three years later he produced a textbook of mechanical physics and published in French an account of some research he had undertaken on the identity of electric and chemical forces.

In 1812–13 Oersted made another tour of France and Germany. This time he settled in Berlin for a while and published there in German the research paper which already had appeared in French. The next year, when he was back in Copenhagen, he married Inger Birgitte Ballum, whose father was pastor of Kjelby, on the small island of Moer. At about the time of his marriage, he associated himself with a movement to introduce the German language into chemical terminology in place of French and Greek. In addition he sought to raise science at the university to the status enjoyed by theology.

Oersted usually devoted five hours a day to teaching and it was his custom each month to present an account of new scientific advances in a special lecture. It was at one of these that, in the year 1820, or thereabouts, he made his historic discovery: that an electric current can displace a compass needle. He assembled the apparatus, ready to try out after his lecture, and asked the audience whether they would like to remain and see what happened. At first the movement was so small, and its direction so unexpected, that Oersted was unsure that it had been caused by the current. So he tried again with a much stronger current and found that the deflection was unmistakable. After the discovery had been announced he received honours from all sides. The Royal Society of London, for example, awarded him the Copley medal, while the Paris Academy elected him a corresponding member.

Of all the congratulations he received from men of science, the most emphatic were those of Thomas Young, who spoke of the marvellous discovery that elevated Denmark to a rank in science it had not held since the days of Tycho Brahe. In Germany it was said that Oersted's experiments in magnetism were the most interesting that had been carried out in that domain of science for a thousand years. Faraday said of him that 'his constancy in the pursuit of his subject, both by reasoning and experiment, was well rewarded in the winter of 1819 by the discovery of a fact that not a single person beside himself [Oersted] had the slightest suspicion, but which, when once known, instantly drew the attention of all who were able to appreciate its importance and value'.

A third state-aided tour in 1822–3 took Oersted to England, where he met Faraday, as well as to France and Germany. On his return he founded the Danish Society for the Promotion of Scientific Knowledge, a major vehicle for his efforts to raise the consciousness of progress in science among the people of Denmark. In 1829 he became director of the Polytechnic Institute in Copenhagen, a position he held for the rest of his life. He also created a special laboratory for research on magnetism following a visit to Göttingen, where he took advice from Carl Friedrich Gauss.

Towards the close of his life Oersted wrote to a friend that

> in my family I am as happy as a man can be. I have a wife whom I love, and children who are dear to me and who prosper. I have three sons – of whom one is of age and is employed in the forestry service of the king – and four daughters, of whom the eldest three are either married or betrothed. My brother, who for some time was commissioner of the king in our provincial parliament, has recently become a Minister of State. As for me, I am still a professor and director of the polytechnic school and secretary of the Royal Society of Sciences.

Oersted's eldest daughter Karen married the professor of chemistry at the university; another, Marie, married the pastor of a place in Zealand. The youngest, Matilda, remained to cherish her father's old friend Hans

Christian Andersen, who bequeathed to her the manuscripts of his famous stories. The husband of Oersted's sister, Barbara Albertine, became president of the Supreme Court of Norway. His brother Anders became prime minister of Denmark. Another brother, Niels Randulph, an officer in the Russian army, fell at the battle of Leipzig in 1813. They had come a long way from the impecunious pharmacist of a generation or two before.

Oersted's study of Danish literature was unceasing, and he published a great deal of non-scientific work himself, usually related to his metaphysical interests; he also wrote for the newspapers. In his later years he was described as of open countenance, of florid complexion, somewhat stout, in manners kindly, by nature gracious, loyal to the king, devoted to his country and to the cause of humanity. In his scientific work he was often baffled but never discouraged, his perseverance helped him to the end. The jubilee of his association with the university in 1850 was celebrated by a torchlight procession of his past and present students. The Danish government presented him with a country house near Copenhagen, but he did not live to enjoy it for long; within a few months he became ill and died peacefully in the Danish capital on March 9, 1851.

4   From Ohm to Helmholtz

Our next five remarkable physicists were born in the thirty-two years from 1789 to 1821. Two came from England, two from Germany and one from America.

Georg Ohm (1789–1854)

Two centuries ago the science and practice of electrical measurement hardly existed. With a few exceptions, ill-defined expressions relating to quantity and intensity retarded the progress of electrical investigations. Until well into the nineteenth century there was no branch of physics in which there were so many differences of opinion and uncertainties as in those related to *Galvanismus*, and scarcely a physicist whose views did not differ from those of every other upon important principles. Yet amidst this confusion a discovery was made that was destined to create order out of chaos. This discovery resulted from the experimental work of Georg Simon Ohm.

The future physicist, born on March 16, 1789 in the Bavarian university town of Erlangen, was the eldest son of Johann Wolfgang Ohm, master locksmith, and his wife Maria Elisabeth (née Beck), daughter of a master tailor. Of the Protestant couple's seven children, only two others reached maturity: Martin, born in 1792, and Elisabetha Barbara, born two years later. Their mother died in 1799, when Georg was scarcely ten years old, but their father lived until 1822. He was a remarkable autodidact, who gave his sons a solid education in mathematics, physics, chemistry and philosophy, while insisting that they also learn the locksmith's craft. It is to the younger son Martin that we owe an account of their early life in the (unpublished) autobiography he left after his death in 1872.

According to this, the mathematical ability of Georg and Martin was recognized at an early age by a professor of mathematics at the University of Erlangen named Karl Christian von Langsdorff, who gave them advice and encouragement. The gymnasium, which was associated with the university, provided the usual classical instruction with just a little history, geography and mathematics. After attending the school from 1800 to 1805, Georg Ohm went on to the university and studied mathematics, physics and philosophy before lack of means and his self-sacrificing

father's disapproval at his supposed overindulgence in the pleasures of dancing, billiards and ice-skating forced him to withdraw after just three semesters.

So in 1806 the seventeen-year-old Georg Ohm began what proved to be a long struggle to earn a living. He started teaching mathematics at a private school in Gottstadt bei Neydau, a village in the Swiss Canton of Bern. In his spare time he studied scientific works, such as those of Euler, Laplace and Lacroix. He was hoping to be able to continue his university education at Heidelberg but when he sought Langsdorff's advice, it was that he would do better to continue reading on his own. After two and a half years at the school he obtained a position as private tutor in Neuchâtel for another two years. During this period he had some contact with the nearby Pestalozzi Institute, which had an important influence on his thinking about school education, as we shall see.

In 1811, in accordance with his father's wishes, Georg Ohm returned to the University of Erlangen to obtain his degree. He subsequently taught mathematics there as a *Privatdozent* for the next three semesters. Being a fine draughtsman, he supplemented the meagre income this provided by making architectural plans of the university hospital and library and other works. Unfortunately there was little prospect of advancement at the university, and it was to be many years before he obtained a proper university post. Soon lack of money obliged him to seek employment from the Bavarian government as a schoolteacher; but the only post he could obtain was one teaching mathematics and physics at the low-prestige and poorly attended Realschule in Bamberg, where he worked with great dissatisfaction until the school's dissolution in 1816. The next year he was assigned, in the capacity of auxiliary instructor, to teach a section of mathematics at the overcrowded Bamberg Oberprimärschule. Meanwhile he wrote an elementary textbook of geometry, dedicated it to his father and published it in 1817. It was not a success.

Georg Ohm always lived simply. He was a man of marked energy, of middle height, compactly built, sturdy and strong, what Germans describe as the Martin Luther type of physiognomy. His eyes were large and penetrating; his mouth revealed wit, satire and good humour. In diction and phrase he excelled; moreover, his voice was full, and far into his life it retained its attractive quality. He was a good, conscientious teacher, but the conditions under which he worked were unfavourable. For example, the Bamberg students had benches to sit on, but no desks at which to write. Their mathematical knowledge at entry was so slight that physics to them was at first unintelligible. He had progressive ideas about the role of mathematics in education. The student, he believed, should learn mathematics as if it were a free product of his own mind, not as a finished product imposed from without. Ideally, by fostering the idea that the highest life is devoted to pure knowledge, education should create a self-reliance and self-respect capable of withstanding all vicissitudes. These views reflected not only his own early education but also the years of isolation in Switzerland and of personal and intellectual deprivation in Bamberg.

In 1817, the year his geometry textbook appeared, Georg Ohm was appointed to the position of *Oberlehrer* of mathematics and physics at the recently reformed Jesuit gymnasium in Cologne. He found there a library and facilities for experimental work. Equally importantly he found congenial colleagues who were enthusiastic about learning and teaching. Encouraged by this, Ohm was able to combine research in the well-equipped laboratory with his duties as a teacher; his students found him inspiring. He prepared himself by studying the French classics – at first Lagrange, Legendre, Biot and Poisson, later Fourier and Fresnel. However, it was Oersted's discovery of electromagnetism in 1820 that led him towards experimental work in electricity and magnetism, particularly the elucidation of the galvanic circuit.

Meanwhile Georg Ohm's younger brother Martin was getting on well. He too had spent some years as a schoolteacher, in the course of which he married a Swiss lady. He had written a textbook on number theory and, perhaps as a result of this, had secured the position of Privatdozent at Berlin University in 1821, which he combined with some teaching at the prestigious Friedrich Wilhelm Gymnasium. In Berlin he made a good impression on some of the influential members of the Academy of Sciences, which led to his appointment as associate professor at the university in 1824, followed fifteen years later by promotion to full professor. By contrast, his brother Georg was overburdened with teaching and had become convinced that his life had run into a dead end, that he must extricate himself from what had become a stultifying situation in Cologne. Fearing that otherwise he would never marry, he was determined to prove himself to the world and to have something solid to support his efforts to obtain a position in a more stimulating environment. He decided that he must devote himself full-time to research and persuaded the authorities to grant him leave of absence at half-pay for a whole year to pursue it in Berlin, alongside his more successful brother Martin. The outcome was his masterpiece, the treatise *Die galvanische Kette, mathematisch bearbeitet*, known in English as *The Galvanic Circuit*, which appeared in 1827. This describes how resistance, strength of current and potential difference are related. Among other results this contains the fundamental law which bears his name.

Georg Ohm structured his theories in conscious imitation of Fourier's *Théorie analytique de chaleur* of 1822, a fact that may have led him to de-emphasize its experimental side in favour of an abstract deductive rigour, in striking contrast to the inductivist tone of his earlier writings. Although he did not spell out how, he wished the analogy between electricity and heat to be taken seriously, not as something coincidental but as revealing some underlying relationship. While Ohm's law was independently confirmed by other scientists, his theories came under attack, mainly because they were based on experiment and observation rather than philosophical speculation in the spirit of Hegel, whose romantic ideas were much in vogue at the time. The criticism to which he was subjected demoralized Ohm, who resigned from his position in Cologne in order to remain in Berlin with his brother.

For the next six years he was without a regular appointment, earning a pittance by teaching cadets at the institution which later became the Military Academy. A university post remained his goal, but in Prussia the higher academic doors were closed to him. In influential quarters his theories were rejected, if they were understood at all. Moreover, his brother

Martin, with whom he was living, had acquired the reputation of being a dangerous revolutionary. So in 1833 Georg decided to return to Bavaria, hoping for a position at the University of Munich. In this he was unsuccessful; instead, he was appointed professor of physics at the Polytechnic Institute in Nuremberg, where he remained for the next fifteen years. Unfortunately, apart from the title of professor, the position was no improvement over the one he had held in Cologne.

After the publication of *Die galvanische Kette*, Georg Ohm turned away from electrical research and directed his attention towards molecular physics. His aim was to investigate, with the aid of analytical mechanics, the form, magnitude and mode of operation of atoms, but he never had enough free time to undertake such an ambitious project. Nevertheless, he continued scientific research in other areas, notably he anticipated Helmholtz when he discovered in 1843 that the human ear recognizes only sinusoidal waves as pure tones, automatically performing an analysis of any periodic sound into its component tones. *Die galvanische Kette* was translated into English in 1841, fourteen years after its original publication, into Italian in 1847 and into French in 1860, although parts of the book were translated earlier.

Increasingly, however, Georg Ohm was given other responsibilities, in addition to his teaching duties, and these effectively put an end to his researches into molecular physics. For some years he was Rector of the Polytechnic Institute, while he also served as Inspector of Scientific Education for the Bavarian State. It was not until 1849 that he achieved his ambition of a senior post at the University of Munich, initially as associate professor, then as full professor three years later. He was also given a seat in the Senate and made curator of the mathematical–physical collection of the Bavarian Academy. His last important piece of research was in 1852–3, when he investigated interference phenomena in uni-axial crystals. By this time, however, he was over sixty and almost blind, although unable to retire because in Bavaria there were no pensions for university professors. Following a stroke he died in Munich on July 6, 1854, having given what proved to be his last lecture the previous day, and was buried in the Sudliche Friedhof cemetery.

Georg Ohm had to wait a long time before his scientific work was properly appreciated in his homeland, and for some years it remained largely unknown anywhere else. However, by the early 1830s it was beginning to be used by at least the younger German physicists working in electrical science. British and French physicists seem not to have become aware of its profound implications until the late 1830s, but the situation improved when the Royal Society of London awarded him its prestigious Copley medal, particularly referring to his research on the conductivity of metals and on galvanometers. It was acknowledged that it would have been of great value to British investigators if they had known of his work earlier. Specifically, Faraday, whose profile follows, was referring to Georg when he wrote in 1830 that 'Not understanding German, it is with extreme regret I confess I have not access and cannot do justice to the many valuable papers in experimental electricity published in that language.' Even ten years later the Manchester physicist James Joule announced the law we know as Ohm's law as if it were a new discovery in a short paper submitted to the Royal Society.

As a Copley medallist and foreign member of the Royal Society, a corresponding member of the Berlin and Turin Academies, a full member of the Bavarian Academy and Knight of the Order of St Michael, Ohm was not lacking in honours. Although at times he was miserably poor, it was said that he displayed no bitterness as to his lot. Germany had been destabilized by the Napoleonic wars at a crucial stage in his career, and, although Ohm was not involved in the fighting, it was a miserable period for civilians as well as soldiers. There can be little doubt that he could have contributed far more to science if he had had better opportunity to do so. The Polytechnic Institute in Nuremberg where he spent the major part of his career is now the Ohm Institute; in front of it is a statue of Georg Ohm by Wilhelm von Rümann, which was unveiled in 1895.

MICHAEL FARADAY (1791–1867)

James, the father of Michael Faraday, was the blacksmith of Outhgill, a village near Kirkby Stephen in Westmorland. In 1786 he married Margaret Hastwell, a farmer's daughter, and soon afterwards they moved south and settled in Newington Butts, south of the Thames near London Bridge. The future scientist was born there on September 22, 1791, the third of four children, of whom his sister Elizabeth and brother Robert were a few years older and his sister Margaret a few years younger. The letters of Faraday's parents display intelligence and great religious earnestness. The father died in 1810 after years of poor health, leaving his impoverished widow to support herself and the children by taking in lodgers. Her influence on her son Michael was profound; she lived until 1838, by which time he had been recognized as the greatest experimental physicist in the world.

From Newington the Faraday family moved house several times, ending up near Manchester Square, on what was then the western fringe of

London. Up to the age of thirteen Michael's education 'consisted of little more than the rudiments of reading, writing and arithmetic at a common day school', he recalled, 'My hours out of school were passed at home and in the streets.' In the same area there was a bookbinder and stationer's shop kept by a French *émigré* named Riebau, to whom the youth was apprenticed. Faraday lived at this establishment for eight years, working as a bookbinder and acquiring manual skills that later stood him in good stead in his experimental work. In these years he was strongly influenced by a book by the eighteenth-century divine Isaac Watts entitled *On the Improvement of the Mind*. He followed its suggestions for self-improvement, such as the formation of a discussion group with others of similar age who were interested in the exchange of ideas. When his apprenticeship with Riebeau expired, he went on to work for another French *émigré*, with whom he was less happy.

Faraday's contemporaries describe him as proud, kind and gentle, simple both in manner and in attitude and with extraordinary animation of countenance. His voice was pleasant and his laugh hearty. In height, he was somewhat below average. His head from forehead to back was so long that his hats had to be specially made for him. In youth his hair was brown, curling naturally; later in life it approached white and was always parted at the centre. It was from his mother that he acquired much of his character, especially the rejection of all social and political distinctions. His parents were both Sandemanians, as he became himself. After the Restoration John Sandeman, with his father-in-law John Glas, had seceded from the Presbyterians and founded this small religious sect, which stressed the love of the Creator. Sandemanians believe in the literal truth of the Scripture, pledging themselves to live according to the Bible and in imitation of Christ.

Faraday's interest in science was first aroused by a chance reading of the article 'electricity' in a copy of the *Encyclopaedia Britannica* that he was rebinding. The article, which was written by one James Tytler, was somewhat heretical in its views, but it stimulated Faraday to try to verify the statements in it, and much later the influence of Tytler's unorthodox theories regarding the nature of electricity was still apparent. Meanwhile he took advantage of some of the lectures on scientific subjects being given in London, especially those given by a certain John Tatum, who gathered together a group of young men at his home every Wednesday night to discuss scientific matters and make use of his library. They called themselves the City Philosophical Society; this group seems to have played a part in the foundation of Birkbeck College. In this way Faraday obtained a basic scientific education, covering electricity, galvanism, hydrostatics, optics, geology, theoretical mechanics, chemistry, astronomy and meteorology.

It was chemistry that interested Faraday most at this stage in his career. Here he set great store by Jane Marcet's book *Conversations on Chemistry, Intended more Especially for the Female Sex*. This popular work had been much influenced by the lectures of Davy which she had attended at the Royal Institution. A customer of Riebau's gave Faraday tickets to attend Davy's lectures himself. He went and from the gallery took careful notes; thereupon he adopted Davy as his role model. Soon an accident opened up an opportunity for him to leave the craft of bookbinding behind him and begin a scientific career. Davy had temporarily been blinded as a result of an explosion during a chemical experiment, so he needed an amanuensis. Faraday was recommended to Davy, who was much impressed by his work. When Davy's assistant at the Royal Institution was dismissed for misconduct in 1813, Faraday was appointed in his place.

Davy was a close friend of Coleridge, who brought to England the philosophical ideas of Kant and his followers. These ideas helped to provide a unity to Davy's work, and even more strongly that of Faraday, because it was in tune with his religious beliefs. His science was characterized by brilliant flashes of insight soundly supported by experimental evidence. Davy exerted a most important influence on Faraday's intellectual development, which took place gradually. In the autumn of 1813 Davy and his wife went abroad on an extended honeymoon, taking Faraday with them as factotum. The party went to France, Switzerland and Italy, calling on eminent men of science, such as Ampère and Volta, wherever possible. Unfortunately Davy's wife, who was something of a snob, insisted on treating Faraday as a personal servant. On their return to London, after eighteen months on the continent, Faraday began to try his hand at experimental work; encouraged by Davy, he published his conclusions in the *Quarterly Journal of Physics*, until in 1820 he had a paper accepted for the more prestigious *Philosophical Transactions* of the Royal Society.

About this time Faraday made the acquaintance of Sarah Barnard, the sister of one of the friends he had made at the City Philosophical Society, and they were married in 1821; she was twenty-one, he twenty-nine. Like him she came of a Sandemanian family; Sandemanians tended to marry, and to find most of their social life, within the sect. Soon after marriage Faraday became a full member of the sect, which played such an important part in his life, and made his confession of faith, later becoming a Deacon and later still an Elder, one of three responsible for the administration of the church in London. 'A just and faithful knight of God', said his fellow-scientist John Tyndall, 'I think that a good deal of Faraday's week-day strength and persistency might be referred to his Sunday Exercises. He drinks from a fount on Sunday which refreshes his soul for a week.' Exclusion from the church was a severe punishment, which he suffered on at least one occasion.

By this time Faraday hardly needed Davy's patronage any more. Davy was never able to accept Faraday as a social equal, whereas John Herschel, son of the great astronomer, readily did so. When the two men first met Faraday was still a lowly assistant at the Royal Institution, whereas Herschel, his junior by six months, was a graduate of Cambridge, where he had been senior Wrangler and Smith's prizeman, and was already a fellow of the Royal Society. They became firm friends and Herschel never failed to provide Faraday with encouragement and support when necessary. For example, he was one of the first to sign the certificate nominating Faraday for the fellowship of the Royal Society, whereas Davy, who was president at the time, made strenuous efforts to have the nomination withdrawn. Exactly why Davy did so is unknown, but Faraday was nevertheless elected fellow in 1823. Davy nominated Faraday for the position of secretary of the newly founded club, the Athenaeum; however, as soon as the club had become well established he resigned the office of secretary while remaining an ordinary member.

The next twenty years saw Faraday make one scientific discovery after another, initially in chemistry and later in electricity, where his classic investigations laid the foundations for the science we know today. His earliest scientific work was on the liquification of gases, in 1823; his first major contribution to science, the discovery of benzene, followed two years later. However, it is with electricity and especially electrochemistry that his name is permanently linked. After discovering the process of electrolysis in 1832, he went on to work out the laws which control it, no mean feat for a scientist without mathematical training. He was thirty when he discovered electromagnetic rotations, forty when he discovered induction, using it to produce the first electrical generator and transformer, and fifty-four when he discovered the magneto-optical effect and diamagnetism. Not many scientists have begun their creative lives so late or continued so long. Faraday's most famous work is his *Experimental Researches in Electricity*, in three volumes, made up of papers that had mostly originally appeared in the *Philosophical Transactions* or the *Philosophical Magazine*. This was followed by his similar *Experimental Researches in Chemistry and Physics*.

By 1825 the financial position of the Royal Institution left much to be desired. Faraday helped to strengthen it by instituting the celebrated series of Friday evening discourses, of which he gave over a hundred himself. These served to educate the English upper class in science, particularly those of its members with influence in government and the educational establishment. To prepare himself for lecturing, at which he excelled, Faraday took lessons in elocution, and he had some experience as a preacher, but these were of little importance compared with the care he took over preparation and presentation. He thought that, even though a lecture might be written out, it should not be read to the audience. His influence as a lecturer consisted less in the logical and lucid arrangement of his materials than in the grace, earnestness and refinement of his whole demeanour. 'Except by those well acquainted with his subjects, his Friday evening discourses were sometimes difficult to follow', said Tyndall, 'but he exercised a magic on his hearers which often sent them away persuaded that they knew all about a subject of which they knew but little.' One of his auditors spoke of his gleaming eyes, the hair streaming out from his head, his moving hands and his irresistible eloquence: his audience took fire with him and every face was flushed. Another commented that 'no attentive listener ever came away from one of Faraday's lectures without having the limits of his spiritual vision enlarged, or without feeling that his imagination had been stimulated to something beyond the mere exposition of physical facts'.

The Faradays lived in rooms over the Royal Institution, known as the Upper Chambers, from which a back staircase led directly to the laboratory in the basement, where he enjoyed good facilities for his experimental work. He gave regular courses, in addition to the routine lectures, and was heavily involved in the organization of the various activities of the Institution, of which he was effectively administrator. From 1829 to 1852 he also held the position of professor of chemistry at the Royal Military Academy at Woolwich. He was also greatly in demand to give practical advice, commercial analysis, expert testimony and public service generally. In 1836, for example, he became chief scientific adviser to Trinity House, the ancient foundation responsible for the erection and maintenance of coastal installations such as lighthouses and buoys; the electrification of lighthouses was just in its early stages. He gave advice on the prosecution of the war against Russia. In 1841, faced with too many calls on his time and energy, Faraday experienced a nervous breakdown; he had previously sometimes suffered from loss of memory and dizziness. Memoranda written by Faraday at this time show that his mind was seriously disturbed; there is a theory that his condition might have been due to mercury poisoning. He went to Switzerland accompanied by his wife, who was also in poor health, and his brother-in-law the watercolourist George Barnard. As soon as his health permitted he resumed scientific work, this time on magnetism, but the important discoveries all precede the breakdown. There is a parallel with Isaac Newton, who also experienced a nervous breakdown, but, unlike Faraday, Newton never recovered his scientific drive.

In early days Faraday had added to his unduly modest salary from the Royal Institution a supplementary income from what he called 'commercial work'. This supplement might have become important to him, but just as it showed signs of expansion Faraday abandoned it. The fall in his commercial income was correlated with his discovery of magneto-electricity, when worldly gains became contemptible compared with the rich scientific landscape which opened up before him. In 1835 the Tory Prime Minister Sir Robert Peel wished to offer Faraday a Civil List pension, but, following a change of government, it fell to the Whig Lord Melbourne to make the offer, which he did with such ill-grace that the proud Faraday felt insulted and turned it down. When this became public knowledge there was uproar; the King intervened, Melbourne apologised and Faraday accepted the pension of £300. Among the honours he received from other countries was membership of the Legion of Honour and a knighthood from the King of Prussia; also he was a foreign member of the Paris Academy.

Faraday maintained a strict separation between his religion and his science, but both were of the utmost importance to him. The tenderness of his nature made it difficult for him to resist the appeal of distress, but he preferred to distribute his charitable gifts through some organization that assured him they would be well bestowed. Faraday had been interested in the visual arts from an early age and was personally acquainted with some of the leading artists of his day. He gave advice on the conservation of works of art to institutions such as the British Museum, the National Gallery and Westminster Abbey. He collected drawings, engravings, lithographs and photographs of scientists and other notables of his day. Although himself not one of the pioneers of photography, he was much interested in this new development.

By the middle of the 1850s Faraday had gone as far in research as he could, and, as we shall see, his work was taken up by the young Scotsman Clerk Maxwell, whose theory of the electromagnetic field built on the foundations that Faraday had laid. In fact he, before others, noticed the failing strength of his brain and declined to impose on it a weight greater than it could bear. His mind deteriorated rapidly, and, even if he had been able to understand Maxwell's mathematics, it is doubtful whether he would have been able to follow Maxwell's reasoning and appreciate these new developments.

As his mental faculties declined, Faraday retreated gracefully from the scientific world. He turned down invitations to accept the presidency of the Royal Society and that of the Royal Institution; he found the idea of the latter position so upsetting that it brought on another nervous crisis. He concentrated what remained of his energies on his teaching functions at the Royal Institution. It was more in his Christmas lectures to audiences of children, rather than in those he addressed to adults, that his unequalled ability as a teacher was most in evidence. His Christmas lectures for a juvenile audience for 1859/60, on the various forces of matter, and for 1860/61, on the chemical history of a candle, were edited by William Crookes and have become classics. However, even his lecturing abilities began to fade, and he was forced to abandon the lectern in 1861, after experiencing fits of giddiness and loss of memory. He arranged for Tyndall to deputise for him at the Royal Institution. In the closing years of his life the Faradays lived at Hampton Court, not in the palace but in a house on the Green placed at his disposal by Queen Victoria at the suggestion of Albert, the Prince Consort. Michael Faraday died in his seventy-fifth year on August 25, 1857 from 'decay of nature'; at his own request he was buried without pomp or ceremony in a simple grave in Highgate cemetery. 'A mighty investigator,' said Tyndall at his funeral, 'nothing could equal his power and sweetness as a lecturer.' His wife Sarah died in 1879; they left no descendants.

GEORGE GREEN (1793–1841)

In 1828 an essay entitled *The Application of Mathematical Analysis to the Theories of Electricity and Magnetism* was published. The author was a largely self-educated thirty-five-year-old miller named George Green. The *Essay* attracted little attention at the time but is now regarded as one of the landmarks of modern mathematical physics. The story of his life is one of the most remarkable in this collection.

A baker named George Green lived in Nottingham, a historic town in the Midlands of England. In 1791 he married a woman named Sarah Butler. Her father helped him set up a bakery in the town centre. Their only son, also named George, was born on July 14, 1793. Two years later their daughter Ann was born; she later married a cousin, William Tomlin, who is the source of most of what is known about the life of the future mathematical physicist.

George Green junior was eight years old in 1801 when he went to Robert Goodacre's Academy, the best school in the town. Goodacre wrote textbooks on arithmetic and in later years would travel in Britain and the USA giving lectures on popular science. On the roof of the building he constructed an astronomical observatory. The school was well equipped with scientific instruments, and probably the boy learned to use them in the short time he was at the school. We know that, when it was enlarged some years later, it offered teaching in reading, English grammar, penmanship and arithmetic. Young Green was probably taught these. It also offered geography, the use of globes, mathematics, book-keeping, English composition, natural philosophy, astronomy, history, Latin and Greek. French was taught by a native; there seems no doubt that the boy learned some French at some stage. Later Tomlin wrote that his 'profound knowledge of mathematics' soon exceeded his schoolmaster's.

Not many of the pupils at the Academy stayed more than a year or two, so it was unremarkable that the boy was withdrawn after four terms. Then, at the age of nine, he started work in his father's bakery. The business was prospering and George Green senior, who already owned some houses in Nottingham itself, was able in 1807 to purchase an attractive plot of land in the nearby village of Sneinton, only a mile from the town centre. The land stood on a hill, and on this site he built a 'brick wind corn-mill'. The mill itself, which stood fifty feet high, was surrounded by auxiliary buildings, including a house for the miller. This was occupied by a manager, while the family continued to live in Nottingham and run the bakery.

Some ten years later George Green senior built a substantial family house alongside the mill, with two front rooms and five bedrooms, and moved there with his wife and son. As Tomlin tells us, the son 'lived with his parents until the termination of their lives and duly rendered assistance to his father in the prosecution of his businesses'. When George Green senior died in 1829 at the age of seventy, his wife having predeceased him by four years, he left money and his Nottingham property to his daughter Ann Tomlin and the milling business to his son. There was nothing romantic about being a miller. It was exhausting work; moreover, the dust from the operation accumulated in the lungs and caused a disease similar to silicosis. Tomlin goes on to say that the work was 'irksome to the son who at a very early age and with in youth a frail constitution pursued with undeviating constancy, the same as in his more mature years, an intense application to mathematics or whatever other acquirements might become necessary thereto'.

We must now go back some years to 1823, when George Green senior was still alive. This was a period when the industrial and scientific revolution had led to the formation of various intellectual and cultural institutions in some of the more enterprising English towns. In Nottingham such an institution was located in a building called Bromley House, and known by that name. It had been founded in 1816 and soon became the popular venue for the wealthier citizens, the leisured, the cultivated and the philanthropic. It was a meeting-place for the committees of the newly formed Nottingham School of Medicine and of the Ladies' Bible Society, the Literary and Debating Society, the Geological, the Natural History, the Geographical and the Astronomical Societies. The Amateur Musical Society also met there; artists exhibited there. However, most importantly, there was a library, called the Nottingham Subscription Library, which non-members could use for a fee. When Green became a member there were about a hundred subscribers, who were mainly members of the leading town and county families or the more prominent local firms. There were also professional men, doctors, lawyers and so on.

Originally £600 had been spent on books for the library, but further purchases were made, and members could ask for particular books to be ordered. They were mostly in the categories of literature, history and biography, with some travel and natural history. When Green was a member, the catalogue listed no more than a dozen serious scientific and mathematical works, including treatises by Lagrange and Laplace, but none of them are the particular books to which Green refers or makes use of in the *Essay*. There was, however, a set of the *Philosophical Transactions* of the Royal Society. In its pages Green would have been able to follow the course of discoveries in science for the previous 160 years.

At the same time, one has to ask what was happening between the time he left school, at the age of nine, and the time he joined the library, at the age of thirty? We know that he worked in the bakery and the mill, but it seems he used to retreat to a room on the top floor of the mill to study whenever he could. If someone was guiding his studies, the most likely candidate seems to be Toplis, headmaster of the Free Grammar School in Nottingham. Toplis was the second head who was a Cambridge graduate in mathematics. He was in his twenties when he came to Nottingham and stayed until 1819, by which time Green was twenty-six. After that he returned to Cambridge as a fellow of Queens College. (The college library contained a good collection of books on mathematics in which the French mathematicians of the time were particularly well represented.) The methods Green used are distinctively French. As well as with Laplace's magnum opus, he seems to have become familiar with papers by Biot, Coulomb and Poisson, among others.

Toplis was among the few who recognized that Britain was being left far behind in the field of mathematical research after more than a century of steady progress on the continent, particularly in France. He was an enthusiast for the work of the French school and had translated the first two books of Laplace's *Mécanique céleste* into English. His translation was published in Nottingham in 1814, when Green was twenty-one, and Toplis presented a copy to the Nottingham Subscription Library when he became a member. Until about 1817 the Green family lived very close to the Free Grammar School, where Toplis was resident. Although there is no certainty that Toplis was Green's mentor, the circumstantial evidence is persuasive. Five years after Green joined the library his first and greatest work, the *Essay on Electricity and Magnetism*, was published in Nottingham by private subscription in 1828. There were 51 subscribers, half of whom were fellow members of Bromley House. It was one of the first attempts to apply mathematical analysis to electrical phenomena and was of such importance that it has been described as the beginning of modern mathematical physics in Britain. In it Green gave the name *potential* to Laplace's analytical device and used it in relation to electricity, where it proved an invaluable tool in the development of electromagnetic theory. The *Essay* is also important for the introduction of new mathematical techniques, known nowadays as *Green functions*, which are regularly used throughout mathematical physics, and *Green's theorem*, which is fundamental in differential geometry as well. Such a seminal work should have been published by one of the learned societies, rather than in such an obscure way, but Green seems to have thought it would not be accepted, 'coming from an unknown individual'. Unfortunately the result was that the *Essay* attracted little attention, and, greatly discouraged, Green seems to have given up his investigations for the next few years.

Green now had full responsibility for running the family business, which he had inherited. Like his father, however, he relied on a manager, named Smith, who had a daughter, Jane, who was nine years younger than Green. When she was twenty-two she bore him a daughter, Jane, who was known as Jane Smith, like her mother. Later two sons and three daughters were born and known by the surname of Green, although the parents never married; only the first daughter was known by the other surname.

One day in January 1830 Green had a conversation with a certain Mr Kidd from the city of Lincoln, which had highly significant consequences. Green had dedicated his *Essay* to the Duke of Newcastle, who controlled the representation of Nottingham in Parliament. He had also sent a copy on publication to a subscriber living near Lincoln, Sir Edward Ffrench Bromhead, who replied most encouragingly and apparently offered to send any further mathematical work by Green to one of the learned societies. Unfortunately the diffident Green did not take up this offer – someone apparently told him it was just a polite gesture – but Kidd assured him that it was intended seriously. So, nearly two years after publication of the *Essay*, Green wrote a long, apologetic letter to Sir Edward explaining the reason for the delay and promising to take up the offer if it still held good. This started a correspondence between the Lincolnshire baronet, himself a mathematician, and Green, who seems to have visited his seat of Thurlby Hall on several occasions.

Sir Edward had been one of the Cambridge students who founded the Analytical Society in 1819. This body, which later became the Cambridge Philosophical Society, helped to rescue Cambridge science from its torpid condition. After graduating from Cambridge in 1812, he studied law at the Inner Temple and, on returning to Lincolnshire, began playing a prominent part in city and county affairs. However, he retained his Cambridge connections and invited Green to accompany him on a visit to his *alma mater*. Green declined: 'You were kind enough to mention a journey to Cambridge on June 24th to see your friends Herschel, Babbage and others who constitute the chivalry of British science. Being as yet only a beginner I think I have no right to go there and must defer that pleasure until I shall have become tolerably respectable as a man of science should that day ever arrive.' The possibility of going up to Cambridge as an undergraduate had already been mentioned by Green in April of that year. He naturally turned to Sir Edward for advice as to 'which college would be most suitable for a person of my age and imperfect classical attainments'. Unsurprisingly, Sir Edward recommended his own college, Gonville and Caius. In preparation for his move to Cambridge, Green, having leased the mill after his father's death, let the family house: we have no information on the whereabouts of Jane, who by this time had four children. He resigned from the Nottingham Subscription Library, having had the satisfaction of presenting a copy of his first memoir published in Cambridge the previous year. By the end of 1833 Green had written three memoirs. Two were published by the Cambridge Philosophical Society and one by the Royal Society of Edinburgh.

Green arrived in Cambridge early in October 1833 and was admitted as an undergraduate to the college usually known as Caius (pronounced keys), a community of some 170 students and twenty fellows. He was much older than the normal age of admission. On his departure for Cambridge, Sir Edward had given him letters of introduction 'to some of the most distinguished characters of the university that he might keep his object steadily in view under some awe of their names and look upwards, not of course with any view of trespassing on the social distinctions of the university, in my time more marked than at present, but that he might venture to ask for advice under any emergency'. Green managed to pass the general examination in Latin, Greek and ecclesiastical history. After the first eight months he wrote to Sir Edward: 'I am very happy here and am I fear too much pleased with Cambridge. This takes me in some measures from those pursuits which ought to be my proper business, but I hope on my return to lay aside my freshness and become a regular Second Year Man.'

The Senate House (or Tripos) examination dominated the undergraduate course. It consisted of 200 questions to be answered over seven days in midwinter in the unheated building. The first three days were on book work, the books being the *Elements* of Euclid and the *Principia*, and passing on these was sufficient for most students. For the better students it was just a question of speed. However, for those aspiring to honours there were four more days of more difficult problems over a wide range of subjects. The lectures were optional; for most students the bulk of the instruction came from a private tutor or coach, but it is unlikely that Green could afford this extra expense.

Green achieved fourth place on the order of merit in the notorious examination and thereby became eligible for a college fellowship. As soon as a vacancy occurred, two years later, he was elected. Fellowships carried the right to rooms in college and meals at the common table, free of charge, and a small stipend. His duties were far from onerous. He is known to have set and supervised examinations; his manner on such occasions was recorded as gentle and pleasant. As a Cambridge graduate he could attend meetings of the Philosophical Society and present his papers there in person. Three papers of his were published in 1838 and three more the next year. Two were on hydrodynamics, two were on the propagation of light and two on the propagation of sound. Thus Green had altogether eight papers published in Cambridge, one in Edinburgh and, of course, the original *Essay* in Nottingham. The *Essay*, on electricity and magnetism, and the last memoir on the propagation of sound are considered to be his most important works. During his six years at the university he had achieved a considerable reputation. 'He stood head and shoulders above all his contemporaries inside and outside the university', wrote someone who was there at the time.

Yet only six months after his admission to the fellowship at the end of October 1839 he took the coach home to Nottingham for the last time. In the words of his cousin Tomlin: 'He returned, indisposed after enjoying many years of excellent health in Sneinton, Alas! With the opinion that he should never recover from his illness and which became verified in little more than a year's time by his decease on May 31, 1841.' For all his six years in Cambridge and his reputation there, Green came home to die in relative obscurity. All the local newspaper had to say was the following: 'we believe he was the son of a miller, residing near to Nottingham, but having a taste for study, he applied his gifted mind to the study of mathematics, in which he made rapid progress. In Sir Edward Ffrench Bromhead, Bart., he found a warm friend, and to his influence he owed much, while studying at Cambridge. Had his life been prolonged, he might have stood eminently high as a mathematician.'

Green died of influenza aged forty-seven, in a house in Nottingham occupied by his partner Jane Smith and her family of seven children. Clara, the youngest, had been born the previous year, only a few weeks after his final return from Cambridge. Jane Smith was the one who reported his death; and she was with him at the end. He was buried with his parents in St Stephen's churchyard in Sneinton. On the gravestone he is described as 'Fellow of Caius College, Cambridge'. In the will he had made some ten months previously, he describes himself as 'late of Sneinton in the County of Nottingham and now of Caius College, Cambridge, Fellow of such College'. Whatever the immediate cause of death, it seems most likely that the 'indisposition' which caused him to return home was the miller's version of silicosis, although Felix Klein, in his account of the development of mathematics in the nineteenth century, attributed it to alcoholism; perhaps he heard this from Green's contemporary the mathematician James Joseph Sylvester.

In time the mill at Sneinton fell into disrepair. Any papers that Green may have left seem to have been destroyed. His descendants were unaware of their genius of an ancestor. Recently, however, the city has begun to take pride in one of its most distinguished citizens. The mill has been restored and a little museum added, in which one may learn about Green's work and its consequences, and perhaps purchase some flour ground at the mill. When the bicentenary of Green's birth occurred in 1993, scientific conferences were held to discuss the influence of his work, and a ceremony was held at Westminster Abbey, where a simple plaque was unveiled in his memory. It can be found near the base of the monument to Newton.

Joseph Henry (1797–1878)

The American physicist Joseph Henry made important contributions to the investigation of electromagnetism. He built the largest electromagnet then known, which could lift 300 kilograms; he also discovered electromagnetic induction, independently of Faraday, and invented an early form of the telegraph and the electrical relay. On principle he never patented any of his inventions, which when exploited commercially made others rich but not the inventor. The unit of inductance is named after him. Sadly, his recognition as a great experimental physicist arrived too late to afford him any satisfaction. Instead it was rather his wise administration of the newly established Smithsonian Institution that made him famous.

Joseph Henry was born in Albany, the state capital of New York, on December 17, 1797. His father William Henry was a sometime day labourer from Argyle, distantly related to the earls of Stirling, while his mother Ann (née Alexander) was the daughter of a miller. William Henry died young, and it was chiefly his widow Ann who brought up her son. She was a small woman with delicate, rather beautiful features, who lived

to an advanced age. She was a devout and strict member of the Scottish
Presbyterian Church, the principles of which she passed on to her son. Before
he had turned six she sent him to nearby Galway to live with her stepmother
and her twin brother John. After three years of elementary school Joseph got
a job in a general store, where the storekeeper, an educated man, encouraged him to continue with his education after work. When the boy was
approaching fourteen he returned home to Albany, where he was apprenticed to a watchmaker and silversmith. After the business had failed he
was released from his apprenticeship, but not before he had acquired some
practical skills that were to be useful to him later.

Although Albany was not a large place in those days, it boasted an
unusually good theatre, which for a year occupied most of the young man's
leisure hours. He belonged to an amateur group for which he acted, produced
and wrote plays. This experience may have helped in later years when he
needed to be an effective public speaker. During this period Joseph Henry
is described as being remarkable for his good looks, delicate complexion,
slight figure and vivacious nature. His lively temperament made him a general favourite. The town had an excellent school, almost a college, called the
Albany Academy, where Joseph Henry continued his education by attending
night classes in geometry and mechanics, followed by calculus and chemistry, while supporting himself by working as a private tutor and doing a

little teaching. He was inclined towards science, but in the USA the only career open to a scientist at this time was in medicine.

Albany also possessed a scientific society, the Albany Institute, with a miscellaneous library of some 300 volumes. Henry obtained the post of librarian and gave a few lectures to the members, complete with experimental demonstrations. Meanwhile he was studying the classic *Mécanique analytique* of Lagrange. The versatile young man also led a grading party for a new highway running from the Hudson River to Lake Erie. His success in this work brought him offers of similar positions elsewhere, but he turned them down in favour of the post of professor of mathematics and natural philosophy at Albany Academy in 1826. This meant that he had to renounce his earlier idea of becoming a physician, but he had not got very far with that.

Henry was now on the threshold of the scientific career which was to bring him such distinction. He would have read Priestley's *History and Present State of Electricity* and might also have been aware of Coulomb's work. Perhaps he also knew of Ampère's work and of Volta's invention of the battery, as a source of electric current. In any case, Joseph Henry spent a good deal of his spare time in the useful exercise of repeating the classic experiments.

A maternal uncle of Henry's, a successful businessman in Schenectady, had died and his widow had brought her family to live in Albany. Her son Stephen was a delicate and sensitive youth, who at the age of eighteen had graduated from Union College with high honours in mathematics and astronomy. Since they had similar interests the two cousins saw much of each other; soon Stephen was also on the faculty of the Academy and later he would follow Henry to Princeton. Meanwhile Henry had been courting Stephen's sister Harriet, and they were married by the end of 1829. They had six children, of whom two died in infancy and the only son on the threshold of manhood. The three survivors, Helen, Mary and Caroline, were to brighten their parents' declining years. Modest, retiring and understanding, Harriet furnished a home life into which Henry could retreat from the pressures of the outside world. She seems to have been an ideal wife for him.

In his thirty-sixth year Henry was appointed to the chair of natural philosophy at the College of New Jersey, which would become Princeton University many years later. He started work there in 1832. The college was at a low ebb, with only about seventy students, but the situation was beginning to improve and later Henry was to say that the fourteen years he spent at Princeton were the happiest of his life. When he first arrived

there was general amazement at his appearance: 'His clear and delicate complexion, flushed with perfect health, bloomed with hues that maidenhood
might envy. Upon his splendid front, neither time nor corroding care, nor
blear-eyed envy, had written a wrinkle nor left a cloud; it was fair and pure
as monumental alabaster. His erect and noble form, firmly and gracefully
poised, would have afforded to an artist an ideal model of Apollo.'

One advantage of the move to Princeton was that it brought him into
contact with a wider circle of men engaged in scientific research. No longer
was he obliged to work in an intellectual vacuum. For example, a short
journey on the newly opened railroad brought him to Philadelphia, seat
of another university and home of several learned societies. One of them
was the American Philosophical Society, to which he was elected in 1835.
It was to the members of this society that he communicated the results
of his Princeton researches. Unfortunately the society was rather slow in
publishing papers presented to it, so Henry was continually at a disadvantage
in the competition with Faraday, who took no interest in what was being
done in America.

It is interesting to compare the lives of Henry and Faraday. The
former was born in 1791 and died in 1863, the latter was born in 1797
and died in 1878; but their periods of peak productivity were almost identical. They both sprang from the same working class that had no traditions
of study and that did not possess the means of educating its children. Both
were influenced to follow a scientific career by the chance reading of books
that revealed the works of nature to the imagination. They were both deeply
religious men who regarded nature as the handiwork of their Creator. However, no comparison of the two would be complete without reference to
their relative situations at this stage of their careers. We have seen Henry
at work in a small town, with only very little spare time, hampered for
want of funds and facilities, with very few technical companions. Faraday
was much more favourably situated at the Royal Institution, living in the
world centre of intellectual activity, with as much time as he wanted for
research, in constant communication with some of the most brilliant technical minds of the day. Henry found that his discoveries were anticipated by
Faraday over and over again; he became understandably disappointed and
discouraged.

In 1836, in recognition of his work during his first four years at Princeton, the trustees of the college granted Henry a year's leave of absence on
full salary. It was a good time to escape from the USA, for the nation was just
entering upon a financial depression. England by then was in many respects

a different country from the one against which the American colonists had rebelled. Reform was in the air, and the spirit of industry had taken a firm grip on the people. Although London was his destination, Henry began his journey by first going to Washington, to collect some letters of introduction; he was not impressed by the federal capital, where he was destined to spend so much of his life.

Once in London Henry lost no time in visiting the Royal Society and the Royal Institution. Unfortunately it was the Easter vacation and Faraday had just left town, but they met before long, and Henry heard Faraday lecture. After two useful and enjoyable months, he moved on to Paris for two more weeks. Observing the passing scene, he found many Gallic customs curious and was struck by the way women were engaged in many occupations followed only by men in America. He deplored the prevailing military spirit, particularly the obligation for every able-bodied male to serve one day each month on military duty. After brushing up his French, he met a few of the French men of science and attended a meeting of the Paris Academy, but otherwise he seems to have made little real contact with the leading French scientists. Finally, after a short visit to the Netherlands, he returned to London to stay with his former student Henry James, father of the psychologist William and of the well-known novelist.

Both Joseph and Harriet being of Scottish descent, like so many New Englanders, they took the packet to Edinburgh and called on many of the notables there as well as elsewhere in Scotland. He ended his tour in Liverpool, where the meeting of the British Association was in progress. On his return to America, Henry felt well informed about the state of physics in Europe, especially experimental physics and the teaching of science. He came back in excellent form, strong and fervent, charged with enthusiasm to resume his own investigations into electromagnetism. For ingenuity, completeness and novelty in the examination of new phenomena, these next researches serve as a model for the experimental physicist. Their significance lies not so much in the material which he added to science as in the new territories he opened to the view of those who followed in his footsteps.

Henry had now reached his forty-ninth year. All the high promises that had accompanied him to Princeton had been fulfilled in the fourteen years he spent there. He had found congenial companions and duties well suited to his powers. He had been valued and honoured by members of the faculty, while the students held him in reverence. Although far from exhausted scientifically, rather at the height of his powers, he chose to withdraw from active scientific work and become an organizer and

administrator. The occasion which brought about the change was his decision to accept the post of secretary to the Smithsonian Institution. With
complete philosophical detachment he had evaluated the needs of American
science and threw in his lot with the new venture to promote the satisfaction of those needs.

To understand how it came about that a man who was not American
and had no connection with the USA should have founded this great
American institution, it is necessary to say something about the life of the
founder. James Smithson, known in his youth as James Lewis or Luis Macé,
was the illegitimate son of Hugh Smithson, first Duke of Northumberland
of the third creation, and the wealthy and high-born Elizabeth Macé. He
was born in France in 1765, matriculated at Oxford University as a member
of Pembroke College at the age of seventeen and early displayed a taste for
science that never left him. After having been admitted as a fellow of the
Royal Society in 1787, he read papers to that body and published a number
of them, chiefly on chemistry and mineralogy. He already knew most of the
notable men of science of his time, such as Arago in Paris and Cavendish in
London. He spent most of his later life on the continent of Europe, associating with the men of science in Berlin, Paris, Rome, Florence and Geneva,
but eventually settled in the French capital, although he was in Genoa when
he died. Apparently the reason he left England was that, because of the ille-
gitimacy and the refusal of his father to acknowledge their relationship, he
never received the social recognition to which he believed he was entitled.
Also perhaps he had too high an opinion of his own abilities and a tendency
to flaunt his wealth.

In his will, made when he turned sixty, Smithson left a life interest
in his considerable fortune to a nephew, after which it should pass to the
'United States of America, to found at Washington, under the name of the
Smithsonian Institution, an Establishment for the increase and diffusion of
knowledge among men'. Previously he had intended to leave his fortune to
the Royal Society, and it is something of a mystery just why he changed
his mind. After some discussion the bequest was accepted by Congress and
before long over 100 000 gold sovereigns were delivered to New York. Thereupon there was great debate, extending over nine years, on how precisely
this princely legacy should be used. Eventually it was agreed that there
should be a library, a museum and an art gallery.

At the end of 1845 Henry was appointed secretary and first director of the new institution. He took up office within days. Most of his
friends advised him against moving to Washington, pointing out his lack of

experience in administration and organization, and Henry himself felt no great enthusiasm about accepting the appointment, recalling the case of Newton, who made no discoveries after becoming Warden of the Royal Mint. The Princeton authorities promised him that they would welcome his return if he decided he had made a mistake. His intentions were of the best, but it soon became all too clear that members of Congress were going to interfere and create problems. His immediate problem was to prevent the capital of the endowment being plundered, for example on a grandiose building, through the over-ambitious plans of Congressmen. Almost at once he was offered a chance to escape when he was offered a chair at the University of Pennsylvania; although this was probably the most desirable scientific position the country had to offer, ideal for his needs, he turned it down, greatly to the loss of science, as he did subsequent offers. Henry hoped that he would be able to return to experimental work once the Smithsonian got under way, but that was never a real possibility.

Before long the many-turretted red sandstone building in vaguely Romanesque style known as the Castle was under construction, on a site in the Washington Mall close to malodorous water and far from the residential area of the city. Henry began to organize a series of publications and a system of exchanges with other institutions that enabled the library to develop. The subjects in which he took an interest are bewildering in their variety. For some, such as meteorology, he was enough of an authority himself, but for others, such as zoology, he consulted outside experts. Increasingly the federal government used the Smithsonian as an all-purpose institution, for example in designing the land surveys and expeditions which were so necessary for the exploration of the American continent. Advice was freely given when telegraph lines were being laid or railroads being planned. Neither was the work confined to the USA. Today the Smithsonian Institution comprises the largest complex of museums and art galleries in the world.

The family, consisting of Joseph, his wife and the three daughters, lived in the Castle itself, where they entertained liberally. The building was badly damaged by fire in 1865 and much valuable material destroyed, including Smithson's voluminous papers and mineralogical collection, as well as Henry's correspondence and research notes. After the end of the American Civil War he made another visit to Europe with his daughter Mary. He continued in his various offices until the age of eighty and was still at work at the end of 1877 when he found one morning that his right hand was paralysed. Nephritis, in those days incurable, was diagnosed, and he knew that he did not have long to live. Joseph Henry died in his sleep in Washington on May 13, 1878 and was buried in the Rock Creek cemetery in Georgetown. A bronze statue by William Wetmore Story, which stands at the entrance to the Castle, was unveiled in 1883.

Henry played an active part in establishing the American Association for the Advancement of Science, which dates from 1848. Fifteen years later he became one of the original members of the National Academy of Sciences, an exclusive body like the European academies, which was founded by President Lincoln to provide technological advice during the Civil War. Although Henry was regarded as the foremost physicist in his own country, he found it difficult to understand why his work met with so little approbation abroad. He missed the fame his discoveries warranted through not publishing the results of his experiments in time; and when he reformed his dilatory habit and became punctual in announcing his results, he still remained inconspicuous because the vehicle he chose was relatively unknown in Europe. By the time the international scientific world had awakened to a recognition of his merit, he was already dead.

HERMANN VON HELMHOLTZ (1821–1894)

Helmholtz was among the most versatile of nineteenth-century scientists. He was both a physiologist and a physicist. He is famous for his epoch-making researches on the physiology of the eye and ear, and he is also famous for his definitive statement of the first law of thermodynamics. Hermann Ludwig Ferdinand Helmholtz, to give him his full name, was

born in Potsdam on August 31, 1821. His father, August Ferdinand Julius Helmholtz, had served with distinction in Prussia's war of liberation against Napoleon and, after studying at the new University of Berlin, became a senior schoolmaster at the Potsdam gymnasium, teaching German, classics, philosophy, mathematics and physics. His mother Caroline (née Penne), the daughter of a Hanoverian artillery officer, was descended in the male line from William Penn, the founder of Pennsylvania, and on her mother's side from a family of French refugees. She was described as being profoundly emotional and of sharp intellect while excessively simple in appearance. The future scientist was their eldest son; they had two younger daughters and another son, Otto, born in 1833, as well as two other sons who died in infancy.

Helmholtz was a delicate child who early displayed a passion for understanding how things worked, but otherwise developed slowly. At the age of seven he began his school education, moving up to the gymnasium in the spring of 1832. There he made good progress generally but particularly in the exact sciences, for which there was an excellent teacher. Although he was keen on physics, he was persuaded by his father, who could not afford university fees, to take up the study of medicine, entering the Friedrich Wilhelm Medical Institute of Berlin in 1838. Students at the institute were entitled to attend lectures at the university; he took full advantage of this and also studied a great deal on his own. They also received some financial support in return for a commitment to serve for five years as military surgeons after they qualified. By 1842 Helmholtz had progressed sufficiently to be appointed house-surgeon at a hospital and completed his doctoral thesis on the structure of the nervous system in invertebrates, the histological basis of nervous physiology and pathology.

The next step in his career was to discharge his obligation to serve as a medical officer in the army. In 1843 he was appointed assistant surgeon to the Royal Hussars at Potsdam. His duties with the squadron left him plenty of spare time, which he devoted to science, for example by studying the mathematician Jacobi's great treatise *Fundamenta nova functionem ellipticarum* and by continuing research in physiology. By 1847 he was fully qualified in medicine and became engaged to a young lady of the distinguished von Velten family named Olga; however, marriage could not take place until he had secured a permanent appointment.

In July 1847 he revealed himself as a master of mathematical physics when he unveiled his theory of the conservation of energy at a meeting of the Physical Society. He was perfectly willing to concede that others had

had somewhat the same idea, but he was the first to formulate the principle clearly and demonstrate it conclusively by scientific methods. While it was enthusiastically welcomed by the younger physicists and physiologists of Berlin, the older scientists almost without exception rejected it. Among the mathematicians Jacobi was one of those who unhesitatingly proclaimed the significance of Helmholtz' work.

In 1848, through the influence of Baron von Humboldt, Helmholtz was released from the remainder of his military service to become lecturer at the Academy of Arts and assistant in the Anatomical Museum in Berlin. He held these positions only briefly, since the following year he was appointed associate professor of physiology and director of the physiological institute at the Albertina University of Königsberg, enabling him to marry. The next year he published the first part of his classic work on measurements of the time relations in the contraction of animal muscles and the rate of propagation in the nerve; the second part followed two years later.

Early in 1851 Helmholtz invented the ophthalmoscope, a turning-point in the way his work was regarded. This little invention, which took him only two months to design and construct, made him famous. Being back in favour with the authorities, from then on he was left free to follow the promptings of his scientific curiosity. He went off on a tour through the Swiss Alps to northern Italy, where he was enchanted by Venice, returning via Trieste and Vienna. En route he took the opportunity to call on the leading scientists at the places he passed through. On his return he was promoted to full professor at Königsberg; his inaugural lecture 'On the Nature of Human Sense Perceptions' was a splendid example of his gift for making difficult scientific problems, close to the frontiers of research, intelligible to a general audience. This led to his famous physiological theory of colour vision, based on ideas of Young.

Helmholtz and his wife were happy and settled in Königsberg. They were cheerful and contented, serious and industrious yet averse to no social pleasures, so they gradually acquired an agreeable circle of friends, who shared the interests of both wife and husband. Helmholtz' wife was a faithful help-meet and true comrade, who worked and wrote for him. He read aloud to her the lectures he was going to deliver, so that she might judge how they would appeal to an educated audience. However, her health, which had long been a cause of concern, steadily grew worse. The doctors thought that the cold climate of Königsberg was one cause of her frequent illnesses, so, when the professorship of physiology in Bonn fell vacant, Helmholtz enlisted the support of von Humboldt in order to have himself transferred there.

In 1855, when he moved to Bonn, Helmholtz' great treatise, the *Handbook of Physiological Optics*, was completed. It was the culmination of five fruitful years spent in Königsberg, years of prolific academic work and high achievement in various branches of science. Just before leaving for Bonn, he received an invitation from William Thomson, the future Lord Kelvin, to address the forthcoming meeting of the British Association in Hull. On the way to Hull he passed through London, where he saw Faraday at the Royal Institution and George Biddell Airy, the Astronomer Royal, at the Royal Observatory. 'Faraday is as simple, charming and unaffected as a child', he noted, 'I have never seen a man with such winning ways.' At the meeting he was particularly impressed by the fact that his audience of 850 included 236 ladies. 'Here in England', he wrote home, 'the ladies seem to be very well up in science.'

Helmholtz soon accustomed himself to life in Bonn, where his wife's health improved considerably, thanks to the milder climate. Officially his duties were to teach anatomy, for which he was certainly qualified but felt the lack of any previous teaching experience. He also lectured on his main subject of physiology. In his experimental work he was at first handicapped by the lack of scientific instruments; he had been able to take only very few of them with him from Königsberg and he found practically nothing useful when he arrived at Bonn. His interests were now turning from the theory of vision to that of sound. He applied the mathematics of Fourier series, which led to another classic work, the *Theory of the Sensations of Tone*.

However, Helmholtz had hardly settled down in Bonn when he received an invitation to move again, this time to Heidelberg, where he would be appointed professor of physiology, rather than anatomy. Bonn attempted to retain him by increasing his salary and promising to improve the facilities provided for anatomy. He and his wife already had made many good friends in Bonn. Against this, Heidelberg boasted Robert Bunsen and Gustav Kirchhoff, two of the leading scientists of the period. Although the state of his wife's health made another move undesirable, in the end he decided on Heidelberg. The Prussians, who governed Bonn, made every effort to dissuade him, but they succeeded only in delaying the transfer until 1858, when he joined Bunsen and Kirchhoff to inaugurate the most glorious age in Heidelberg science.

The Helmholtz family, which now included two small children, Käthe and Richard, settled down well in Heidelberg. He was elected a corresponding member of the Academy of Sciences in Munich and attended its festival in March 1859, meeting some of the other academicians, as well as the composer Richard Wagner and King Ludwig II. Meanwhile he received the news from Potsdam that his father, to whom he was closely attached, had died following a stroke, while at home it was increasingly clear that the health of his beloved wife was rapidly declining. She died at the end of the year, leaving him despondent. He was already suffering from migraine and fainting fits, brought on by the strain. He sought comfort and distraction in his work. Further recognition came with corresponding membership of the academies of Vienna and Göttingen and the award of the Order of the Golden Lion from the Netherlands.

The subject of the next profile will be Lord Kelvin. When he was still just William Thompson, he met Helmholtz for the first time when he was in Germany in 1855. Afterwards Helmholtz wrote to his wife that 'I expected to find the man, who is one of the first mathematical physicists of Europe, somewhat older than myself, and was not a little astonished when a very juvenile and exceedingly fair youth, who looked quite girlish, came forward . . . he far exceeds all the great men of science with whom I have made personal acquaintance, in intelligence and lucidity and mobility of thought, so that I felt quite wooden besides him sometimes.'

A remarkably close friendship and scientific collaboration sprang up between Helmholtz and Thompson, which lasted until the death of Helmholtz, nearly forty years later. In the summer of 1860 he went to stay with Thomson on the Isle of Arran; and the following February wrote to him about his plans for the future:

> I had seriously to think of introducing a new order of things, and if this had to be done, it was better it should be soon. In the end it came about more rapidly than I expected, for when love has once obtained permission to germinate, it grows without further appeal to reason. My fiancée is a gifted maiden, young in comparison with myself, and is I think one of the beauties of Heidelberg. She is very keen-witted and intelligent, also accustomed to society, as she received a good deal of her education in Paris and London . . . I should never have presumed, as a widower with two children, and no longer in my first youth, to seek the hand of so young a lady, who had every qualification for playing a prominent part in society. However it all came about very quickly and now I can once more face the future happily. The wedding is to be at Whitsuntide.

Helmholtz' second wife, Anna von Mohl, was a woman of great force of character, talented, with wide views and high aspirations, clever in society and brought up in a circle in which intelligence and character were equally esteemed. In 1862 a son, Robert Julius, was born, but this nearly cost Anna her life and the son soon became fatally ill. Then in 1864 she gave birth to a daughter, Ellen Ada Elizabeth.

At Easter 1861 Helmholtz made another visit to England, to lecture on the physiological theory of music, and was prevailed upon to give an evening discourse at the Royal Institution on the application of the law of the conservation of energy to organic nature. In 1864 he was back in England for six weeks, mostly in London but also in Oxford, Glasgow and Manchester: he lectured at the Royal Society and the Royal Institution and met the young Clerk Maxwell for the first time. The following year he was in Paris, returning for the Ophthalmological Congress held there during the Great Exhibition of 1867.

By 1868 Helmholtz had been in Heidelberg for ten years, and the Prussians decided to make another effort to secure his return. The chair of physics and mathematics at Bonn had been left vacant by the death of Julius Plücker, which led Bonn to enquire whether Helmholtz would be interested in returning as professor of physics (Plücker was unusual in being well-qualified in both disciplines). In his reply Helmholtz explained that physics had originally been his principal scientific interest and that he had been led to medicine, particularly physiology, by force of circumstances. He went on to say that 'what I have accomplished in physiology rests mainly upon a physical foundation. The young people whose studies I now direct are for the most part medical students, and most of them are not sufficiently grounded in mathematics and physics to take up what I should consider the best of the subjects that I could teach . . . Lectures in pure mathematics I could not well undertake; in those on mathematical physics I should treat mathematics as the means and not as the end.' Interestingly, in 1868 he surprised the scientific and mathematical world by publishing an essay 'On the Facts that Underlie Geometry', which had much in common with Bernhard Riemann's masterly paper 'On the Hypotheses that Underlie Geometry', published fourteen years earlier, of which he was unaware, but there were some differences and even errors in Helmholtz' work.

In the end Helmholtz decided to remain at Heidelberg. Another son, Friedrich Julius, was born, but did not thrive. At the beginning of 1870 Helmholtz was elected an external member of the Berlin Academy, and efforts were now made to interest him in moving to the Prussian capital. The faculty of philosophy at the University of Berlin had a vacancy to fill, following the death of Magnus, and proposed to offer it to either Kirchhoff or Helmholtz, with a preference for the former: 'if Helmholtz is the more gifted and universal in research, Kirchhoff is the more practised physicist and successful teacher. While Helmholtz is more productive, and is always occupied with new problems, Kirchhoff has more inclination to teaching; his lectures are a model of lucidity and polish; also from what I hear he is better able to superintend the works of elementary students than Helmholtz.' By all accounts Helmholtz was a poor lecturer, who simply read out verbatim passages from the books he had written. Despite this, he attracted outstanding students.

When it became clear that Kirchhoff would not leave Heidelberg, Helmholtz was invited to state his requirements for taking the Berlin post. Although the university had been founded in 1810, science was not taught until later and even then only at an elementary level. Helmholtz was determined to make Berlin a major centre for physics. He asked for an institute of physics, with the necessary equipment, of which he would be director, and an official residence. The Minister of Education agreed to his terms and the Helmholtz family prepared to move. Confirmation of the appointment was held up until the end of the year, because of the Franco-Prussian War. Almost immediately afterwards he received a letter from Sir William Thomson asking whether he would be willing to accept the chair of experimental physics in Cambridge, but by then Helmholtz felt committed to Berlin.

Helmholtz and his wife Anna entertained in great style. Their parties brought together intellectuals, scientists, artists and leaders of both government and industry. Their daughter Ellen married the son of the great industrialist Werner von Siemens. The two children by Helmholtz' first marriage had been, since his second marriage, in the care of their grandmother. The daughter Käthe grew up a serious woman, greatly loved and admired but never satisfied, a gifted artist who painted in the ateliers of Berlin and Paris. She spent a year in France and England. After she had married in 1872 a daughter was born, but then her health declined and she died five years later. Helmholtz' eldest son Robert volunteered for military service and was sent to the front where he was accidentally injured; after discharge from the army he became a student at the Munich polytechnic before embarking on a successful career in mechanical engineering.

In 1871 Helmholtz was back in Britain again for the meeting of the British Association at Edinburgh, meeting beforehand the mathematical physicist P.G. Tait and the naturalist T.H. Huxley at St Andrews. Later he joined Sir William Thomson, by this time Lord Kelvin, for a cruise on his yacht. Helmholtz was struck by Kelvin's easy relationships with students, engineers, seafarers and aristocrats. The sea was rough, but this was not allowed to interfere with scientific work.

In 1876, although he had only been there for five years, he was elected rector of the university; in his inaugural address he particularly praised the academic freedom enjoyed by German students and faculty, 'which amazes all foreigners'. At the end of his year of office he went on a tour of Italy, seeing various scientists en route, and in 1880 made a similar tour of Spain. The next year he went to London, accompanied by his wife Anna, to give the Faraday lecture at the Chemical Society and then on to Cambridge to receive an honorary degree. In Germany he was elevated to the ranks of the hereditary nobility and so became entitled to use the prefix 'von'.

In 1887 he was appointed president of the new physical–technical institute at Berlin Charlottenburg, which had been established through the munificence of his friend Werner von Siemens. During the next few years Helmholtz was much involved in the work of setting up the new institute. Although he retained his position at the university, he was relieved of many of the routine duties attached to it. He crossed the Atlantic for the first time on the occasion of the Chicago World's Fair; his wife, who went with him, had serious misgivings about the wisdom of making the journey. They were impressed by the cities of the east and by Niagara Falls, but not by the prairies; they went as far west as the Rockies. Felix Klein, who was with them, wrote an account of what happened on the voyage back to Europe: 'we were sitting in the smoking room till about 10 p.m. with a perfectly calm sea . . . when Helmholtz, remarking that it was time to go to bed, went down a fairly steep stairway leading to the saloon. Then we heard a heavy fall . . . on which we all hurried below, and were in time to see Helmholtz lifted by a number of stewards at the foot of the gangway, and carried into his cabin; there was a pool of blood on the floor.' His wife Anna, who was suffering from sickness at the time, wrote to the children:

> Professor Klein came in, and broke to me that your father had fallen down the companion way, and was bleeding from the forehead and nose, and that two doctors were with him; and he led me into the ship's doctor's cabin. There lay your father covered with blood, but he appeared to be conscious and was able to answer all questions. At first they feared an apoplectic stroke, which I never believed for a moment, but I think one of his old and long-forgotten swoons must have suddenly come over him. Evidently he had become unconscious before the fall, since he did not put out his hands to protect himself but fell heavily on his face.

Helmholtz gradually recovered from the accident, but he was weakened by it and suffered from double vision and vertigo. There were other shocks in store; the death of his beloved son Robert, the constant illness in mind and body of his invalid son Fritz and then the untimely death of his disciple Heinrich Hertz, a great loss to science. Nevertheless, Helmholtz was able to carry on both his scientific research and his administrative duties at the institute until, on July 12, 1894, he experienced a stroke, which left him confused and partially paralysed. His condition gradually deteriorated until the end came in his seventy-third year on September 8, 1894. His wife survived him by five years; their son Fritz died in 1901, at the age of 33.

The enormous influence Helmholtz wielded derived from his magnetic personality. The notoriously reticent and undemonstrative physicist Max Planck gives us some idea of how impressive this was:

> In his whole personality, his incorruptible judgement and in his modest manner he represented the dignity and truth of science. I was deeply touched by his human kindness. When in conversation he looked at me with his quiet, searching but benevolent eyes, I was seized by a feeling of boundless, childlike devotion. I would have been prepared to confide in him anything which affected me deeply in the certainty of finding in him a just and mild counsellor, and an appreciative or even praising word from his mouth gave me greater happiness than all the success I could achieve in this world.

5   From Kelvin to Boltzmann

Our next five remarkable physicists were born in the twenty-one years from 1824 to 1844. Two came from Scotland and one from each of Austria, England and America.

WILLIAM THOMSON (LORD KELVIN OF LARGS) (1824–1907)
In 1894 William Thomson had held the chair of natural philosophy at Glasgow University for fifty years; there was a great gathering of scientists from all over the world to celebrate the occasion. Together with his friend Helmholtz in Germany, he had been the foremost figure in transforming the science of physics as it was known at that time. Three years later he retired from the chair; the periodical *Vanity Fair* published a caricature of him, with the following note:

> He has been President of the Royal Society once, and of the Royal Society of Edinburgh three times. He has been honoured by nine universities – from Oxford to Bologna. He is the modest wearer of German, Belgian, French and Italian orders, and he has been twice married. He knows all there is to know about heat, all that is yet known about magnetism, and all he can find out about electricity. He is a very great, honest and humble scientist who has written much and done more.

William Thomson, the future Lord Kelvin, was born on June 26, 1824 in a comfortable house on the outskirts of the Irish city of Belfast. He was the fourth of seven children, four sons and three daughters. Their mother, Margaret (née Gardner), came from a Scottish mercantile family. Their father James Thomson, an Ulster Scot, was professor of mathematics at the non-sectarian Belfast Academical Institution. Margaret Thomson died in May 1830, shortly after giving birth to their youngest son Robert, so the upbringing of the children devolved entirely on the father. In 1832 the family, reduced to six children by the death of one in infancy, moved to Glasgow, where James Thomson became professor of mathematics at Glasgow College, the educational heart of the ancient University of

Glasgow, where he had earlier been a student. He was already known as the author of several mathematical textbooks, the royalties from which helped to supplement his meagre salary.

Initially William and his elder brother James were taught at home. In 1834 both boys began their formal education by matriculating at Glasgow, where the academic environment was one characteristic of Scottish universities at the time, which differed greatly from those of Oxford and Cambridge. They were much more universities of the people, where untutored boys were sent to train for the professions. In Scotland natural philosophy was an essential part of an all-round course that began with philosophy and ended with theology. Whereas at Cambridge there was not even a chair of natural philosophy, at Glasgow there were professorships of astronomy and chemistry as well as natural philosophy.

William Meikleham, then holder of the chair of natural philosophy, had a great respect for the French approach to physical science, as exemplified by Legendre, Lagrange and Laplace. Encouraged by Meikleham, William Thomson read Laplace's *Mécanique céleste* and Fourier's *Théorie analytique de la chaleur* in French during a family continental tour in 1839. His brother James, who had a passionate interest in all things mechanical, had decided to become an engineer. The third son John was embarking on a career in medicine, but succumbed to the epidemic of typhus in 1847 which followed the famine in Ireland; the youngest brother Robert went into the insurance business and emigrated to Australia. Since William was most attracted by mathematics, his father sent him to Cambridge.

Like Cavendish before him, William Thomson was admitted to Peter-house, where he was coached by the famous William Hopkins. Because of the exaggerated importance attached to finishing in the first rank of the Tripos examinations, Thomson's studies at Cambridge did not influence him as deeply as had those of his years at Glasgow. Much time and effort was devoted to learning how to deal with the particular kinds of mathematical problems set, which were only rarely related to any physical questions that were not considered in Newton's *Principia*.

His father was not surprised to find that, compared with Glasgow, putting a young man through Cambridge was not cheap. He regularly admonished his son not to be extravagant and to avoid gaining a reputation for idleness and dissipation, advising him especially to avoid the rowdy rowing men who would waste too much of his time. He was dismayed when his high-spirited son, who was inclined to be impetuous, wrote that he had bought a share in a small boat for himself, having been carried away by the excitement of the boat races. Sculling, he told his father, was excellent exercise and much more enjoyable than walking in the dull Cambridgeshire countryside. Increasingly he participated fully in student life, winning trophies for sculling, and became one of the most popular undergraduates of his year.

When the time came for him to sit the Tripos examination in January 1845, Thomson came out second on the list. He was surprised not to be senior Wrangler; no-one was in any doubt that he was the best mathematician of his year. In the highly competitive examination for the Smith's prize, which followed, he came out first, having found the questions more to his liking. He was elected a foundation fellow of Peterhouse and college lecturer in mathematics later in the year, although still barely twenty-one. He had done very well, but his future career was to be in Glasgow, not Cambridge.

At Cambridge there was still little interest in the work of the French analysts, but during his student years Thomson did not neglect to expand his knowledge of French mathematical techniques and theories. Soon after graduation, encouraged by his father, he travelled again to France. Quite by chance he came across a reference to Green's *Essay*, and was able to borrow two copies just before he set off to spend the summer in Paris. He read it on the journey and when he arrived showed it to Liouville, Sturm and other French scientists. They were astonished to find that the unknown Green had been at least twenty years ahead of anyone else. Thomson gave a copy of the *Essay* to August Leopold Crelle for publication in his journal, and later it was published again in a German translation. When Crelle asked Thomson to provide a biographical introduction, he wrote to Caius for information, and the college forwarded his letter to Sir Edward Bromhead. Sir Edward replied and also contacted Tomlin. Without the letters they sent Thomson, we would know even less about George Green. For the rest of his life Thomson never failed to express his admiration for Green's achievements.

Thomson's studies in Paris were crucial for the subsequent development of British physical science, but it was not so much the lectures as the practice in new experimental work which he found so important later. He said he was particularly indebted to his teacher Victor Regnault for 'a faultless technique, a love of precision in all things, and the highest virtue of an experimenter – patience'. During this period he developed the technique of electrical images, first read about Carnot's theory of the motive power of heat and formulated a methodology of scientific explanation that strongly influenced Maxwell.

After his 1845 visit to Paris, Thomson returned to Scotland. The next year, following the death of Meikleham, he was elected to the professorship of natural philosophy at the University of Glasgow, the post he held for the remaining fifty-three years of his life. The election was hotly contested, but in the end he was selected unanimously. He was also elected a fellow of the Royal Society of Edinburgh. In 1849 Thomson's first great memoir, *A Mathematical Theory of Magnetism*, was published, which led to his election to the Royal Society of London. In 1851 a second great memoir appeared, *On the Dynamical Theory of Heat*. In this work he postulated the existence of a state of complete rest, which he called absolute zero, the base for the temperature scale to which the name Kelvin is nowadays attached. In these years he used to spend some time each summer in Cambridge, to keep in touch with the scientists there and in London, and once or twice he revisited Paris as well.

At that time neither in Scotland nor in England was there a research laboratory in a university or anywhere else in which students could work. Thomson, having enjoyed such facilities in Paris, was keen to establish similar opportunities for his students, so he extracted a small sum from the university for that purpose. It made possible the first teaching laboratory in Britain, albeit of a modest sort. He was also greatly interested in developing measuring instruments of high accuracy, and the facilities of his new laboratory made that possible also. The new professor soon became popular through his radically new professionalism, marked by a clear research orientation.

For an impression of Thomson in the lecture room we have these lines of one of his students, written years afterwards: 'Lord Kelvin [as Thomson later became] possessed the gift of lucid exposition in ordinary language remarkably free from technicalities. Occasionally he got out of range with the majority of his class, but there was no obscurity in his statement, it was simply beyond their grasp . . . he had no syllabus of lectures and used no notes in lecturing. He had his subject clearly before him and dealt with it in logical order. He was not dictating a manual of natural philosophy to his students . . . he considered it unnecessary for him to teach what could be got in an ordinary textbook and that his province was to supplement this.' On one occasion he described his method as follows: 'the object of a university is teaching, not testing . . . the object of the examination is to promote the teaching. The examination should, in the first place, be daily. No professor should meet his class without talking with them. He should talk to them and they to him. The French call a lecture a *conference*, and I admire the idea involved in the name. Every lecture should be a conference of teaching and students.'

After his beloved father had died in an outbreak of cholera in 1849, Thomson began actively seeking a wife. Disappointed in his overtures to one young lady, he married, on the rebound, the highly cultured and intellectual Margaret Crum in September 1852. She was three years younger than he was and came from one of the newer commercial families of Glasgow, with whom he had been friendly for many years. Unfortunately her health declined rapidly following an exceptionally strenuous honeymoon tour around the Mediterranean, and attempts at treatment were largely unsuccessful. It is not known what was wrong with her, but, after prolonged suffering and many setbacks, she died in June 1870, after seventeen years of marriage. Another misfortune in this period was an accident when Thomson was engaged in the Scottish sport of 'curling' on ice, in which he fractured a thigh-bone; this continued to be troublesome for the rest of his life.

It was Thomson who drew the attention of the scientific world to the theories of Sadi Carnot, the pioneer of thermodynamics, and his English follower James Joule. For a number of years Thomson and Joule collaborated on research in this area, and, in a posthumous tribute to Joule, Thomson said that 'the genius to plan, the courage to undertake, the marvellous ability to execute and the keen perseverance to carry through to the end the great series of experimental investigations by which Joule discovered and proved the conservation of energy in electric, electromagnetic and electrochemical actions, and in the friction and impact of solids, and measured accurately, by means of the friction of fluids, the mechanical equivalent of heat, cannot be generally and thoroughly understood at present. Indeed it is all the scientific world can do just now in this subject to learn gradually the new knowledge gained.'

The role that James Thomson played in the life of his younger brother must not be underestimated. After working in industry for some years, where he gained much valuable practical experience, James had settled in Belfast and become professor of civil engineering at Queen's College, until in 1873 he moved to be professor of engineering in Glasgow. The Cambridge mathematical physicist Sir Joseph Larmor referred to James as 'the philosopher, who plagued his pragmatic brother' to obtain a comprehensive understanding of the problems he dealt with. As Helmholtz explained in 1863, 'James was a level-headed fellow, full of good ideas, but cares for nothing except engineering, and talks about it ceaselessly all day and all night, so that nothing can be got in when he is present. It is really comic to see how the two brothers talk at one another, and neither listens, and each holds forth about quite different matters. But the engineer is the more stubborn, and generally gets through with his subject.' At dinner parties where both brothers were present, it was considered advisable to seat them as far apart as possible.

Some of Thomson's many scientific friends have already been mentioned, but not all. One of his most important and long-standing friendships was with the mathematical physicist George Stokes, later Sir George, whose career was almost exclusively in Cambridge. From the time of their first meeting in 1845 until the death of Stokes in 1903 they were in frequent correspondence; Thomson seems to have consulted Stokes at every opportunity. Another of Thomson's disciples was Tait, who had become professor of natural philosophy in Edinburgh; they collaborated on the classic *Treatise on Natural Philosophy*, first published in 1867, which for many years was the bible of theoretical physicists in Britain and other lands. Tait, who thrived on controversy, was a protégé of the Irish mathematician Sir William Rowan Hamilton and, after the latter's death, took over the mission of promoting the use of quaternions; Thomson firmly resisted his attempts to use them in their joint work.

Telegraphy by land lines dates from about 1837; by 1850 there was a successful submarine cable between England and France. However, it was the Atlantic telegraph of 1866 which most caught the popular imagination, and Thomson played a leading part in that enterprise. His interest and reputation brought him to the attention of a consortium of British industrialists who, in the mid 1850s, proposed to lay a submarine telegraph cable between Ireland and Newfoundland, to improve communication between Europe and North America. Telegraphy was by then a well-developed and extremely profitable business; the idea of laying such a cable was not new. The undertaking provided perhaps the first instance of the complex interaction between large-scale industrial enterprise and theoretical electricity.

Thomson was brought in early as a member of the board of directors of the project and he played a central role in its execution. The directors had entrusted the technical details of the project to an industrial electrician named Whitehouse, and many of the difficulties which plagued it from the outset resulted from Whitehouse's insistence on employing his own system of electric signalling. Thomson had developed a very sensitive apparatus, the mirror galvanometer, to detect the minuscule currents transmitted through miles of cable, but Whitehouse refused to use it. Thomson had asserted that the length of the cable would, by a process of statical charging of its insulation, substantially reduce the rate at which signals could be sent unless low voltages were used, so low that only his galvanometer could detect the currents. However, the Whitehouse–Thomson controversy stemmed primarily from Whitehouse's jealousy of Thomson's reputation.

The first attempt to lay such a cable in 1857 ended when it snapped and was lost. The second attempt, a year later, was successful, but the high voltages required by the Whitehouse method reduced the ability of the cable to transmit signals rapidly, just as Thomson had predicted. Whitehouse privately recognized the inadequacy of his own instruments and surreptitiously substituted Thomson's galvanometer while claiming success for his own methods. This deception was soon discovered, and the ensuing controversy among Whitehouse, the board of directors and Thomson combined theoretical science, professional vanity and financial ignominy. A third cable was laid in 1865 and, with the use of Thomson's instruments, it proved capable of rapid sustained transmission. Thomson's role as the man who saved a substantial investment made him a hero to the British financial community and to the Victorian public in general; indeed he was knighted for it by Queen Victoria in 1866.

Sir William's success with the Atlantic cable, and the close relation thereby established between testing in the laboratory and application to the electrical industry, opened the way for his ambitious marketing of scientific knowledge. Through a carefully developed and cleverly exploited system of patents and partnerships, his financial returns on his scientific capital were such that he soon became a wealthy man. This kind of success was deplored by other scientists, especially those of France and Germany, who maintained an almost religious belief in the importance of keeping science pure, uncontaminated by industrial applications.

Some of his most useful and profitable inventions were related to navigation. One of the most successful was the patent magnetic compass, of which no fewer than 10 000 were sold; this gradually superseded the older type which was unreliable in ironclad ships. Sir William purchased a schooner of 126 tons partly for his own pleasure but also so that he could try out such inventions at sea. It also provided a way he could escape from the pressures of his many different responsibilities. One cruise took him to the island of Madeira, where he was entertained by the Blandy family of wine shippers. It was then that he met Frances Anna Blandy, who was about fourteen years younger, and they were married in 1874. Neither of his marriages resulted in children.

Throughout the latter part of the nineteenth century the progressive University of Glasgow was becoming increasingly cramped on its ancient site among the slums in the High Street. Eventually enough money was raised by public subscription, by government grant and by the philanthropy of local industrialists to build a grand cathedral of learning on a new site, which opened in 1870. It included a new physical laboratory, vastly superior to the old one. Two years later Sir William was elected a life fellow of Peterhouse, which had the result that he visited Cambridge more frequently and was able to play a leading role in the efforts being made to reform the university syllabus by introducing more natural philosophy. In 1876 and again in 1900 the Mastership of Peterhouse became vacant and each time he was asked whether he would be interested in the position. Each time he declined decisively, as he did when the Cavendish chair became vacant in 1879. He was such a grand figure in Glasgow with his electrical engineering business as well as his academic position that it is hardly surprising that he did not wish to leave. In 1891 the students elected him Lord Rector of the University of Glasgow, with which he was so closely associated, and in 1904 he was appointed Chancellor.

In 1876 the science journal *Nature* printed an appreciation of Sir William's achievements by Helmholtz, which concludes by saying that 'British science may be congratulated on the fact that in Sir William Thomson the most brilliant genius of the investigator is associated with the most lovable qualities of the man. His single-minded enthusiasm for the promotion of knowledge, his wealth of kindliness for younger men and fellow-workers, and his splendid modesty, are among the qualities for which those who know him best admire him most.'

Although he was no longer young, Sir William still seemed to have boundless energy and a list of all his varied activities would take up far more space than is available here. However, one that should certainly be mentioned is his first visit to America in 1884, to give a course of twenty lectures on molecular dynamics at the new Johns Hopkins University in Baltimore. The invitation originally came from the mathematician J.J. Sylvester, who was then at Johns Hopkins but had moved to Oxford by the time the lectures were delivered. The audience included many of the American scientists of that time. Although on this occasion he was too busy with his other interests to see much of the USA, he returned in 1897 for the meeting of the British Association in Toronto and took the opportunity to visit the Canadian and American West.

In 1892 Sir William was created Baron Kelvin of Largs in the County of Ayr, the first 'scientific' peer of the realm (the Kelvin is the stream that runs through the gardens of the university). This might not have happened, however, had he not also shown his support for the Conservative Party, particularly in relation to Irish questions. He always laid stress on his Ulster Scottish background and thought of himself as more Irish than Scottish. From his father he inherited the enlightened liberal tradition which owed its origin to Ulster of the eighteenth century and which strove for the establishment of a non-sectarian framework of government and institution, free of party rivalry and based on merit alone.

In 1896, the jubilee year of his professorship, there were great celebrations and he was decorated with the Grand Cross of the Royal Victorian Order. In 1899, the year in which he retired from his chair, he became the first foreigner to be decorated with the Grand Cross of the Legion of Honour. Academic jubilees are regularly celebrated in continental Europe but less so in Britain; Kelvin's was a particularly splendid occasion, where messages of congratulation reached him from all over the world, and it was remarked that there was an unusual element of spontaneity to the celebrations.

In addition to Netherhall, his country seat near Largs, the Kelvins had a London home at 15 Eaton Place, in Belgravia. In the House of Lords he made some fourteen speeches, six of which related to maritime affairs. During the same period he published almost 130 scientific papers, including six in the year of his death and two published posthumously. His eldest brother James died in 1892, at the age of seventy, and Elizabeth, the eldest sibling and last surviving sister, also died in the same year. Thus he outlived all the other members of his family and many of his closest scientific colleagues as well.

For a glimpse of the Kelvins at the turn of the century, an old friend of theirs recalled the following: 'The dear Kelvins arrived. He is a great source of anxiety. He insists on doing everything but is not well and once or twice in the last week he has turned faint at dinner. He did so last night but gulped down some champagne & came around & went to a party in the evening. His troubles are connected with digestion & Maimie is distracted and hoping she is providing the right food. We love having them but shall be thankful when they are safely away.'

In September 1907 Lady Kelvin suffered a major stroke. Her illness rendered her aged husband anxious yet optimistic. Until then his own health had been fairly robust, but at the end of November he too became gravely ill: his doctor diagnosed the complaint as 'a severe chill of the liver (duodenal catarrh)'. The accompanying fever continued and Lord Kelvin died in Netherhall on December 17, 1907, at the age of eighty-three. The funeral took place at Westminster Abbey, two days before Christmas.

JAMES CLERK MAXWELL (1831–1879)

Clerk Maxwell is regarded generally as standing at the high-water mark of classical physics. One of his greatest admirers was Albert Einstein, who commented that 'it was Maxwell who fully comprehended the significance of the field concept; he made the fundamental discovery that the laws of electrodynamics found their natural expression in the differential equations for the electromagnetic fields. This led to the electromagnetic theory of light, one of the greatest triumphs in the grand attempt to find a unity in physics.'

James Clerk Maxwell was born on June 13, 1831 at 14 India Street, Edinburgh, today the home of the International Centre for Mathematical Sciences. John Clerk Maxwell, his father, was a land- owner of scientific bent. He had built to his own design a modest house named Glenlair, on an estate of 1500 acres near Dalbeattie in Galloway owned by the family. He was an advocate (the Scottish equivalent of a barrister) although more interested in technology than the law. Attending the meetings of the old-established Royal Society of Edinburgh was one of his chief pleasures. Another regular was John Cay, whose daughter Frances he married when he was thirty-nine and she thirty-four. Apart from a daughter who died in infancy, James was their only child. He was a delicate boy, like Helmholtz, an odd and eccentric child of an unusually inquisitive nature. His mother

saw to his early education; after she died in 1839 her unmarried sister Jane helped the boy's father with his upbringing. At the age of eight he was entrusted to a private tutor for a few years, until he was ready to begin his formal education at the new Edinburgh Academy, which provided its pupils with a good background in mathematics and some understanding of physics, as well as the classical languages. When he left he was not only top of the class in mathematics and English but very nearly so in Latin as well. Although he had been shy as a young child, he lost the shyness at school, where he came in for a lot of bullying, mainly because of his sensible clothes, designed by his father, and his rustic Galloway accent, but he also stuttered and, throughout his life, when ill at ease he could lapse into 'chaotic statements'. In class he began to enjoy classical literature and mathematics and formed some enduring friendships, notably with Tait. As soon as he was old enough, his father took him to the lectures at the Royal Society of Edinburgh.

After six years at the academy the young man went on to the University of Edinburgh, where he attended courses in natural philosophy and mathematics; he was most impressive by Sir William Hamilton's class in logic. During the long summer vacations, when he was at Glenlair, he pursued his own lines of investigation, feeling increasingly confident of his mathematical and scientific powers. After three useful years at Edinburgh the next step was to be Cambridge, although his father had misgivings about this.

Following in the footsteps of Thomson a few years earlier, Maxwell was admitted to Peterhouse, but, because there were rather too many able students in his year, he was advised to migrate to Trinity, where the chances of a fellowship after graduation would be better. There he was coached by the legendary Hopkins, tutor of Cayley, Stokes and other well-known scientists. In his obituary of Maxwell his friend Tait wrote that 'He brought to Cambridge in the autumn of 1850 a mass of knowledge which was really immense for so young a man, but in a state of disorder appalling for his methodical private tutor. Though that tutor was Hopkins the pupil to a great extent took his own way; and it may safely be said that no high wrangler ever entered the Senate-house more imperfectly trained to produce "paying" work than did Clerk Maxwell.' Hopkins described Maxwell as 'the most extraordinary man he had met within his whole range of experience; it appeared impossible for him to think incorrectly on physical subjects.' When the Tripos results were announced, Maxwell was surprised to be listed only as second to the Canadian-born Edward Routh, but he came equal first with Routh in the separate examination for the Smith's prize. He was already recognized as having genius. He was elected to the select essay society called the Apostles, where he read an essay on the philosophy of science, especially the history of the development of scientific ideas. Following a period of 'brain fever', Maxwell became active in the Christian Socialist movement, which tempered the Scottish Calvinism of his upbringing. At this time Oxford University was wracked with religious controversy, but Cambridge was less affected.

As an adult Maxwell was about five foot four in height, strong and athletic. 'He was possessed of dark eyes, jet black hair and beard, and in complexion somewhat pale. His mirth was real, but never boisterous. In disposition he was genial and patient, and he had great power of concentration even amidst distractions. He had considerable knowledge and discrimination in literature, he was a rapid reader, and he had a retentive memory.' Another description of him at the age of eighteen is intriguing:

> James Clerk Maxwell still occasioned some concern to the more conventional amongst his friends by the originality and simplicity of his ways. His replies in ordinary conversation were indirect and enigmatical, often uttered with hesitation and in a monotonous key.

While extremely neat in person, he had a rooted objection to the
vanities of starch and gloves. He had a pious horror of destroying
anything – even a scrap of writing paper. He preferred travelling by the
third class in railway journeys, saying he liked a hard seat. When at
table he often seemed abstracted from what was going on . . .

Already Maxwell was acquainted personally with many of the British men of
science. Even before he went to Cambridge he had created a stir at a meeting
of the British Association in Edinburgh by an intervention in his broad
Galloway accent. Other men of science he met at Cambridge, especially
after he had become a fellow of Trinity in October 1855.

When he learned that the professor of natural philosophy at Aberdeen
had died, he decided to please his father, whose health was causing him
some anxiety, by applying for the post. Maxwell's attachment to Glenlair,
and to the continuity of family tradition, was strong. In the winter of 1855/6
Maxwell returned to Scotland to be with his father and was at his side
in April 1856 when he suddenly died. Since it had been his father's wish,
Maxwell accepted the Aberdeen chair when he was offered it. He had been
publishing work of steadily increasing quality ever since he was a schoolboy and now was on the brink of his greatest achievement. At Cambridge
he had published his first paper on electromagnetism, the branch of science
in which he was to become supreme. He also took up the theory of gases,
another life-long interest. At Aberdeen he was responsible for the course on
natural philosophy given to students whose knowledge was fairly rudimen-
tary. There were fifteen lectures a week, plus demonstrations and examinations. He introduced the idea of having students conduct experiments
themselves, rather than just observe them being done, and later introduced
the same practice in London and Cambridge. In addition he gave evening
lectures to artisans, perhaps reflecting the influence of Christian Socialism.

Maxwell had resigned his Trinity fellowship on appointment to the
Aberdeen chair but, as a Cambridge graduate, he could still compete for the
Adams prize. He proved to be the only candidate in a year when the subject
was the structure of the rings of the planet Saturn; he showed that they
could not be solid but must be made up of particles. Airy, the Astronomer
Royal, described his essay as 'one of the most remarkable applications to
mechanical astronomy that has appeared for many years'. Principal Dewar of
Marischal College, one of the two that made up the university, occasionally
invited Maxwell to his house, where he met Dewar's daughter Katherine,
who was seven years older than he was. They were married in June 1858.

She was not popular with his students or colleagues and became increasingly neurotic as the years went on. She was jealous of his scientific friends to such an extent that he could never invite them to his home.

At this period the existence of two separate colleges in Aberdeen, Marischal and King's, gave students a choice, since there were usually two lecturers in each subject, but Maxwell and others were in favour of a merger. When this was agreed in 1860 many redundancies were inevitable, and Maxwell's post was one of them. He was not much upset by this, because there was a vacancy for a professor of natural philosophy in Edinburgh. Both Maxwell and Tait applied for this. Tait had been runner-up for the Aberdeen chair, but this time it was Tait who was successful, being considered a much better teacher even if Maxwell was a far more powerful researcher. However, it turned out that Scotland's loss was England's gain. The professorship of natural philosophy at King's College, London was also vacant and Maxwell was appointed in the summer of 1860. Election to the Royal Society of London soon followed.

King's College had been founded in 1828 as an Anglican answer to University College, itself a non-conformist answer to Oxford and Cambridge. In London the Maxwells lived at number 8 (now 16) Palace Gardens Terrace, a newly built house in Kensington. The attic of the house was converted into a private laboratory. Maxwell taught and researched in the mornings as a rule, the afternoons being devoted to riding in Hyde Park, and sometimes carried out experiments at home in the evening; more usually he was out giving lectures to artisans, as at Aberdeen, which he seems to have regarded as more worthwhile than his college lectures. The ground floor was given up to his brother-in-law, who was seriously ill; despite being very busy, Maxwell helped his wife look after him.

Scientifically, Maxwell's stay in London was the most fertile period of his career. The precise formulation of the space-time laws for electromagnetic fields was his greatest achievement. The differential equations he formulated showed that the fields spread in the form of polarized waves and with the speed of light. However, in addition to his work on electromagnetism, he carried out both theoretical and experimental work on the viscosity of gases, continued research into colour vision and took the world's first colour photograph. All this against a heavy teaching load. An added demand on his time was membership of various national committees, such as the British Association's committee on electrical units; their recommendation formed a basis for the electrical side of the international system adopted in 1881.

While Maxwell was at King's he got to know other men of science, notably Faraday. They had begun to correspond while Maxwell was at Cambridge but now Maxwell took the opportunity of meeting him personally and attending his famous Friday evening lectures at the Royal Institution. They got on well together. After one of the discourses, when Maxwell was caught up in the crowd of people all pressing to get out, Faraday, appreciating the similarity of the situation to that of molecules in gas, called to him 'Ho Maxwell, you cannot get out. If any man can find a way through a crowd it should be you.'

After five years in London, Maxwell left to return to Glenlair. One of the reasons was to enlarge the family seat, a project of his father's that the son regarded as a sacred trust; in any case he enjoyed being laird and took his responsibilities very seriously. The other was to collect his scientific thoughts together in his masterpiece, *A Treatise on Electricity and Magnetism*, which finally appeared in 1873. During this period he also wrote a textbook on heat, published in 1870, in which the famous Maxwell demon makes his first appearance.

Maxwell was a wealthy man, who had no need to earn his living as an academic. Of course he was isolated from the scientific community at Glenlair, but kept in touch with his main associates by mail, particularly with Stokes, Thomson and Tait; there was so much correspondence that the post office provided him with a private mail box. Also the Maxwells usually wintered in London, where he could participate in the scientific life of the capital. At Cambridge, where he was several times an external examiner, he had some success in extending the scope of the Tripos syllabus, for example to include electricity and magnetism. In 1868 the position of Principal of St Andrews University became vacant; Maxwell applied, but, despite strong support from the faculty, he was not appointed, perhaps for political reasons; the Principal is appointed by the Crown on the recommendation of the prime minister of the day.

As we have seen, it was the universities of Scotland, rather than England, that were in the vanguard of scientific education in the first half of the nineteenth century; in physics, at least, Oxford and Cambridge were in the doldrums. The situation began to improve after Royal Commissions to enquire into the two ancient English universities were instituted in 1850. In that year Oxford established the honours school in natural sciences. A year later Cambridge followed suit with the natural-science Tripos; this excluded physics, although the mathematical Tripos already included a good deal of mathematical physics. When the enlightened Prince Albert was Chancellor of the latter university, he wished to see some of the principles and practices of the increasingly successful German universities adopted. Not a great deal happened in this direction, but the need for reform was obvious. Although the use of apparatus for class demonstrations of the principles of natural philosophy, electricity and magnetism went back to the eighteenth century, there was no professorship of experimental physics until 1870, when one was established. For the initial appointment an attempt was made to interest Thomson, as we know, and then Helmholtz. The electors then offered the chair to Maxwell. His first impulse was also to say no, but in his letter of refusal he asked a number of detailed questions about the position, and the outcome was that he was elected on March 8, 1871.

In 1872 another Royal Commission, chaired by the eighth Duke of Devonshire, William Cavendish, was charged with the task of looking into the relation of the state with science, to assess whether more state support might be needed. It uncovered some remarkable facts. At Oxford University there were nine fellowships in natural science out of 165, at Cambridge three out of 105. The duke, a second Wrangler and Smith's prizeman, was the Chancellor of Cambridge at the time. He decided to reform the university off his own bat, and in 1874 the Cavendish laboratory was erected through his munificence. The duke not only paid for the building, but also for the apparatus needed to equip it initially. Since there had been no previous physics laboratory in Cambridge, unlike in Oxford, Maxwell needed to explain to the architect what was needed, on the basis of his own experience and visits to laboratories elsewhere. The building served the needs of Cambridge well for nearly a century. Not much else followed from the report of the Devonshire Commission.

By this time Maxwell was an establishment figure, traditionalist and conservative. Although he was an enthusiast for the popularization of science, his lecturing style had never been good. In Cambridge the audiences he drew were pitifully small, normally two or three, half a dozen at most. He had always been interested in history of science, and had it in mind to write a history of dynamics, in particular:

> The cultivation and popularization of correct dynamical ideas since the time of Galileo and Newton have effected an immense change in the language and ideas of common life, but it is only within recent times, and in consequence of the increasing importance of machinery, that the ideas of force, energy and power have become accurately distinguished from each other. Very few, however, even of scientific men, are careful to observe these distinctions; hence we often hear of

the force of a cannon ball, when either its energy or its momentum is meant, and the force of an electrified body when the quantity of its electrification is meant.

Instead he settled down to editing the papers of Henry Cavendish, a task he could easily have passed on to someone else. What he discovered, in that mass of unpublished material, has already been described in the profile of Cavendish. Tending to the needs of his invalid wife also took up a lot of Maxwell's time; at one point Maxwell did not sleep in bed for three weeks, lecturing and running the laboratory by day, and sitting by his wife all night. He transcribed all the Cavendish papers into legible form, longhand, by himself, by candlelight in the long nights sitting at her bedside. He had contracted smallpox at Glenlair soon after their marriage and often asserted that her ministrations had saved his life. It seems strange that, quite apart from Röntgen, whose wife was tubercular, both Clerk Maxwell and William Thomson (in the case of his first marriage), married women who soon became chronic invalids, as did the mathematician Sir William Rowan Hamilton.

Of his own collected papers, fifty-eight out of 101 were published during this Cambridge period, but of these twelve are book reviews, six are lectures, eight are articles he wrote for the *Encyclopaedia Britannica*, nine are review papers and twenty-one are short (though often not unimportant) notes. Only in 1878 does Maxwell seem to have returned to full production, with the two powerful late papers on gas theory. However, from the beginning of 1879 it had been observed that there had been 'some failure of the old superabundant energy', and in the Easter term he had been unable to lecture. During the summer he returned to Glenlair, improved in health initially, but soon regressed. He knew that he had only months to live, and returned to Cambridge in great pain to have Sir George Paget, his trusted doctor, within call and 'to be with friends'. While still in the prime of life, he died on November 5, 1879, from abdominal cancer, like his mother, leaving no descendants. He was buried in a cemetery near Glenlair next to the graves of his parents and other Clerk Maxwells. A fire in 1929 left the house which had meant so much to him in a ruinous condition.

J. WILLARD GIBBS (1839–1903)

In nineteenth-century America there was a growing interest in science, but few scientists of distinction. Willard Gibbs, the outstanding American scientist of his day, was little known to the general public in his own country. He sprang from a family with long and distinguished academic, but not

scientific, connections. As one of them said, 'Though but a few of them, perhaps, could be termed of exceptional distinction, yet we find that for generations back they were, without exception, men notable for their intellect, education and integrity, who held positions of public responsibility; while their wives, as far as the more limited records show, were often women of an intellectual character.'

The subject of this profile was born in New Haven, Connecticut, on February 11, 1839, the fourth child and only son of Josiah Willard Gibbs the elder and of his wife Mary Anna (née Van Cleve). Of his three older sisters, Anna, Eliza and Julia, only the first and third survived to maturity, while a fourth sister Emily died when only twenty-three. Their father, born in 1790, was a graduate of Yale College, the old university which gives the town of New Haven its claim to fame, although previously the family had sent its sons to Harvard. He was a good all-rounder and, although he was outstanding in mathematics, he became professor of sacred literature at his *alma mater*, where he taught philology. Their mother was a woman of unusual character and attainments, who took a particular interest in ornithology.

The boy's education began at home and continued at private schools in New Haven, where he had a traditional classical education. As an undergraduate at Yale from 1854 to 1858, Willard Gibbs excelled in mathematics and Latin, gaining prizes in both subjects and winning scholarships. After graduation he entered the Yale advanced programme in engineering and earned the first American Ph.D. in that subject in 1863 with a thesis *On the Form of the Teeth of Wheels in Spur Gearing*; he explained that the subject reduces to an exercise in plane geometry. His mother died in 1855, perhaps of tuberculosis, and his father in 1861. The son also was suspected of being tubercular and suffered from astigmatism; probably these were among the reasons why he did not volunteer to serve in the armed forces in the American Civil War, which began just after he had graduated. Instead he was appointed to a three-year tutorship at his *alma mater*. At Yale, such tutors were expected to teach as required any of the prescribed subjects of the first two years of the college course, so Gibbs found himself teaching mainly Latin texts while continuing to work on applied science. He patented 'An Improved Railway Car Brake', a mechanism allowing American trains to dispense with the services of a brakeman.

The close-knit family was now reduced to Willard and his two older sisters Anna and Julia. The three of them set sail for Europe in 1866. They went first to Paris, enjoying the life of the city, while Willard Gibbs attended lectures at the Sorbonne and the Collège de France, mainly on mathematics but also on physics, and studied the classic memoirs of the great French scientists. His health was continuing to give cause for concern and so, on medical advice, they spent the later part of the winter on the Mediterranean, at the end of which he was pronounced free of tuberculosis. They then resumed their itinerary and travelled in leisurely fashion to Berlin, where they spent two semesters. Here Gibbs' younger sister Julia, who for some years had been engaged to one of his classmates at Yale, married and returned with her husband to New Haven, leaving the older sister Anna to keep Willard company. Again he attended lectures on mathematics and physics at the university and studied the German scientific literature. Since Willard Gibbs senior was a German scholar, it may be assumed that his son was sufficiently familiar with the language to derive real benefit from his time in Germany, which was completed by two semesters in Heidelberg, where he must have overlapped with the famous Russian woman mathematician Sonya Kovalevskaya. In physics he would have had the opportunity to hear Helmholtz and his distinguished colleagues Bunsen and Kirchhoff. Although detailed information is lacking, these three years of post-doctoral study in France and Germany seem to have been crucial to

his later development as a creative scientist. Britain was not included in
the itinerary. On his return to America in 1869 he brought back with him
a far more comprehensive overview, and a much deeper understanding, of
current research both in mathematics and in physics than he could have
easily obtained anywhere in the USA during this period.

At this time Yale was about to install its first new president in a
quarter-century and some of the senior professors wanted to redirect the
university onto a course more in keeping with the changing times. The
existing system was designed to train men of an active, extrovert disposition for executive positions in politics, law, the church and commerce.
Religion and law were the chief objects of study. Willard Gibbs had absorbed
the distinctive spirit of the university without question. Together with his
sisters and brother-in-law, who was librarian of Yale, he settled down in the
home in which he had been brought up, with Julia taking the place of his
late mother. Yale appointed him professor of mathematical physics in the
graduate school in 1871. He was the first occupant of the chair and retained
it for the rest of his life. Professorial salaries at Yale were not meant to be
sufficient to live on, but Willard Gibbs was unusual in being paid nothing at
all; fortunately he had private means. In 1873 Bowdoin College, in Maine,
attempted to lure him away from Yale, but, although the only response of
the Yale Corporation was to pay him a small salary rather than none at all,
he chose not to leave New Haven: financial considerations did not weigh
heavily with him. In 1884 the new and innovative Johns Hopkins University
also attempted to attract him. Johns Hopkins was the first American university to regard research as part of its mission, and, if he had moved to
Baltimore, Willard Gibbs would have been given the opportunity to create
the first research school of mathematical physics in the USA. However, he
still chose to remain at Yale.

Willard Gibbs had published nothing in mathematical physics at the
time he was appointed professor; in fact, his only real research had been
in engineering, not in physics at all. Nevertheless, by 1873 he had published a contribution to the mathematical theory of thermodynamics in the
Transactions of the Connecticut Academy of Arts and Sciences. He went
on to revolutionize the study of physical chemistry in a two-part paper
'On the Equilibrium of Heterogeneous Substances' published in the same
journal between 1875 and 1878. Meanwhile he became an enthusiast for
Grassmann's *Ausdehnungslehre* and Hamilton's theory of quaternions: 'if
Gibbs cannot be given credit for originality of methods yet he deserves praise

for the sensitivity of his judgement as to what deletions and alterations should be made in the quaternionic system in order to make a viable system', wrote a colleague tactfully.

In his mature years Gibbs was a man of striking appearance:

> A little over medium height, with a good figure, he carried himself well and walked rather rapidly and with a purposeful stride. He was always neatly dressed, usually wore a soft felt hat on the street, and never exhibited any of the physical mannerisms or eccentricities sometimes thought to be inseparable from genius. His hair and full beard were grey and his complexion clear and ruddy – almost florid. His eyes were blue and could twinkle amazingly on occasion. His forehead was high, his nose well formed and of a good length, and his mouth capable of a very sweet and intimate smile. In repose his expression was rather grave and abstracted, but how it could light up and become animated in greeting a friend or in pointing up a humorous turn to the discussion! His countenance was very mobile in conversation, every thought revealing itself in his changing expression. His manner was cordial without being effusive and conveyed clearly the innate simplicity and sincerity of his nature.

Being of a retiring disposition, Gibbs was not considered a good teacher. His lecturing style, based on the logical exposition of abstract principles, was clearly heavily influenced by the courses he had attended in Germany, in contrast to the more-pragmatic British style based rather on the development of skills relevant to the solution of practical problems. Few students could understand what he was trying to convey. He worked in almost total isolation, without the stimulus of colleagues or research students, although he was in correspondence with men of science, particularly in Britain. So far as is known, he never carried out any experimental work. He was abnormally modest and reticent about his scientific achievements, so his reputation is almost wholly due to his publications. His great importance rests on his comprehensive application of mathematics to chemical subjects, and, although some parts of his work were unwittingly duplicated in Europe, his overall grasp of chemical thermodynamics, a branch of science he can reasonably be said to have founded, was unrivalled in his time. In statistical mechanics, his methods, which were more general and more readily applicable than Boltzmann's, came to dominate the field.

The honours which came to Gibbs in later years were many and varied, but his work was particularly appreciated in Britain. In 1885 he was elected a corresponding member of the British Association, in 1891 an honorary member of the Cambridge Philosophical Society and of the Royal Institution, the next year an honorary member of the London Mathematical Society. The Royal Society, of which he was a foreign associate, awarded him the Copley and Rumford medals. Despite all these British honours, he never set foot in Britain, although twice invited to address the British Association. Again, although he was a corresponding member of the leading scientific academies of continental Europe and the recipient of honorary degrees from several European universities, he never returned to the continent. In his homeland, amongst other honours he was elected to the National Academy of Sciences at the early age of forty, although he seldom attended its meetings. He was awarded the Rumford medal by the American Academy of Arts and Sciences. He never became a member of the American Physical Society and only joined the American Mathematical Society shortly before he died. Except for a customary summer vacation in the Adirondacks, he never cared to travel far from New Haven.

In 1898 Willard Gibbs lost his older sister Anna, the constant companion of so many years. Towards the end of his life he was thinking about the possibility of another visit to Europe, but nothing came of it. Early in 1903 he began to experience some health problems, which were not thought to be serious, but suddenly in New Haven on April 28, 1903 he died aged sixty-four. After his death not much was found in the way of unpublished material, since it was his custom to work everything out in his head. He did not read exhaustively in the literature of the subjects of his research. He usually preferred to work out for himself the results obtained by others; he said that he found that easier than trying to follow the reasoning of someone else. It is significant in this connection that few succeeded in learning about the discoveries he made from his own publications. His writings demanded 'extraordinary attentiveness and devotion', it was said, his mode of exposition was 'abstract and often hard to understand'. Einstein rated Gibbs' book on statistical mechanics a masterpiece but added 'it is hard to read and the main points have to be read between the lines'. At a different level the vector notation, so useful in physics, was invented by Gibbs; although the advocates of quaternions stoutly defended Hamilton's theory, they gradually had to concede defeat. 'If I have had any success in mathematical physics', he told one of his students, 'it is, I think, because I have been able to dodge mathematical difficulties.'

JOHN WILLIAM STRUTT (LORD RAYLEIGH) (1842–1919)

During the course of the nineteenth century the world of science became increasingly institutionalized. Nevertheless, there were still first-class scientists who, like Cavendish in the previous century, were not associated with any particular institution, but preferred to work at home, at least for the greater part of their career. Our next subject is one of the last of these. He was born on November 12, 1842 as John William Strutt, the eldest son of the second Baron Rayleigh. The peerage, to which he later succeeded, was of Terling Place, Witham, Essex; his birthplace was Langford Grove, near Maldon, in the same neighbourhood. His immediate ancestors on his father's side were landed gentry with little or no interest in science. His mother Clara Elizabeth la Touche (née Vicars) came from an Irish family with a distant relationship to that of the scientist Robert Boyle. In his boyhood Strutt displayed no unusual talent, just the normal boyish interest in the natural world. Owing to occasional bouts of ill-health, his schooling consisted of short periods at Eton and Harrow followed by four years at a small boarding school near Torquay, where he showed no interest in classics but began to develop decided competence in mathematics.

In 1861, being almost twenty, young Strutt went up to Cambridge and entered Trinity College as a fellow-commoner like Thomas Young. Here he became a pupil of Routh, the senior Wrangler of Maxwell's year, who was developing a reputation as an outstanding coach. Under Routh's guidance Strutt acquired the grasp of mathematics which stood him in good stead in later years. He also benefited from the lectures of Stokes, the Lucasian professor of mathematics, who was greatly interested in experimental physics and performed many demonstrations in front of his classes using apparatus of his own. In the mathematical Tripos of 1865 Strutt came out as senior Wrangler and also came first in the Smith's prize examination. By that time he had clearly decided on a scientific career, though some members of his family doubted the propriety of this in view of the social obligations inherent in his eventual succession to his father's title and position. However, Strutt seems to have considered that such obligations should not be allowed to interfere with scientific work.

In 1866 Strutt was elected a fellow of Trinity, thus further emphasizing his scholarly leanings. By this time the traditional lengthy grand tour of the continent by members of the aristocracy was going out of fashion. However, with some friends he made a short visit to Italy, and the following year he crossed the Atlantic, visiting Canada briefly but mainly the eastern USA, then in the throes of reconstruction after the Civil War. Because Strutt had a keen interest in politics, it was suggested that he might stand for Parliament, as a representative of the University of Cambridge, but he decided that, although he was in general a supporter of the Conservative Party, he disagreed with its policy on certain questions; in any case, he wanted to concentrate on a career in science. Immediately after his return to England, he purchased an outfit of experimental equipment.

At Cambridge, because of the lack of a university physical laboratory, students received little or no direct encouragement to embark on experimental investigations for themselves. However, the young scientist's first serious research was experimental in character and the results were presented at the Norwich meeting of the British Association in 1868. This marked the beginning of a lifetime of research, stimulated by his own curiosity and by early careful reading of the current scientific literature, which provided many suggestions for independent investigations into the puzzles and questions left by previous researchers. In those early days he was much encouraged by correspondence with Maxwell, who was always eager to help a youthful colleague.

Strutt had become acquainted with Arthur Balfour, the future prime minister, as a fellow student at Cambridge, and through him met his sister Evelyn. They shared a common interest in music; he persuaded her to read Helmholtz's *Theory of Sensations of Tone*. They married in 1871; college statutes meant that he had to vacate his fellowship on marriage. Not long after this, a serious attack of rheumatic fever threatened for a time to cut short his career and left him much weakened in health. On medical advice he decided not to winter in England, so an expedition to Egypt was planned as a recuperative measure. This took the form of a journey up the Nile by houseboat as far as Nubia, during which he started work on his famous textbook on the theory of sound. Shortly after returning to England in the spring of 1873, Strutt succeeded to the title on the death of his father and took up residence at the family seat of Terling; we should now refer to him as Rayleigh, or the third Rayleigh, rather than Strutt. It was then that he installed the laboratory where so much of his later experimental work was done. The installation was by no means an elaborate one, and visitors were inclined to be surprised that such obviously important results could be obtained with what was considered even in his day to be rather crude equipment. However, Rayleigh early manifested an economical turn of mind and took real pleasure in making extensively improvised apparatus yield precise results.

In 1873 Rayleigh was elected to the Royal Society. Six years later the untimely death of Clerk Maxwell left the Cavendish professorship vacant. Pressed by many scientific friends to let his name go forward for the post, Rayleigh finally consented, being partly influenced by the loss of income from the Terling estate due to the agricultural depression of the late 1870s. It does not appear that he ever contemplated retaining the professorship for an indefinite period and, indeed, he gave it up after five years. The teaching duties of the Cavendish professor were not onerous; he was required to be in residence for eighteen weeks of the academic year and to deliver at least forty lectures during that period. The creation of a modern physical research laboratory, he concluded, depends on a number of factors. Of course a stimulating leader is essential. However, there have been many such who failed to create schools because the other factors were not present; the leader must be 'one who is not only abounding in energy and ideas, but also one who can without too great an effort throw himself into the difficulties of others. This requires a peculiar kind of versatility not always easily combined with great powers of concentration on any one line of thought. It is also necessary to have a productive line of investigation opening up. Finally it is necessary to have the right kind of pupils. They must be men of the not very common kind of ability which makes a scientific investigator: they must not be too young: and they must be provided with the means of subsistence while the work goes on.'

Rayleigh embarked vigorously on a programme of developing elementary laboratory instruction. It is difficult to appreciate what a task such a programme involved; collegiate instruction in practical physics was almost a new thing, and there was little to go on save the teacher's imagination. Under his direction laboratory courses were developed for large classes in heat, electricity and magnetism, properties of matter, optics and acoustics. This was pioneering work of a high order and a beneficial influence on the teaching of physics throughout England and elsewhere. Previously research had been carried on, by and large, outside the universities, which thus remained quite out of touch with real progress in physics until well into the second half of the nineteenth century. At the same time the publication of his textbook on the *Theory of Sound* in 1877 also made an important contribution to the teaching of theoretical physics; Helmholtz and Klein had a very high opinion of it. It was a two-volume work; a third volume was projected but never written. Because it was on the Nile that the book had its genesis, the first part was written without access to a large library.

The stay in Cambridge was a period of broadening professional and social relations. Through his relations by marriage, Rayleigh came into contact with British politicians, especially those of the Conservative Party. In particular, Lord Salisbury, several times prime minister, was a personal friend. During the Cambridge period Rayleigh became increasingly involved in the affairs of the British Association and was elected to preside over the Montréal meeting of 1884, the first outside the United Kingdom. This meant another visit to North America; he took advantage of this to spend two months touring Canada and the USA, where he was hospitably entertained by various American and Canadian physical scientists.

By this time Rayleigh's financial situation had improved, so he did not feel any further need for an income from his scientific work. Therefore, immediately after the return from the USA, he resigned from his Cambridge post and went back to Terling, which remained his scientific headquarters for the rest of his life. Probably many contemporaries in the peerage as well as some tenants on his estate thought his preference for scientific activities rather peculiar, but Rayleigh went his own way with typical British imperturbability. Terling is close enough to London to permit frequent visits to the metropolis, so Rayleigh could, without too much inconvenience, spend time on committee work for the government or for bodies like the Royal Society. Though he clearly loved his laboratory, he was no recluse and gave freely of his time to endeavours for the advancement of science. He also kept up a voluminous correspondence with scientific colleagues.

In the later stages of his career Rayleigh continued to take a lively interest not only in the British Association but also in the Royal Society, which he served from 1885 to 1896 as secretary and then from 1905 to 1908 as president. From 1882 to 1905 he held the chair of natural philosophy at the Royal Institution, where he gave over 110 lectures in the afternoon series, on a great variety of topics, and many of the evening lectures as well. In 1896, like Faraday before him, he was appointed scientific adviser to Trinity House, involving numerous visits of inspection to lighthouses. Other public services included serving as vice-chairman of the National Physical Laboratory, as Lord Lieutenant of the County of Essex and as Chancellor of the University of Cambridge, where he founded the prestigious Rayleigh prize. Because of these commitments, the Rayleighs did not travel overseas a great deal, but in 1897/8 they spent some time in India and in 1908 visited East and South Africa.

Early in 1919 Rayleigh's health began to give cause for concern, and he died in his seventy-sixth year following a heart attack at Terling on June 30, 1919. He was buried in the village churchyard; two years later a memorial was erected to him in Westminster Abbey. Public recognition of his achievements came to him in full measure. In 1904 he received the Nobel prize in physics for his part in the discovery of the inert gas argon, a previously unknown constituent of the atmosphere; he donated the cash award which went with it to Cambridge University to improve the facilities of the Cavendish Laboratory and the University Library. The next year he was appointed to the Privy Council. In the course of his life Rayleigh received numerous honorary degrees from universities at home and abroad and many other distinctions. As one of the original recipients of the Order of Merit in 1902, at the investiture he declared that 'the only merit of which I personally am conscious is that of having pleased myself by my studies, and any results that may have been due to my researches are owing to the fact that it has been a pleasure to me to become a physicist'. Meticulous experimental work, characterized by extreme precision, was Rayleigh's *forte*, particularly during the Cambridge period when he helped to establish standard electrical units for resistance, current and electromotive force. However, he was also a very able applied mathematician, who invented what became known as the WKB method of steepest descents. Like the majority of British scientists of his period, the revolution in theoretical physics set in motion by Planck was something he found difficult to accept.

Ludwig Boltzmann (1844–1906)

Although it would be going too far to say that there was a Viennese school of physics, several remarkable physicists came from the imperial city. Under the Habsburgs, Vienna was the hub of *Mitteleuropa*, but other cities such as Breslau, Budapest, Cracow and Prague were also of major importance. Cultural links with Germany were strong; knowledge of the German language was taken for granted. There was a remarkable freedom of movement between different universities in the area. Austrian universities followed the German pattern, so that students had to face only one examination, either for the certificate required for entry into one of the professions, or for a doctorate. The first degree in physics was the doctorate of philosophy, for which the candidate was required to present an original dissertation to the faculty. Once the dissertation had been accepted the process was completed by examinations called *Rigorosa*, one of which was a topic in philosophy, as a concession to the title of the degree.

In 1837 Ludwig Georg Boltzmann, an Austrian tax official, married Katharina Pauernfeind. He was a Viennese of Prussian ancestry; she was the daughter of a businessman from a wealthy and old-established Salzburg family. Their eldest son Ludwig Eduard was born in Vienna on February 20,

1844. Two years later there was born a second son, who died in childhood, and then a sister Hedwig. Their father was of the Protestant faith, but Ludwig and Hedwig were brought up as Roman Catholics, like their mother. Young Ludwig began his formal education when the family moved from Vienna first to Wels and then to Linz, the chief city of Upper Austria, where he began attending the local *akademisches Gymnasium*. He was an outstanding scholar, showing particular aptitude in mathematics and science. Outside school he took piano lessons from the composer Anton Bruckner and throughout his life he maintained a strong interest in music. In appearance he was short and stout, with curly hair and blue eyes; he suffered severely from myopia.

In 1859 Ludwig Georg died from tuberculosis, a loss his son felt deeply. In 1863 he enrolled at the ancient University of Vienna as a student in mathematics and physics, receiving his doctorate three years later; the Austrian Dr.Phil. was at about the level of a master's degree; there was no equivalent of a bachelor's degree. The director of the institute of physics was the young scientist Josef Stefan, later to become famous for the experimental discovery of the relation between radiant heat and temperature. When Boltzmann began research he particularly appreciated the close contact between Stefan and his students: 'When I deepened my contacts with Stefan, and I was still a university student at the time, the first thing he did was to hand me a copy of Maxwell's papers and since at that time I did not understand a word of English he also gave me an English grammar.' Maxwell's work did not become generally known outside Britain until considerably later. In 1869 Boltzmann was awarded the *venia legendi*, the right to lecture, having been appointed assistant professor the previous year. He began working at a small laboratory in a house on Erdbergstrasse. The facilities were minimal, but the physicists who worked there were full of ideas.

Two years later he was appointed to the chair of mathematical physics at the University of Graz, which was rapidly rising in importance at that time and soon ranked with the best in Central Europe. The director of the institute of physics, August Toepler, had been appointed just before Boltzmann and was working hard to develop the institute, with a new building, new apparatus and much larger research funds. During the next few years Boltzmann completed his theory of the statistical properties of gases, in which the celebrated equation named after him makes its first appearance. This was written up in a paper published in the Proceedings of the Imperial Academy of Sciences of Vienna in 1872, under the title 'Further Researches on the Equilibrium of Gas Molecules'. In a relatively short time his approach to kinetic theory became widely known, especially in Great Britain. Although it was ridiculed by many at the time, his belief that thermodynamic phenomena were the macroscopic reflection of atomic phenomena, regulated by mechanical laws and the laws of probability, was his most original contribution to science. Later his attempts to prove the second law of thermodynamics were to be of decisive importance in the development of statistical mechanics.

Thanks to funds obtained by Toepler, Boltzmann was able to make short visits to other centres of research, notably Heidelberg and Berlin. At the former university he worked with the chemist Bunsen and the mathematician Königsberger, and at the latter with the physicists Kirchhoff and Helmholtz. He was particularly impressed by the last of these, although put off by the Prussian Geheimrat's chilly manner. 'Yesterday I spoke at the Berlin Physical Society', he wrote to his mother, 'You can imagine how hard I tried to do my best not to put our homeland in a bad light. Thus in the previous days my head was full of integrals. Incidentally there was no need for such an effort, because most of my listeners would not have understood my talk anyway. However, Helmholtz was also present and an interesting discussion developed between the two of us. Since you know how much I like scientific discussions, you can imagine my happiness. Especially as Helmholtz is not so accessible otherwise. Although he has always worked in the laboratory nearby, I have not had a chance of talking much to him before.'

In 1873 Boltzmann was offered a full professorship at his *alma mater*. Although it was a chair in mathematics, not physics, it was considered that 'although his researches originated in physics they were also excellent as mathematical works, containing solutions of very difficult problems of analytical mechanics and especially of probability calculus.' Boltzmann accepted the offer despite the fact that in Graz the new institute of physics had just been completed and transformed into an ideal centre for high-quality research in the most advanced physics of the time. The ambition of every Austrian academic then, as now, was to become a professor at the University of Vienna.

Before leaving Graz, Boltzmann had met his future wife, Henrietta von Ailgentler, a young woman with long blonde hair and blue eyes, ten years his junior. Having lost both her parents, she was making a living as a schoolteacher. Boltzmann proposed marriage by letter, in which he said 'Although rigorous frugality and care for his family are essential for a husband whose only capital is his own work, it seems to me that permanent love cannot exist if [a wife] has no understanding and enthusiasm for her husband's efforts, and is just his maid and not the companion who struggles alongside him.' After this, she decided to study mathematics at the university. Although she was allowed to attend lectures in the first semester, at the start of the second the faculty passed a regulation to exclude women students. Henrietta presented a petition to the Minister of Public Education, a former colleague of her late father's, who then exempted her from the regulation, but the problem recurred, so to avoid it she replaced mathematics in her studies by a course in cookery.

Boltzmann was not entirely happy in Vienna. For one thing he had difficulty in finding an apartment to live in. For another he found that he was expected to do a lot of administration. The main problem, however, was caused by his teaching duties, which were those of a professor of mathematics, not physics. So, when Toepler decided to leave Graz and move to Dresden, Boltzmann applied for the post which he had relinquished. Ernst Mach, then in Prague, also applied. Mach was a charming, unassuming man, of whom the psychologist William James said that he had never had such a strong impression of pure intellectual genius. Einstein admired Mach for his independence, incorruptibility and ability to see the natural world as through the eyes of a curious child. Like many other scientists of the time, Mach believed that Newtonian physics needed revision. The Ministry of Research and Education took a long time to decide between them, but in the end Boltzmann was preferred. The young couple were then able to get married and settle down to raise a family.

Of the fourteen years the Boltzmanns spent in Graz, at least the first twelve were happy. They had two sons, Ludwig Hugo and Arthur, and two daughters, Henrietta and Ida; a third daughter, Elsa, was born after the family had left Graz. Professional recognition was not lacking. He was elected to the Imperial Academy and received many honours from foreign scientific academies. He was even offered noble rank, but declined, saying 'Our middle-class name was good enough for my ancestors, and it will be for my children and grandchildren as well.' Occasionally he was invited to dine with the Emperor Franz-Josef. Unfortunately he was a slow eater, whereas the Emperor barely touched his food. Court etiquette did not allow guests to continue after the Emperor had finished, so Boltzmann's plate was removed almost before he had started.

The nature-loving Boltzmann used to take long walks in the country, during which he taught the children all about plants and butterflies. These walks, coupled with ice-skating in winter, were his main form of exercise, but he also enjoyed swimming and installed some gymnastic equipment in his house for the family to use. He also bought a farmhouse near Oberkrois-bach, with a commanding view over a large part of Styria, and enjoyed country life. He purchased a cow and consulted his colleague, the professor of zoology, as to how to milk it. In Graz he entertained frequently, including students among the guests. He was familiar with classical German literature and liked to quote from Schiller and other standard authors. He played the piano regularly, especially the works of Beethoven and Mozart. He also played in chamber recitals and attended concerts and operas.

However, Boltzmann, lacking the stimulus he might have received in one of the major centres of modern science, started to feel dissatisfied and began to neglect his duties in the institute. On the experimental side, he spent a considerable amount of effort on repeating the experiments of Hertz and verifying Maxwell's theory of the electromagnetic nature of light. Scientific distinctions and awards kept arriving, but problems were on the way as well.

In 1888 Boltzmann accepted the offer of the chair in Berlin made vacant on the death of Kirchhoff, but, after Frau Helmholtz at dinner warned him that he would not be happy in the Prussian capital because of his informal style and well-known sense of humour, he created some annoyance by withdrawing his acceptance. When, almost immediately afterwards, he decided, for reasons that are unknown, that he must leave Graz anyway, Munich seized the opportunity and appointed him to a newly created chair in theoretical physics. At a farewell ceremony, his former colleagues expressed the hope that he would return to Austria some day; the country could not afford to lose one of its leading scientists. In Munich Boltzmann was finally able to teach the subjects dearest to his heart. Although compared with Vienna there was 'wonderful equipment, but far fewer ideas', he added 'we must not let the Austrian education ministry know that good work can sometimes be done with inferior equipment'. Looking back to the start of his career, he recalled that

> Erdberg has remained through all my life a symbol of honest and inspired experimental work. When I managed to bring some life into the physics institute of Graz I called it jokingly 'Little Erdberg'. By this I did not mean that the space was small, it was probably twice as big as Stefan's institute; but I had not yet achieved the spirit of Erdberg. Still, in Munich, when the young graduate students came and did not know what to work on, I thought how different we were at Erdberg!

Today there is nice experimental equipment, and people are looking
for ideas on how to use it. We always had enough ideas; our only worry
was the experimental apparatus.

Boltzmann spent four peaceful years in the Bavarian capital, during
which students came from all parts of Europe and even from as far away as
Japan to study under him. As well as lectures on physics, he gave lectures
on mathematics, especially the theory of numbers. He used to meet his
colleagues at the *Hofbräuhaus* to discuss academic questions over glasses of
beer. At this time university professorships in Bavaria were not pensionable.
Recalling the case of the blind Georg Simon Ohm, who died without a
pension in the most miserable circumstances, he started to worry about the
future of his family. His own sight started to deteriorate; he feared that he
was going blind. To spare his eyes, his wife Henrietta read the scientific
literature to him.

For some idea of his personality we can turn to some letters written
by a visiting Japanese physicist: 'I think no-one can be as competent as he
[Boltzmann] is, except for Helmholtz. His lectures are extremely transparent; he speaks lucidly, not like Helmholtz who speaks rather awkwardly.
But he is an odd little fellow and sometimes ends up doing unintelligent
things.' Others recalled that 'He never exhibited his superiority. Anybody
was free to put questions to him and even to criticize him. The discussion
would take place quietly and the student was treated as an equal. Only later
would the student realize how much he had learned from it.' 'He was not
upset even if a student disturbed him at home when he was working. The
great scientist remained available for hours, always keeping his patience and
good temper.' As for his lectures: 'He gave a course that lasted four years. It
included classical mechanics, hydrodynamics, elasticity theory, electrodynamics and the kinetic theory of gases. He used to write the main equations
on a very large blackboard. By the side he had two smaller blackboards on
which he wrote the intermediate steps. Everything was written in a clear
and well-organized form. I frequently had the impression that one might
reconstruct the entire lecture from what was on the blackboards. After each
lecture it seemed to us as if we had been introduced to a new and wonderful
world, such was the enthusiasm that he put into what he taught.'

After a time Boltzmann began to feel a strong desire to return to
Vienna, and, when this became known, friends in Austria tried to find a
way to achieve this. A suitable opportunity arose when his former teacher
Stefan died in 1892, and there was a move to install Boltzmann as his

successor. Boltzmann was interested, but, when the conditions of his appointment in Munich were improved, he felt that he ought to stay. Then Vienna came back with another offer, including satisfactory pension rights. For months he remained undecided, but he finally came down in favour of Vienna and returned there in June 1894.

Boltzmann soon realized that moving back to Vienna was a mistake. He made no secret of the fact that in Austria he found far fewer students properly prepared for scientific work than in Germany. It was more like secondary-school education, wasting his talents and aspirations, said his wife. The pleasant social circle to which he had belonged eighteen years before no longer existed. His university colleagues were not particularly welcoming. The situation worsened when Mach, who rejected atomism, moved to Vienna as well. For Boltzmann it was unbearable to have a 'malevolent' colleague who openly fought the very theory to which he had devoted his entire professional life. 'Boltzmann is not malicious but incredibly naive and casual', Mach wrote to a friend, 'he simply does not know where to draw the line. This applies to other things too, which are important to him.'

In 1900 Boltzmann decided to accept an appointment as professor of theoretical physics at the University of Leipzig, where the physical chemist Wilhelm Ostwald had built up a leading research centre. The Leipzig faculty recommended Boltzmann as the 'most important physicist in Germany and beyond'. Unfortunately, the strain of reaching this decision led to another nervous breakdown. Although Ostwald was a personal friend of Boltzmann, he was also an ally of Mach and opposed Boltzmann's theories violently, so Boltzmann found himself in a worse situation than before. In addition he felt homesick for Austria and made an attempt at suicide. He began to negotiate for a return to Vienna, where the chair he had vacated had not been filled and his adversary Mach had retired, after suffering a stroke. However, the Minister for Research and Education did not find it easy to explain Boltzmann's personality problems to the Emperor Franz-Josef, or to still rumours that he was mentally ill and would not be able to perform his duties properly. Nevertheless, Franz-Josef agreed to his reappointment on condition that Boltzmann gave his word of honour never to accept a position outside the Empire again.

Fellow scientists from many lands contributed to a notable *Festschrift* in honour of his sixtieth birthday; he had already received numerous academic and scientific honours, including an honorary degree from Oxford. Among the various countries Boltzmann visited for scientific purposes was the USA. He first went there in 1899, accompanied by his wife, to lecture and receive an honorary degree from Clark University in Worcester, Massachusetts. After a promising first decade, Clark was about to be eclipsed by the even newer University of Chicago, but at the time Clark was an exciting place to visit. They made a tour of the major cities of the eastern USA. Five years later he was in St Louis for an International Congress, accompanied by his son Arthur Ludwig, and the next year he gave a course at a summer school held at the University of California at Berkeley. Unfortunately, the audience found his English, of which he was very proud, almost incomprehensible and were sorry he was not prepared to lecture in his native language. He wrote an entertaining account of his experiences.

By all accounts, Boltzmann was an outstanding teacher on his home ground. His mastery of his science and his love for it were combined with a zest for lecturing and an attention to detail and procedure. He always lectured completely without notes, but his blackboard technique was carefully developed; the sequence of equations clearly inscribed on the board gave an exact account of the structure of ideas he was expounding. The exposition was not so formal and so highly polished as to smooth over the difficulties or make the audience lose interest. Boltzmann never hesitated to point out and correct his own errors; he welcomed questions and discussion, and his lectures drew crowds of eager students. In 1902, for example, he began his mechanics course by offering his students 'everything I have: myself, my entire way of thinking and feeling' and asking the same of them: 'strict attention, iron discipline, tireless strength of mind'. One of those who heard him was Lise Meitner, who recalled that

> Boltzmann had no inhibitions whatsoever about showing his enthusiasm when he spoke, and carried his listeners along naturally. He was fond of introducing remarks of an entirely personal character into his lectures . . . His relationship to students was very personal. He not only saw to their knowledge of physics but tried to understand their character. Formalities meant nothing to him, and he had no reservations about expressing his feelings. The few students who took part in the advanced seminar were invited to his house from time to time. There he would play the piano for us – he was a very good pianist – and tell us all sorts of personal experiences.

The return to Vienna was not without its problems. For example, he had resigned from the Imperial Academy when he went to Munich but the Emperor was against his immediate reappointment, so he had to wait to be re-elected; Boltzmann thought this offensive. He led an active social life in the city and yet was often still working in the early hours of the morning. However, his health was becoming a major concern. His sight was steadily deteriorating. He suffered from asthma, headaches and chest pains. He began to worry that his wits and his memory would suddenly desert him in the middle of a lecture. He suffered from the characteristic mood-swings of manic-depression.

At least partly for health reasons, Boltzmann, with his wife and youngest daughter, spent a few days at Duino, a village near Trieste, famous for its castle perched on a rocky promontory of the Adriatic coast, having the sea on one side and deep forests of cork trees on the other. It was there, on September 5, 1906, at the age of sixty-two, that he took his own life. It seems most likely that he hanged himself from a crossbar of the window in his hotel room, while his wife and daughter were out swimming. The following day he had been due to return to Vienna to start his lectures. The dean of the faculty of philosophy at the university had written to the Ministry the previous May that Boltzmann was suffering from a severe form of neurasthenia and should abstain from any scientific activity. Boltzmann had announced lectures for the summer semester but had to cancel them because of his nervous condition. Mach wrote: 'In informed circles one knew that Boltzmann would most probably never be able to exercise his professorship again. One spoke of how necessary it was to keep him under constant medical surveillance, for he had already made earlier attempts at suicide.'

6   From Röntgen to Marie Curie

Our next five remarkable physicists were born in the twenty-three years from 1845 to 1867. Two came from England, two from Germany and one from Poland.

Wilhelm Conrad Röntgen (1845–1923)

Wilhelm Conrad Röntgen, the discoverer of X-rays, was born on March 27, 1845 in Lennep im Bergischen, a small town in the part of the Rhineland which then belonged to Prussia. His father Friedrich Conrad Röntgen was a cloth manufacturer and textile merchant. The family of his mother Charlotte Constanze (née Frowein) were in the same line of business, and, although they too came from Lennep, her parents lived in the Netherlands. In 1848, when their son was three, the family moved to the town of Apeldoorn in Holland, about a hundred miles distant, and as a result they became Dutch citizens rather than Prussian. After elementary school, Wilhelm attended a nearby private boarding school until at the age of seventeen he entered the technical school in the provincial capital Utrecht. There he lived with one of the teachers, and later recalled that

> the father of this family was a fine scholar, a solid character and a really splendid man who understood superbly the task of guiding young persons along the correct path in life. The mother was a loving, cultured and kindly woman who provided the proper atmosphere for a full life, one of happiness and at the same time pleasant stimulation. There was no time for foolish and stupid things, but much for creative activities. Self-created amusements were offered at community celebrations, but otherwise diligent work was demanded in serious learning. That was a happy and equally rewarding time, and looking back on the years of my youth, I have to add that I also went horse-riding and ice-skating and generally busied myself in many wholesome outdoor activities.

The young man was aiming to enter the famous old University of Utrecht, but a problem arose, which was to cause him considerable

difficulty. After an incident in which he refused to tell on another boy who was at fault, Röntgen was expelled from the technical school. He then went to another school where he was prepared for university entrance but failed to qualify. Even so, he attended lectures at the university for a short while, but, because he was not a matriculated student, he would not have been able to proceed to a degree. Röntgen then decided to try instead for a place at the Polytechnikum of Zürich (since 1911 part of the Eidgenössische Technische Hochschule, or ETH), which it was possible to enter by passing a special examination. However, when he was all ready to sit this he went down with an eye infection, but was admitted anyway.

'Röntgen was a tall, handsome youth who, with penetrating brown eyes, looked soberly at the world', wrote someone who knew him in Zürich:

> He had a well spread nose and a large mouth. He had thick black wavy hair, was clean shaven and always dressed well. His father supplied him with sufficient money for clothes and extravagant fun and good living. Usually his attire was a modest dark coat, grey trousers, a soft wing collar with a large bow cravat, and a gold watch chain dangling over his waistcoat. He was conservative and modest. Although he disliked blatant conviviality, he was fun-loving, a popular young man

in a city where students were not always as serious as its sober citizens. He often went riding along the lake-shore, but when opportunity presented itself he would hire a fancy two-horse carriage to ride in style through the city streets. He did not join one of the *Burschenschaften*, which went in for carousing and duelling, but enjoyed the company of other students from the Netherlands who had their own informal club.

In Zürich at that time there was a scholarly innkeeper named Johann Gottfried Ludwig who had been forced to leave his native Jena because of suspected revolutionary activities. He settled in Zürich, where he purchased an inn and married a Swiss girl, Elsabeth Gschwend. They had four children, Lina Barbara, Anna Bertha, Hans Otto and Maria Johanna. Röntgen was a regular customer of the inn and became attracted by Bertha, the second daughter of the innkeeper. She was a charming woman, tall and slender, with a healthy and wholesome attitude towards life. When she became ill and was hospitalized in a sanatorium he hired, on her thirtieth birthday, a carriage drawn by four perfectly matched horses and driven by a top-hatted coachman in splendid livery, and drove up to the sanatorium to present her with a huge bouquet of red roses. It was agreed that they would marry once his future was more secure.

Röntgen's studies were going well. He received his diploma as a mechanical engineer in 1868 and then moved over to physics, in which field he wrote his doctoral dissertation on thermodynamics, after taking a course from the great Rudolf Clausius. When this was accepted his parents came to Zürich to inspect his fiancée; they found her well-educated, of good family, intelligent, of good character and very agreeable. However, they felt that, although over thirty, she was not yet fully prepared for matrimony, so it was agreed that she would go to Apeldoorn to learn cooking and other domestic skills. At the same time Röntgen obtained the position of assistant to the physicist August Kundt, from whom he learnt the importance of taking the utmost care in experimental work, often repeating an experiment over and over again until the results were absolutely certain. When Kundt moved to Würzburg he took his assistant with him. However, Röntgen found the laboratory facilities there disappointing, a regular complaint throughout his life.

At that time Friedrich Kohlrausch was one of the most eminent German physicists. Röntgen, in the course of some experiments on specific heat, found some errors in the published work of Kohlrausch and sent a report to the prestigious *Annalen der Physik und Chemie*. When this first scientific paper of his was accepted, he felt he could look forward to his future with some confidence, and therefore get married. The wedding took place in Apeldoorn at the beginning of 1872. A few months later Kundt announced that he was moving again, this time to Strasbourg, and asked Röntgen to come with him. This he was glad to do, because his prospects of advancement in Würzburg seemed poor; even the first step of becoming a Privatdozent was thwarted by his unsatisfactory school record.

After the Franco-Prussian war, Strasbourg (or Strassburg) lay in German hands, and the ancient university was being reopened with new and well-furnished buildings. For once he could not find fault with the equipment in the laboratory. His elderly parents moved to Strasbourg to join the young couple. In 1874 Röntgen himself was appointed Privatdozent. The following year he was offered a full professorship in physics at the Agricultural Academy in Holdenheim, close to Stuttgart. After some hesitation he accepted the post, but soon regretted doing so, because the experimental facilities were so inadequate. When Kundt became aware of the situation, he arranged for Röntgen to return to Strasbourg as associate professor of theoretical physics, at the age of thirty-four.

This second Strasbourg period was in many ways the most fruitful of his professional life. His experimental work made good progress and he published his results on a wide range of topics but particularly on the physical properties of crystals. Soon he was ready for a full professorship and in 1879 the offer of one came from the University of Giessen, some 30 miles north of Frankfurt, where he would be director of the institute of physics. A new building for the institute had been agreed; not only could Röntgen choose the equipment for this, but he also had a say in the choice of his own assistants. The Röntgens found the small town of Giessen congenial and were pleased when his parents did so too. In 1880 his mother died, and his father died four years later. In 1887 the Röntgens, realizing that they would never have children of their own, decided to adopt the daughter of Bertha's brother Hans, their six-year-old niece Josephine Bertha.

Two years later Röntgen made one of his most important discoveries when he demonstrated the correctness of a prediction of the Faraday–Maxwell theory of electromagnetism. This discovery, one of the foundations of the modern theory of electricity, made Röntgen famous internationally. By this time he had been at Giessen ten years and was forty-three years old. He began to receive offers from other universities. One was Utrecht, but, although Bertha liked the idea of returning to the Netherlands, he turned it down. Another was to return to Würzburg as professor of physics and director of the new institute of physics, where there was accommodation for the director in the form of a spacious nine-room apartment. He accepted the latter position without hesitation and took office on October 1, 1888.

At this age the bearded Röntgen was impressive in appearance. As a colleague wrote:

> He had a penetrating gaze and unassuming manner which gave him extraordinary character and dignity. His nature was amiable and always courteous, but his reticence and shyness amounted almost to diffidence when he received strangers. In later years his shyness acted as a wall; it protected him from selfish and curious persons, but it also kept many fine and sincere persons away. Occasionally Röntgen would appear distrustful, but actually he concealed a deep sympathy and understanding. After he was convinced of the sincerity of his caller he would extend a warm friendliness toward him. He had close friends among his colleagues at the university who remained close to him for the whole of his life.
>
> Röntgen showed no patience with persons whose behaviour was actuated by selfish personal motives or was noticeably shaded by personal prejudices. His was an intellectual honesty which characterized not only his work but his attitudes as well. He believed that their attitudes interfered with the progress of science and were actually detrimental to the person. Many times he was gruff to people only concerned with their own importance and he was not apologetic for such behaviour. He detested those who would try to cover up superficial academic knowledge with dazzling but utterly meaningless theories.

Röntgen disliked lecturing, speaking in such a low voice that students at the back of the class could hardly hear him at all. As an examiner he was dreaded. Social activities took on a certain routine: four or five families from his group of close friends would gather for dinner at one or other home. Dinners were elegant, with fine white wines and red wines drawn from a cask. For recreation the Röntgens made use of a small log cabin they had bought some five miles from the city as a base for walks. Unfortunately, Bertha did not enjoy good health, and their adopted daughter was not strong either.

Röntgen was elected Rector of the University of Würzburg, for the customary year of office, after which he and his wife took an Italian tour, although normally they always took their summer holidays in the Swiss Alps. In June 1894, just before they left for Italy, Röntgen was conducting a series of experiments with cathode rays. Many other scientists were interested in these, so he was anxious to proceed as rapidly as his meticulous standards would permit. Cathode rays would not penetrate more than a few centimetres in the air but he discovered a new form of radiation, which was much more powerful. Helmholtz had predicted that, on the Maxwell theory, a sufficiently short-wave light-ray should go right through solid materials. He tested this out on various materials. Lead blocked the radiation entirely, but other metals let it through. However, what struck him most was that, when he held up his hand, the screen showed the bones of his hand, encased in darker shadows. He decided that, although the eye might be deceived, the photographic plate could not be, and took photographs of the phenomenon. As was his practice, he repeated the observation over and over again, until he was satisfied, and within a few weeks he was.

So far the work had been kept secret, but now Röntgen felt sufficiently confident to publish his results, albeit in just a small academic journal. The historic paper he wrote, entitled 'A new kind of ray', appeared in time for him to send out copies to many of the leading scientists in the field on New Year's Day 1896. There were sceptics, but Röntgen had a reputation for careful experimentation. The exciting news of his discovery produced a flood of messages of congratulation from all over the world. The potential for applications in medical diagnostics was immediately recognized; the university made him an honorary doctor of medicine, while the students put on a torchlight parade in his honour. The Kaiser, Wilhelm II, summoned him to Berlin, to give a demonstration, and decorated Röntgen with the Prussian Order of the Crown, Second Class. He declined an invitation to speak before the Reichstag. This was perhaps the first time that a scientific discovery had, almost overnight, caused such a sensation and received so much publicity, to Röntgen's horror. He returned to the scientific investigations that had occupied him before his great discovery. At first the X-ray pioneers saw no reason to protect themselves against the effect of the radiation, but before long some of them began to experience loss of hair and burns, particularly on the hands.

In 1900 Röntgen left Würzburg, with considerable reluctance, to become professor of physics and head of the institute of physics at the University of Munich. The next year, when Nobel prizes were awarded for the first time, he received the prize in physics 'for the discovery of the remarkable rays subsequently named after him'. Although he always called them X-rays, others called them Röntgen rays. In his will he left the prize money to the University of Würzburg to be used in the interests of science, but unfortunately the legacy became valueless together with the rest of Röntgen's personal fortune when the German economy collapsed after the First World War. The Prince Regent of Bavaria bestowed upon him the title of 'Geheimrat' in 1908 and soon afterwards the city council of Weilheim, where he owned a hunting lodge, made him an honorary citizen. His wife Bertha, whose health had been failing for some years, died in 1919; the following year he retired from his position at the university while retaining facilities in the laboratory of the physics institute. A staunch patriot, he was deeply depressed by the German defeat and its consequences. Röntgen died of rectal carcinoma at the age of seventy-seven on February 10, 1923, in his Munich apartment; his ashes were buried with those of his wife and parents. Röntgen was of the opinion that his discoveries and inventions belonged to humanity and that they should not in any way be hampered by patents, licences or contracts or controlled by any one group. His altruism made a very favourable impression on scientists both in Germany and abroad. A statue of him was erected on the Potsdam Bridge in Berlin.

Joseph John Thomson (1856–1940)

Manchester in the middle of the nineteenth century was a prosperous industrial town with an active cultural life. Many of the northern businessmen were devout nonconformists. Their faith often impelled them to live austere

personal lives and frequently to initiate schemes of social amelioration and works of personal charity, often at considerable cost to themselves. The advancement of education, public health and science were the main areas which attracted these public-spirited men. One of the most important of these initiatives was Owens College, the predecessor of Manchester University, founded in 1851 by a wealthy bachelor who had been in the textile trade. Its aim was to provide an education modelled on the practices of the Scottish universities, rather than Oxford and Cambridge, and to be of use to young men who intended to enter commerce or industry. The courses were based on traditional subjects: Latin and Greek, mathematics, natural philosophy, English and history, chemistry, foreign languages and natural history. The degrees were those of the University of London, which imposed no residence requirement. There were no faculties of law, medicine or technology.

Joseph John Thomson, later known informally as J.J., was born on December 18, 1856. His father Joseph James Thomson managed a small business selling antiquarian books, occasionally acting as publisher as well. This prospered sufficiently to give him a reasonable living and a house with servants in the middle-class suburb of Cheetham Hill. The Thomsons had lived in Manchester for several generations but originally they came from lowland Scotland – hence the spelling of their name without the letter p. Joseph married Emma Swindells, who also came from Manchester. She has been described as small, with bright dark eyes brimming with kindness, and dark hair hanging in clusters of ringlets over her ears. They had one other child, Frederick, who was four years younger than J.J.

Although not much seems to be on record about the father, who died when J.J. was only sixteen, it is recalled that he was acquainted with some of the scientific and literary people of Manchester, many of whom would probably have patronised his bookshop. When J.J. was a boy he had been introduced to the famous Joule by his father, who had afterwards said 'some day you will be proud to say you have met that gentleman'. J.J. was a shy boy but determined and 'knew which way he was pointing'. When asked what he wanted to do when he grew up, he replied that he wanted to do original research, whereupon one of those present tapped him on the head and said 'don't be such a little prig, Joe'. By the age of fourteen he could be described as a 'rather pallid, bony youth with the air of a serious but happy student, unassuming and modest without diffidence, very approachable and friendly'.

J.J.'s father seems to have decided that his elder son's aptitude for mathematics indicated that engineering would be a more suitable choice of career for him than bookselling, perhaps hoping that the younger son would enter the business instead. Since J.J. had completed his secondary education by the age of fourteen, it was thought that Owens College would fill the gap until he was old enough to be apprenticed to a Manchester engineering firm. However, after the early death of his father there was no money to pay the premium for the apprenticeship; the family business was sold and his widow, with her two sons, moved to a smaller terraced house nearer Owens, where J.J. was in his second year. She could only just afford to keep him there.

After a difficult first twenty years, the college had started to flourish, despite having rough and overcrowded facilities. The range of lectures in engineering, mathematics and natural philosophy was impressive. There was probably nowhere outside Oxford and Cambridge that could match it in England, though Glasgow was perhaps its equal in Scotland. In effect J.J. went through university in three subjects while still a schoolboy. After he left, having thrived on the teaching it provided, the college authorities made a regulation raising the minimum age of admission.

To complete his education J.J. was determined to go to Cambridge, but his mother could not afford to send him there unless he won a scholarship. As there were no entrance scholarships for engineering or physics, the route had to be through mathematics. He succeeded on his second attempt at winning a scholarship to Trinity, the college with the highest reputation and the most difficult to enter as a scholar. There were several reasons why he chose Trinity. The first was his own self-confidence in his ability; he did not hesitate to test himself. The second was that a number of his teachers at Owens had come from the college. Thirdly, Trinity, as we know, was currently the college of Clerk Maxwell, the first Cavendish Professor, whose famous treatise on electricity and magnetism was a starting-point for much of J.J.'s early thoughts about research.

In the course of his long life J.J. never lived anywhere else but Manchester and Cambridge. Moreover, at Cambridge he was living either in Trinity, as undergraduate, college fellow, lecturer, young professor or head of the college, or nearby, and all these homes were within a few minutes walk of each other and of the Cavendish laboratory, which was then in the town centre. He liked Cambridge: the air, he said, seemed to favour him. Not once in that time could he recall a working day lost by sickness. His relations with his mother had always been close; she was a sweet-natured person, proud of her son without, apparently, understanding anything of his work. Up to her death in 1901, he regularly spent part of his summers with her, after his marriage as well as before, either in Manchester or at some seaside resort, usually with his younger brother Frederick, who worked for a firm of calico merchants in Manchester. A lifelong bachelor, Frederick admired his elder brother's success; after he retired early due to ill-health he spent the last three years of his life in Cambridge.

Mathematics was still (at Cambridge, at least) the usual approach to the study of physics, and the best students had to face the peculiar rigours of working for the mathematical Tripos. The examination was held in January in the unheated Senate House; during the midday break J.J. went to the barbershop for a shampoo. Having been coached, like Strutt, by the famous Routh, he was listed as second Wrangler, close behind Joseph Larmor, who spent his life in Cambridge at the adjacent St John's College. The two men became life-long friends, though there is little correspondence between them to show it, mainly because they lived so close and saw each other so often. In J.J.'s final year as an undergraduate Clerk Maxwell died, after years of poor health and periods of absence from the Cavendish. It was a matter of great regret to J.J. that they never met, but was able to acquire Maxwell's armchair, and kept it for the rest of his life in his study at home; much of his physics was done sitting in that chair.

After graduating J.J. stayed on at Trinity to try for a prize fellowship, which would provide him with seven years of leisure for research. His choice of subject for his fellowship thesis, on the transference of kinetic energy, followed on from the inspiring lectures on energy he heard at Owens College. Three years were allowed for the preparation of such a thesis, but, with great concentration and application, J.J. completed his within one year and with characteristic self-confidence submitted it, against the advice of his tutor. He was successful; the following year he won the Adams prize for a dissertation on the motion of vortex rings and became assistant lecturer in the college; in 1883 he became university lecturer as well.

In the three years between his fellowship election in 1881 and 1884, the crucial year of his early academic career, J.J. published a number of important papers on physics. These were both theoretical and to a lesser extent experimental, including – in 1883 – the first paper on the subject he was to pursue throughout his later life 'On the theory of electric discharge in gases', a new area of research at the Cavendish and one that was to lead to the discovery of the electron. They made his reputation in what was then the small world of professional mathematicians and physicists. In the first half of 1884, when still under thirty, he was elected a fellow of the Royal Society. Without that it is doubtful whether a few months later, when Rayleigh retired from the Cavendish chair, J.J. would have been appointed to succeed him. As it was the appointment took everyone by surprise. The fact that he should have put himself forward for the post at all was further proof of his sturdy self-assurance; only eight years previously he had arrived in Cambridge as a freshman.

There were nine electors for the chair and at least four of them, in the previous three years, had directly been involved in adjudicating J.J.'s work for the Trinity fellowship, the Adams prize or the fellowship of the Royal Society. In particular, the external elector was Sir William Thomson, no relation to J.J., who, as we have seen, could himself have taken the post on several occasions if he had so wished. With his commanding reputation and equally formidable manner of speech, it was probably his support that swayed the decision in J.J.'s favour. The electors took a bold decision and backed the original thinker of great promise rather than other candidates with more proven experience of teaching and experimental work, two of whom had actually taught J.J. in Manchester.

After the Cavendish laboratory had been started under Maxwell and consolidated under Rayleigh, J.J. was the right man to develop it further. In the 1880s physics generally was passing through a quiet period; there were no grand themes, but a lot of tidying up and improvement of measurements as well as a very necessary improvement in the methods of teaching and examining. The number of students at the Cavendish was small, and the organization of the laboratory informal. It took some time for J.J. to become established in the post, find promising lines of research and raise the money to enlarge and equip the laboratory. The professor personally conducted the administration, including, it seems, the book-keeping. J.J. wrote his correspondence standing up at a desk with a sloping top, in a neat and legible handwriting. It was the only thing about him that was neat, since he was noticeably casual in organizing his paperwork, his timetable and his dress. Conditions in the laboratory were described as somewhat chaotic.

One of the best-known Cambridge families of the time was that of Sir George Paget. He came up to Caius in the 1830s to read mathematics and, after distinguishing himself in the Tripos, switched to medicine because it would lead to a fellowship at Caius. He had progressed until he had become Regius Professor of physic, in which role he transformed the teaching of medicine in Cambridge. He married and had a family of ten children, of whom seven reached maturity. Among these were the twins Rose and Violet, who were born in 1856. Unlike her more extrovert sister, Rose grew up to be intellectual, precise and calmly organized. Although the twins were almost opposite in character, they remained devoted to each other all their lives.

Rose Paget, who had become a student at the Cavendish shortly after J.J. had become professor, had no formal qualification in physics at all, only an intellectual fascination with science, which had been fostered by her distinguished father, and a determined will. She and J.J. married in 1890; after his mother she was the only woman in his life and the centre of his affection for fifty years in a marriage of much warmth and mutual admiration. In the early years they first lived near the centre of Cambridge and then from 1899 in a large Victorian house with an acre of garden called Holmleigh, on the other side of the river Cam from the colleges. They had two children, a son and a daughter: the son, George Paget Thomson, was to become a physicist of distinction, who shared the Nobel prize in 1937 for an experimental demonstration of the wave nature of the electron.

By the early 1890s J.J. had come a long way from his Manchester roots. A star in the ascendant, he attracted the brightest students to the Cavendish. He had played himself in carefully at the laboratory and, while continuing the earlier lines of work and teaching of Rayleigh, had found himself a major field of research in the conduction of electricity through gases. Within a few years this was to lead him personally to his greatest discoveries and bring a world-wide reputation to the laboratory. The nature of cathode rays was then unknown. In 1897 J.J. succeeded in deflecting them by an electromagnetic field, thus showing that they consisted of negatively charged particles. He also measured the ratio of their charge to their mass and deduced that electrons were about two thousand times lighter than the hydrogen atom. These revolutionary discoveries were announced by J.J. on April 30, 1897, at a historic Friday evening discourse at the Royal Institution.

The growth of the Cavendish was also helped by a wise university regulation allowing graduate students of other universities to enter Cambridge as research students. In a photograph of the research students at the Cavendish taken in 1898, as many as nine out of sixteen were from outside Cambridge. At the same time, a timely change in the regulations of the Commissioners of the 1851 Exhibition allowed that well-endowed body to fund science-research scholarships for overseas students. The first to benefit from this was Ernest Rutherford; while still in New Zealand he had read everything J.J. had written and decided that this was the man under whom he wished to work. It was the start of a long and fruitful collaboration and friendship, further described in the profile of Rutherford.

It was the magnetic quality of J.J.'s greatness that drew so many of these brightest minds, and formed a seed-bed that was continually renewed as established members went off to professorships around the world (ninety-seven of them altogether, at fifty-five universities) and new students arrived. For more than twenty years his group of research workers in experimental physics was easily the most important in Britain, and probably in the world. It included a formidable list of scientists, both British and foreign, who in their turn were to make fundamental contributions to science; at least eight of them won Nobel prizes. Although so much was achieved at the Cavendish under J.J., it is striking what was not. It was Hertz who proved the existence of electromagnetic waves, Röntgen who discovered X-rays, Becquerel who discovered radioactivity, and the Curies who discovered radium, none of whom were directly influenced by J.J.

According to his son George Thomson: 'in all his theories J.J. liked to visualize, and for him the mathematics was always merely the language which described the physical and spatial concepts in his mind. He had no idea of mathematics dictating the theory.' He was not good at the physical handling of apparatus himself; he devised it in his mind and his personal assistant constructed it and got it working. To quote his son again: 'He had the physician's gift of diagnosis, and could often tell a research worker what was really the matter with an apparatus that a man had made and struggled with miserably for weeks.' His own apparatus was simply designed and constructed without unnecessary refinement. The phrase 'sealing wax and string' with which a later generation described the Cavendish apparatus of his day is not a great exaggeration; judged by modern standards there was something amateurish about it . . . yet this rather odd collection of glass and brass did in fact play a major part in producing the revolution in physicists' conceptions of the nature of matter and energy that was about to occur.

J.J.'s own mind, it was said, showed restless mental activity and originality. The subjects he pursued were for him the most exciting things in the world, and he communicated this excitement to researchers. Although most of the people in the Cavendish in his day were working on the conduction of electricity in gases, there was no attempt to direct research and quite a few were working on something else. Everyone at the Cavendish loved his characteristic smile, and felt a certain pleasure on hearing a footstep that could only be his. Niels Bohr described J.J. as 'an excellent man, incredibly clever and full of imagination . . . extremely friendly but it is very difficult to talk to him'. Rutherford's first impression of him is recorded in a letter he wrote to his fiancée: 'you ask me whether JJ is an old man. He is just fifty and looks quite young, small, rather straggling moustache, short, wears his hair (black) rather long, but has a clever-looking face, and a very fine forehead and a radiating smile, or grin as some call it when he is scoring off anyone.'

J.J. could be forceful and determined in getting what he wanted for his own work for the laboratory and for his students, and was always conscious that he possessed rare intellectual power, yet he was modest about his achievements. Throughout his written work he took careful account of what others had done. This showed particularly in the field of cathode rays, where much work had been done by German physicists. In the great upsurge of activity following Röntgen's discovery of X-rays there was a good deal of parallel activity taking place and no doubt J.J.'s discovery that the cathode-ray stream was made up of particles of a smaller order of magnitude than the atom, and which were universal constituents of matter, would have been postulated before long by others who were conducting very similar experiments. The fact is that he was the first to see the profound significance of these experiments and it took several years to convince others that he was right. Although J.J. had no gift for languages, he learned enough German, with the help of his wife who knew it well, to read the German scientific journals and kept closely abreast of work being done elsewhere, not only through the journals but also by direct correspondence. In later years he was out of sympathy with the new physics developed by Bohr, but by that time he was no longer Cavendish Professor and no harm was done. He continued to believe that atoms consisted of electrons embedded in a positively charged sphere, a model long superseded. J.J. was described as 'a curious link between the old and new physics. He opened the door to the new physics but never went through it himself. It was Rutherford who went through the door that he opened.'

The growth of J.J.'s intellect seems to have been almost complete by the time he left Manchester. What Cambridge did for him was to supply the mathematical knowledge, skill and discipline which enabled him to understand Maxwell's writings. These profoundly affected him and the resultant struggle to harmonize the view of the physical universe which he had formed at Manchester with that of Maxwell led him to the discovery of the electron. He was assisted in his labours by a comprehensive memory embracing a great variety of subjects from science to athletic records, although at times it was liable to fail him and he was known to repeat the same story to the same people in a matter of minutes. There were other instances, not uncommon in the laboratory, which were not capable of an easy explanation. A researcher would explain to J.J. what he believed was the theory behind the experimental results he was obtaining: J.J. would counter this by propounding quite a different view, and the argument would continue day after day and would finally cease, both sides being unconvinced. Then, perhaps a month later, J.J. would tell the researcher that he had found the explanation of the results they had discussed, and would give a detailed account of the very same theory as that which the researcher had propounded. The unconsciousness of its origin combined with the more perfected form was peculiarly trying. If generally this was dismissed as an instance of the vagaries of great minds, it did from time to time lead to difficulties and misunderstandings.

Being a man of very varied tastes and interests, often unexpectedly pronounced or unusual, Thomson was ready to talk to almost anyone about almost anything, and seemed to be bored by no subject except philosophy, which he once described as a subject where you spent your time trying to find a shadow in an absolutely dark room. He was one of the founders of the Society for Psychical Research and was also keenly interested in telepathy and water-divining.

The outbreak of the First World War in 1914 followed by his election to the presidency of the Royal Society brought J.J. more into public affairs. As British scientists made their contribution to the war effort, normal scientific research ground to a halt. However, the backwardness of Britain in applied science became all too evident and, as president of the Royal Society, J.J. led the efforts to persuade the government that something had to be done. Within the society itself he led the opposition to those who wanted fellows of German descent to be expelled. In 1918, when the war was over, J.J was appointed to the Mastership of his college, Trinity. At the same time he was persuaded to resign the Cavendish chair, and hence the direction of the laboratory, accepting in lieu a personal research chair with facilities in the Cavendish. He maintained his interest in physics, but no longer contributed to it himself. As Master of Trinity he was an effective chairman of college committees, took a keen interest in the students and particularly enjoyed watching their sporting activities, although he was never at all athletic himself.

He was most careless in his attire and appearance, and behaved as if it was a matter of no interest either to himself or to others. As a ready speaker with a remarkable command of English, he could, when occasion demanded, deal with complex, difficult or delicate situations with precision and tact. Travel as a recreation did not in itself appeal to him much, and in his prime he travelled little except to receive an honorary degree, a medal or a prize, or to deliver a prestigious course of lectures, for which he was quite prepared to cross the Atlantic. In later years J.J. played up his Manchester background, though he rarely revisited the city. He spoke with a noticeable Lancashire accent and liked regional dishes such as Lancashire hot-pot. His robust sense of humour came from the north: in a speech he remarked that there were two kinds of physicist in Cambridge, those who made discoveries and those who received the credit for them.

As for honours, apart from the Nobel prize in physics in 1906 for research into the conduction of electricity by gases, he received the principal awards of the Royal Society, in particular the Copley medal. He received a large number of honorary doctorates and was an honorary member of all the leading scientific academies. He was particularly proud of the Order of Merit. J.J. was knighted in 1908, but later declined the offer of a peerage, partly because he did not think himself wealthy enough to sustain the honour. In fact, he died a wealthy man, but this was through shrewdness in the management of his investments rather than through exploitation of any of his discoveries. He was always interested in commercial applications but did not seek to benefit from them personally. Unlike Lord Kelvin, he never took out any patents, though the cathode-ray oscilloscope and the television tube derive directly from his apparatus. In a broader sense the whole of today's electronic industry descends from his key discovery of the electron; no-one could have foreseen the vast range of practical applications a century later. He died at the age of eighty-three on August 30, 1940, in the Master's lodgings, after a progressive decline over the previous four years, and his ashes were buried in Westminster Abbey near to the graves of other great British scientists. There are portraits of J.J. in most of the institutions with which he was associated.

MAX PLANCK (1858–1947)

Albert Einstein said of Max Planck 'Everything that emanated from his supremely great mind was as clear and beautiful as a great work of art; and one had the impression that it all came out so easily and effortlessly . . . for me personally he meant more than all the others I have met on life's journey.' The creator of quantum theory came from an old-established family of lawyers, public servants and scholars. One of his ancestors was a pastor in the south-German region of Swabia who later became a professor of

theology at the Georgia Augusta and one of whose grandchildren became a jurist, the founder of the German civil code. He and Max Planck's father were cousins, the latter being also a distinguished jurist and professor of law first in Kiel and later in Munich. This ancestry of excellent, reliable, incorruptible, idealistic and generous men, devoted to the service of church and state, must be borne in mind if one is to understand the character of the great physicist.

Max Karl Ernst Ludwig Planck was born in Kiel on April 23, 1858, the fourth child of his parents. His mother, Emma Patzig of Greifswald, his father's second wife, came from a family of pastors. He grew up in a conservative, cultured family in prosperous Wilhelmine Germany. When he was nine the family moved to Munich, where he attended the renowned Maximilian Gymnasium. The mathematics teacher he found particularly inspiring, learning from him such fundamental ideas of physics as the principle of conservation of energy. His teachers did not rate him an outstanding student, rather they praised his personal qualities. When it came to choosing a profession, he considered philology but eventually decided on physics. He also became an excellent pianist and found in playing music deep enjoyment and recreation.

After graduating from the gymnasium, Planck studied for three years at the University of Munich. There was no professor of theoretical physics at the university at the time, but there were lecturers who gave him a good foundation in mathematics and physics. It will be recalled that it was then the normal practice in Germany for students to spend their first two years at more than one university. The absence of any examination before the final one made this system possible. Its obvious advantage was that students had the opportunity to hear the great professors of their day. Planck migrated for a year to Berlin, where he heard Helmholtz and Kirchhoff. Helmholtz, he recalled, did not prepare his lectures properly; the students felt just as bored as he seemed to be himself. Kirchhoff's lectures were meticulously prepared, but his delivery was dry and monotonous.

Although Planck's doctoral thesis of 1879, on thermodynamics, was not particularly well received, he was not discouraged. He continued to study the subject in his *Habilitationschrift*, which qualified him to become Privatdozent in 1880, but found that he had been anticipated by Willard Gibbs. He tried to correspond with Rudolf Clausius on matters related to the second law of thermodynamics, but received no replies. However, Clausius' papers on entropy were a major influence on Planck, who later used the concept of entropy as a bridge into the realm of quantum theory. To make himself better known in the scientific world, Planck competed for a prize at the Georgia Augusta in 1887, on the concept of energy, but before he had completed his essay, for which he was awarded second prize, he accepted an associate professorship at the University of Kiel. After the death of Kirchhoff in 1889, the University of Berlin secured Planck for the chair of theoretical physics, at first as an associate professor but within three years as a full professor. He owed this rapid promotion largely to the support of Helmholtz. Planck's lecturing style was to read verbatim from one of his books; if you had a copy you could follow it line by line.

Meanwhile Planck's research into thermodynamics had led him to the inescapable conclusion that something new was needed; he called it the quantum of action. Matter, he decided, emits radiant energy only in discrete bursts; the quantum of energy grows larger as the wavelength decreases, but the product is constant. Planck had no doubts about the importance of his idea. He told his son that it was either complete nonsense or the greatest discovery since Newton. In public, naturally, he was much more modest, so much so that it was said later that he had not begun to realize its full implications himself. It is generally acknowledged that the year 1900 of Planck's discovery marks the beginning of a new epoch in physics, yet during the first years of the new century it did not make much of an impact. Planck himself returned to thermodynamics. Personally and scientifically he was thoroughly conservative and recoiled from his own findings, which clashed with the tenets of classical physics, and he tried hard to find ways of reconciling them.

Planck was very much a family man. His first marriage to Marie Merck, the daughter of a banker, ended in divorce, but he remained on good terms with her. Their four children lived with their mother until she died in 1909. He then married again, to a niece of hers, Marga von Hoesslin. The Plancks lived in Grunewald, an attractive new suburb at the edge of the pine forest west of Berlin. According to his disciple Lise Meitner, 'Planck loved happy, unaffected company, and his home was a focus for social gatherings. Advanced students were regularly invited to his home . . . if the invitation fell during the summer semester we played tag in the garden, in which Planck participated with almost childish ambition and great agility. It was almost impossible not to be caught by him.' Later she said that, while she had been swept along by Boltzmann's exuberance, she loved and trusted Planck for his depth of character. 'He had an unusually pure disposition and inner rectitude, which corresponded to his utter simplicity and lack of pretension . . . he was such a wonderful person that when he entered a room the atmosphere in the room got better.'

When the 'golden age' of physics began at the turn of the century, Berlin was central to its development, as Planck was to its success. Although he was not in the vanguard of those who accepted Einstein's ideas, in time Planck developed the greatest admiration for what Einstein had achieved. The Berlin Academy, mainly at the instigation of Haber, Nernst and Planck, created a special chair for Einstein, which allowed him to pursue his ideas unhampered by teaching and routine work. For many years Planck and Einstein met at regular intervals; their collaboration made Berlin, in the years preceding the First World War, the leading centre for theoretical physics in the world. A friendship that went far beyond the exchange of scientific ideas developed between them. They shared a fascination with the secrets of nature, similar philosophical convictions and a deep love of music. They often played chamber music together, Planck at the piano and Einstein on the violin.

Planck was a pianist of great technical ability, who could play at sight almost any piece of classical music. He also liked to improvize on a given theme, such as an old German folk-song; at one time he had thought of becoming a composer (he composed songs, conducted an orchestra and

accompanied the famous violinist Joseph Joachim). One year a large harmo-
nium, with numerous keys, built at the suggestion of Helmholtz and tuned
harmonically, was delivered to the department of physics. Planck learned
to play this complicated instrument and compared the just intonation with
the normal equal temperament. Unexpectedly, he found that our ears prefer
the latter.

When Boltzmann died in 1906, Planck was invited to succeed him in
Vienna, but colleagues in Berlin persuaded him to stay. He served as Rector
of the university for 1913/4; when the war began he was one of a number of
prominent German signatories to the chauvinistic 'Manifesto of the 93' or
'appeal to the cultured peoples of the world'. Like others he soon regretted
doing so, explaining that he had not read it when he signed it. Within the
Berlin academy, of which he was an influential member, he succeeded in
preventing the expulsion of members from enemy countries.

Planck was deeply rooted in the traditions of his family and nation,
an ardent patriot, proud of the greatness of German history and typically
Prussian in his attitude to the state. During the war years, however, a change
came over him. It was not only the general suffering, the catastrophic end to
the struggle which hurt his patriotic feelings deeply, but grievous personal
loss. Three of the four children of his first marriage died during the war
period. His eldest son Karl was killed in action on the western front in 1916.
The two daughters, Emma and Margarete, were identical twins; Margarete
died in childbirth; her sister took charge of the baby and then she too died
in childbirth. The surviving children were brought up by their grandfather.
Only one son, Erwin, of the first marriage remained, and a younger son,
Hermann, of the second marriage. Erwin also fought in the war; he was
taken prisoner by the French but survived.

In 1928 Planck, being seventy, retired from his university post, but
continued in office as permanent secretary of the Berlin Academy and as
president of the Kaiser Wilhelm Gesellschaft, the Society for the Advancement of Science. In the Wilhelmine period the prestige of chemistry in
Germany was very high, so, when it was proposed that the society create
a number of specialized research institutes, financed at first by German
industry, later with increasing support from the state, it was decided that
the first two should be for chemistry and for physical chemistry. The physics
institute was not organized until after the First World War, when Einstein
was appointed the first director. The Society fully intended that it should
become a proper institute with a building including a laboratory, but at first
it was no more than an office concerned with dispensing research grants.

The Rockefeller Foundation was prepared to help finance the building of a laboratory if the Germans assumed responsibility for its maintenance. This was a project close to Planck's heart and in 1938 he had the satisfaction of seeing the long drawn out negotiations concluded. However, by then the golden age of German science was over. We must go back over the previous decade to understand what had happened.

In 1929 the worldwide economic collapse was under way; in Germany unemployment began to soar. Reactionary groups in Germany never accepted the reality of the defeat of the imperial military machine and sought to exploit a historical fantasy in which an undefeated army was betrayed by a sinister alliance of socialists and Jews. The reactionary forces included not only remnants of the military, industrialists and large landowners, but also many conservative academics. They all detested the social-democratic government of the Weimar republic and were dedicated to its destruction, the generals and business magnates by overt action and the university mandarins by incessant and insidious propaganda. Against this background the Nazis were gaining more and more adherents, and there were demonstrations on the streets of Berlin, in which students were much involved. The summer of 1932 marked the beginning of the end for the Weimar republic: by January 1933 Hitler was Chancellor and soon effectively dictator. Many Germans wrongly believed that, having achieved power, Hitler would moderate some of the more extreme Nazi policies; they regarded Nazism as a passing phase and Hitler as a puppet in the hands of the Reichswehr and the great industrialists. One has to remember that even many highly educated Germans never took the Nazi movement seriously. When they began to realize its disastrous impact it was too late for any serious opposition.

The first three months of the Third Reich began with an attack on Jewish and left-wing intellectuals and members of the cultural elite, such as musicians, artists and authors. Jewish businesses were boycotted. Within Prussia all Jewish judges were dismissed. This action was followed by much more far-reaching legislation 'for the reconstitution of the professional civil service' to ensure loyalty to the new regime. Its purpose was to exclude from state and municipal service 'unreliable elements', socialists and other political opponents as well as Jews. 'Non-Aryan' included 'descended from non-Aryan, particularly Jewish, parents or grandparents. It suffices if one parent or grandparent is Jewish.' The rules made exceptions for those who had fought in the First World War but in practice the only effect was to delay their dismissals. Similar action was taken against socialists and communists, but in scientific circles it was the Jews who were most affected. The act also applied to Privatdozenten, who were stripped of their *venia legendi*. Four years later people with Jewish spouses were included.

The law applied to the teaching staffs of universities and technical institutes, since in Germany they are state employees. In 1933, as a consequence, approximately 1200 academics were dismissed, without notice. Of course the law also meant that many younger people who were hoping for academic positions were prevented from obtaining them, while older people had to decide what to do. Those who were not affected by the new law felt it useless to protest. Some even took advantage of the resulting vacancies to advance their careers. The students, many of whom were Nazi supporters, contributed by organizing the burning of books by Jewish authors. Ordinary people were astonished at the speed and intensity of it all; the course of events was so bizarre and irrational that they could not believe it was happening. Although anti-Semitism of a less-virulent form had existed in Germany before Hitler, not only under the empire but even under the Weimar republic, nothing like Nazi anti-Semitism had been seen before.

Planck's reputation as diplomat, patriot and conciliator, and his professional standing, meant that he was the man to whom the scientists looked for leadership. Universally respected for his absolute integrity and devotion to German science, Planck went to see Hitler in 1933, to plead for reason and restraint, emphasizing the enormous damage being done to science in Germany by the racist laws. Characteristically, Hitler worked himself into such a rage that Planck could do nothing but listen in silence and take his leave. Afterwards Heisenberg said he looked 'tortured . . . and tired'. Hitler had assured him, Planck said, that the government would do nothing further that could hurt science in Germany. Planck trusted that the violence and oppression would subside in time. 'Take a pleasant trip abroad and carry on some studies', he advised a worried colleague, 'and when you return the unpleasant features of our present government will have disappeared.' The logical process of German thinking inhibited resistance. 'If I protest', the reasoning went, 'I shall be removed from my post where I have influence, then I'll have none. So I had better be quiet and see what happens.'

The Prussian tradition of service to the state and allegiance to the government was deeply rooted in Planck. Whatever his personal feelings, he believed that it was his duty to work with the regime; he did not believe that public protest would achieve anything. Busts of Hitler were installed in the institute. Telegrams were sent to Hitler avowing pride in the 'national resurrection' which was taking place and thanking him for his 'benevolent protection of German science'. When Einstein, under pressure, resigned from the Berlin Academy, Planck wrote to thank him for doing so and hoped that they would remain friends. For the next ten years Planck, like other German scientists, omitted all reference to Einstein when relativity and quantum theory were under discussion, believing that it was more important to promote his work than to credit him with it, which might have led to political interference.

By 1938, far from relaxing, the Nazi persecution of the Jews had reached a new pitch of intensity. The admission of Jews to higher education was already very restricted; now it was forbidden. Thoroughly disillusioned, the eighty-year-old Planck resigned as secretary of the academy in 1938, after twenty-six years of service. He still enjoyed good health, due to the simplicity and regularity of his life and his custom of spending most of his holidays in the Bavarian Alps, usually at a small property he owned near Tegernsee. When he lectured it was usually on science and religion.

After the Second World War Planck was but a shadow of his former self. His house in Grunewald had been destroyed in one of the air raids on Berlin, and he had lost all his possessions. His son Erwin, the only surviving offspring of his first marriage, was marginally involved in the bungled July 1944 plot to assassinate Hitler led by Count von Stauffenberg and was executed by the Nazis. Towards the end of the conflict the Plancks took refuge on the estate of a friend on the west bank of the river Elbe, near Magdeburg. There they found themselves between the lines of the retreating Germans and advancing Allied armies; the battle raged around them for days. Eventually the American troops came and took him to safety in Göttingen, where he remained, leaving on only a few very special occasions. A great celebration was being prepared for his ninetieth birthday but a few months previously his health had begun to fail and he died from a stroke on October 4, 1947. Max Planck had received the Nobel prize for physics in 1919, for his discovery of energy quanta, and other honours too numerous to mention.

WILLIAM HENRY BRAGG (1862–1942)

The early life of William Bragg, the founder of the science and art of X-ray crystallography, was tough and testing. He wrote about it in a short autobiography in 1927, not intended for publication, and as a result we have more information about these years, before he arrived in Australia, than we have for many other scientists of his period. Since his son was also an

R. Schwabe Sept 1932

eminent physicist and his first name was also William, we must be careful
to distinguish between them.

The Bragg family came from West Cumberland, the part of England
that lies between the Lake District and the Solway Firth. William Henry
Bragg was born on July 2, 1862 at a farmhouse called Stoneraise Place near
the market town of Wigton. His mother, Mary Wood, was the daughter of the
local vicar. He did not remember her well, for she died in 1869 at the age of
thirty-six when he was barely seven years old. His father Robert John Bragg
had taken up farming after retiring from the merchant marine. He lived on,
but there is little about him in the autobiography. William, the first-born
child, grew to manhood with hardly any parental love or guidance that he
could recall. Instead his boyhood was dominated by an uncle, also named
William Bragg, with whom he went to live after he had lost his mother.
Uncle William was a pharmacist at Market Harborough in Leicestershire,
a widower with no children of his own. His nephew recalled that 'there
were no parties for children; we never went to other people's houses, and
no children came to ours. I think my uncle was too particular. He used to
lecture us terribly, talking by the hour, and I suspect he was not to be shaken
in his opinions by anyone.'

School offered some outlet from this regime. On the initiative of Uncle William the old grammar school at Market Harborough was reopened the same year as his nephew arrived. The master 'was an able man, I believe, . . . and I got on quickly enough'. In 1873, at the age of eleven, Bragg went up for the Oxford Junior Local Examinations at Leicester and was the youngest boy in the whole country to pass, despite failing church history and Greek. An aptitude for mathematics and modern languages rather than the subjects of the old classical syllabus was already becoming apparent.

The few organized school ball-games were 'a great delight' to him, and there were some happy times with his cousin Fanny, who also lived with Uncle William. Otherwise, whatever enjoyment, satisfaction and contentment the young Bragg found in life were discovered primarily within himself. He was already a solitary child: 'I liked peace and was content to be alone with books or jobs of any sort.' However, he was not without personal ambition; his tough childhood had made him self-reliant, quietly self-confident and self-content, these characteristics would sustain him for the rest of his life.

Uncle William would have liked his nephew to go to Shrewsbury, an old-established public school of good repute, but 'In 1875 my father came to Harborough and demanded me; he wanted to send me to school at King William's College on the Isle of Man, where his brother-in-law was a master. I think he became alarmed lest he should lose me altogether.' The few accounts of this college in the second half of the nineteenth century do not paint an attractive picture. The discipline was strict, the social and psychological pressure severe. Cruelty among the boys was widespread, engendered no doubt by the fearful beatings that the masters meted out to their pupils.

Bragg survived and even prospered in this environment by adhering strictly to the rules of the college, by applying himself diligently to his studies, by enjoying to the full the sporting, social and educational opportunities that the school increasingly provided, and by repressing almost totally the emotions he had already learnt to hide. Bragg found much satisfaction in his school work, especially the mathematics, where his school reports testify to his exceptional ability and achievements. In 1880 he won a school prize for mathematics, which took the form of Clerk Maxwell's two-volume treatise on electricity and magnetism. Outside the classroom, Bragg was first prefect and then captain of the school. He participated in many activities, especially the annual theatricals of the Histrionic Society. The ultimate academic goal for school and boys alike was to win a scholarship at one of the Oxford or Cambridge colleges. In 1880 Bragg won a minor scholarship to Trinity

College, Cambridge. The following year he tried again in the hope of upgrading it to a foundation scholarship, but his academic work had stagnated and he did not succeed: 'The effective cause for my stagnation was the wave of religious experience that swept over the upper classes of the school during that year. The storm passed in time, sheer exhaustion, and the fortunate distraction of other things, work and play.'

Bragg went up to Cambridge in 1881. To begin with he was lonely: 'I had no companions', 'I could not afford, or thought I could not afford, to join the Union or the Boating Club.' His carefulness and reserve held him back. He was coached, like Strutt and J.J. Thomson before him, by the famous Routh, an indication of Bragg's own awareness of the Cambridge scene and of Routh's early appreciation of his abilities. In the college examinations of 1882 he won a prize and his minor scholarship was converted into a foundation scholarship, which brought him various small privileges. When he took the first part of the Tripos in 1884 he came out, to his great joy, as third Wrangler. For the second stage he specialized in physics, gaining some experience of experimental work. When he gained first-class honours in the examination at the beginning of 1885, he might have tried for a fellowship, but there happened to be strong competition that year. Instead he started experimental work in the Cavendish Laboratory and by the end of that year he knew Thomson pretty well. Although earlier he was 'much shut in myself, unventuresome, shy and ignorant', after graduation he 'found Cambridge a lovely place and Trinity something to be very proud to belong to'.

Walking along King's Parade to the Cavendish one morning, Bragg was joined by Thomson, who started to talk about the professorship of mathematical physics at the University of Adelaide in Australia, which had just been advertised; Horace Lamb, the incumbent, was moving to Owens College in Manchester. When Bragg enquired whether he might have a chance, Thomson said that he might. Bragg applied by telegraph, only just in time. In fact there was a strong field of 23 candidates; by far the ablest man was excluded 'on personal grounds' (not safe with the bottle). After the interviews, which were held in England, Lamb reported the conclusion to the Chancellor of Adelaide: 'yesterday the interviews were held and – with some slight hesitation between two of the candidates – we unanimously recommend Mr Bragg of Trinity College, Cambridge. It is evident that his mathematical abilities are of the highest and he has also worked at physics in the Cavendish Laboratory under my coadjutor in the appointment, who says his work is very good . . . as far as I can judge the only possible source of misgiving as to the propriety of our choice is Mr Bragg's youth, he is only 23.'

That evening at Market Harborough a telegram broke the exciting news that he had been chosen. Uncle William broke down and wept (Bragg's father had died shortly before). The position was a professorship, where he would be his own master, with a generous salary, and there was all the excitement of going to a new country. The colony of South Australia had been founded only in 1836, but Adelaide was already a fine city of some 30 000 inhabitants; and the university had already been in existence for ten years.

Soon after his arrival in Australia Bragg found time to explore the country to the east of Adelaide as far as Melbourne and Sydney before settling down to work. The problem in Australian universities during this period was neither shortage of money nor conservatism of thought but rather a shortage of students who wanted to study and who could afford to do so. The university capped the educational pyramid but the base of the pyramid was weak. Bragg found himself heavily overworked; the full extent of the demands on his time and physical stamina was something he had not foreseen. Consequently, although he followed new developments in science, he was not involved in research at all at this stage in his career.

One of the leading citizens of Adelaide at this time was Charles Todd, the government astronomer, postmaster-general and superintendent of telegraphs, famous throughout Australia as the architect and builder of the transcontinental telegraph line, which was one of the epic achievements of Australian history. It linked the eastern cities of the country through Adelaide with Darwin on the north coast and then by submarine cable with the outside world. It was after Todd's wife Alice that the town of Alice Springs was named. Bragg had met Gwendoline, one of the Todd daughters, several times at their home in the Adelaide observatory. When the family went on holiday to Tasmania, he accompanied them, proposed marriage to Gwendoline and was accepted.

Despite the heavy pressure of work at the university there were ample opportunities for recreation in Adelaide. Bragg performed in amateur dramatics, played tennis, golf and lacrosse, and took up the new craze of bicycling. Gwendoline was keen on painting and performed in Gilbert and Sullivan operettas; together they had a busy social life. In 1897 he took her on her first visit to England, and it was after their return that he started to think seriously about scientific research – it had never before entered my head, he once remarked. He was particularly interested in Marie Curie's work on the element radium, of which he was able to purchase a sample, and began experimenting with it. He sent his results to Rutherford, the great authority on radioactivity, and received an encouraging reply. Before long he was being asked whether he might be interested in moving closer to the mainstream of scientific research. Because Rutherford was leaving for Manchester, a possibility opened up at McGill University in Montréal, but nothing came of this.

After Bragg had been elected fellow of the Royal Society in 1907, he received and accepted an invitation to become professor of physics at the University of Leeds, not far from Rutherford at Manchester. So now life, work and interests in Adelaide had to be wound up. He had enjoyed success there, success in university teaching, in popular lectures, in sport, in adult education; he was a highly respected citizen, governor of the public library, museum and art gallery, a pillar of his parish church, on the council of the school of mines and prominent in the Australian Association for the Advancement of Science. The family said goodbye to Adelaide at the beginning of 1909. They now had two adolescent children, William Lawrence, the future physicist, his brother, Robert, and a toddler named Gwendoline like her mother. They soon began to regret what they had left behind in Australia, where they loved the life.

For Bragg himself there was a warm welcome by his new colleagues in Leeds although the students did not appreciate his lectures. He became engaged in a tiresome dispute with Charles Groves Barka, then professor at King's College, London. It boiled down to the old disagreement on the nature of light between the corpuscular theorists on the one hand and the undulatory theorists on the other. In research he continued to study radioactivity, especially the passage of alpha and beta particles and gamma rays through matter. In 1912 he invented the Bragg diffractometer for the measurement of X-ray wavelengths and with his son discovered the law of X-ray diffraction.

For Gwendoline initially the contrast between beautiful Adelaide and the dirty industrial city of Leeds, with its rows of poor little back-to-back houses, was quite painful, with only the wild open country to the north to provide relief. Before long, however, she had a house of her own, a pleasant square house with low-pitched slate roof and large garden, with carriage sweep and lawn. She employed two maids and a cook, and began to build up a social circle, including some of the successful manufacturers of the city, the brewers and the steel-makers, the makers of railway engines and of ready-made clothing. She threw herself into social work, together with the wives of such people.

For Bragg struggling at Leeds there was some compensation in the success of his eldest son Lawrence, who was doing brilliantly at Cambridge. Later a certain tension grew up between them. Unlike his father, Bragg junior held to the undulatory theory of light. Even after they had shared a Nobel prize in 1915, for their analysis of crystal structure by means of X-rays, this tension persisted, although it never seemed to disturb their mutual affection. They collaborated on a book, *X-rays and Crystal Structure*, published in 1915. When their research overlapped, Bragg senior never hesitated to give credit to 'his boy'. Bragg senior was always scrupulously fair in giving credit to his son's contributions to their joint work, yet it was assumed by others that the father was just showing parental generosity and that *he* was really the dominant contributor. Again, when his son first used a Fourier synthesis to calculate the electron density in a crystal in an important paper in 1929, he acknowledged that this work owed much to a suggestion put forward by his father much earlier, but later came to feel that the paper should have been a joint one.

In 1915 William Bragg moved from Leeds to London as Quain Professor of physics at University College and became involved in research on underwater sound propagation for the admiralty. After the war he tried to build up a research group at the college, but he did not care for the way the place was run. However, early in 1923 the death of the aged Sir James Dewar, the director of the Royal Institution, created an opportunity for him to move to somewhere more congenial. According to the then secretary of the Institution, Dewar had hoped that Rutherford would be his successor. 'So one day when Rutherford had come from Cambridge to give a lecture at the Royal Institution we interviewed him. He explained that he was too deeply committed at Cambridge to think of changing. "But", he said, "I know of a man who is as well fitted as I am to fill the billet." We eagerly asked the name of this man: "William Bragg", he replied. I supposed he meant young William Bragg [i.e. the son Lawrence]. "No", he answered, "I mean Sir William Bragg, professor of physics at University College, London; he is a great man of science and also a very great man".

When the invitation arrived, on May 7, 1923, William Bragg knew that he would be given the facilities he needed for his own research in the Davy–Faraday Laboratory. Moreover, he had always worked for an understanding between science and the other disciplines, and he would have plenty of scope to pursue this through the noble tradition of public lecturing. Already in 1920 he had given the Christmas Lectures intended for children on 'The World of Sound'. An excellent lecturer himself, he explained that, in his view, 'the value of a lecture is not to be measured by how much one manages to cram into an hour, how much important information has been referred to, or how completely it covers the ground. It is to be measured by how much a listener can tell his wife about it at breakfast the next morning . . .' He particularly felt that it was quite wrong to read a lecture: 'I think it is a dreadful thing to do, something quite out of keeping with everything a lecture should mean. When a man writes out a lecture he invariably writes it as if it were to be read, not heard. The ideas follow each other too fast. It is easy for the lecturer to deliver well-considered rounded phrases but the audience has to follow and to think.'

However, the Royal Institution was very run down, most of the scientific staff were past their best, and it would take energy and determination to restore it to its former glory. William Bragg set to work energetically, and within a year the place had been transformed. Rutherford as visiting professor was often in the laboratory. Gwendoline played her part by entertaining in the Upper Chambers. While they were at Leeds the Braggs had rented a country cottage in beautiful Wharfedale, above Bolton Abbey. Wharfedale was now too far away but they found a romantic old place called Watlands near Chiddingfold in Surrey, within easy reach of London, to provide a retreat from the official residence of the director.

William Bragg had been knighted in 1920; now a different honour arrived, the prestigious Copley medal of the Royal Society, and a stream of other scientific distinctions. Sadly, Gwendoline's health was giving cause for concern and she died in the autumn of 1929. Sir William was not yet seventy, but he tended to get very tired, although determined to carry on. He would walk to the Royal Society at Burlington House from the Royal Institution in Albemarle Street, just a few hundred yards, with his old slow countryman's walk, making several pauses on the way when he seemed to be taking an interest in some shop window. One evening, opening his study door, his younger daughter noticed how wearily he looked up from his papers: 'Daddy', she cried, 'need you work so hard?' He answered simply 'I must my dear; I am always afraid they'll find out how little I know.' In 1935 he was elected president of the Royal Society; he had to be persuaded that at seventy-three he was not too old for this.

Although Sir William was not attracted by politics – he declined an invitation to stand for Parliament – he made an influential radio broadcast, putting the case for what he called 'moral rearmament'. This phrase was taken over by the American evangelist Dr Frank Buchman, leader of the Oxford Group, but Sir William was no follower of his. When the Second

World War began, Sir William was involved in various kinds of committee work. He had always been keen on broadcasting about science; one of his last contributions was to a radio programme on 'The Problem of the Origin of Life'. A few days after that he took to his bed and on March 12, 1942 he died, at the age of seventy-nine. The memorial service was held in Westminster Abbey.

Lawrence Bragg has already been mentioned briefly; he was scarcely less distinguished than his father, with whom he shared a Nobel prize. He might well have had a profile of his own, but he had a much easier start in life than his father. His story begins in Adelaide, where he was born on March 31, 1890. At the age of eleven he started school there until four years later his father decided he was ready for university. At the University of South Australia he read mainly mathematics, graduating with first-class honours in 1908 at the age of eighteen. That was the year that the family returned to England, and it was to be fifty years before he saw the land of his birth again.

Following his father's example, Lawrence Bragg went up to Trinity College, Cambridge, to read mathematics, but moved to physics at the first opportunity. By 1914 he had been elected fellow and lecturer at Trinity. When the war came he joined up and was commissioned at once, on the strength of some previous military experience. He was soon sent to France where he helped to perfect the French technique of locating enemy guns acoustically. The award of the joint Nobel prize in 1915 came as he was setting up an acoustic-ranging station near the front line; his military service brought him decorations and promotion to the rank of major. His younger brother Robert also joined up but died in the ill-fated Gallipoli operation to try to obtain control of the Dardanelles.

As the war drew to a close, Lawrence Bragg was looking for a professorship. After returning briefly to Cambridge, he was appointed to succeed Rutherford at Manchester. At the age of twenty-nine he had no previous experience of teaching undergraduates, although he had a gift for lecturing. However, the students were mainly ex-servicemen who had no mercy on novices. Also, a junior staff member sent him anonymous letters accusing him, amongst others, of incompetence. However, in 1921 he was elected to the Royal Society and, during the same year, he married Alice Hopkinson, whom he had first met at Cambridge. Had she not been a native Mancunian he would have had misgivings about bringing a lively young wife to grimy Manchester and introducing her to his sober colleagues. However, she already knew the city well because her father had been a much-loved physician there. In all respects it was a successful marriage, with four children, namely two sons and two daughters.

By 1929 he was thinking of moving on. He sounded Rutherford out about the Cavendish chair, with no result, and was offered one at University College, London, which he turned down. Worried that he might find himself remaining in Manchester for the rest of his career, he suffered a nervous breakdown, but soon recovered. In 1935 he spent a term at Cornell University in the State of New York and ended the year by giving the Christmas lectures at the Royal Institution, on the subject of electricity. Then, in 1937, he was appointed director of the National Physical Laboratory, at Teddington, on the outskirts of London. He was just reconciling himself to a life of tedious committee work when, following the death of Rutherford, he was elected the next Cavendish Professor.

So Lawrence Bragg and family moved to Cambridge after all, and he ran the Cavendish from 1938 to 1945; for part of the time he was president of the Institute of Physics as well. Once the Second World War began his main preoccupation was to see what contribution he could make to the war effort, not just in Cambridge but nationally. He was knighted in 1941 and became known as Sir Lawrence, his father being Sir William. He had held a non-resident chair at the Royal Institution since 1938 and in 1953 he accepted the offer of the much more important post of resident professor. In spite of his father's efforts there were still serious financial problems at the institution and a need for further reform. Sir Lawrence set to work to put matters right, especially on the research side, and to continue the tradition of first-class public lectures. In 1966, when he retired at the age of seventy-six, he received the Copley medal from the Royal Society and was made a Companion of Honour. After retirement he continued to live in London most of the year, giving lectures at the Royal Institution and elsewhere. He died in hospital near his home in Waldringfield on July 1, 1971, at the age of eighty-one.

MARIE CURIE (1867–1934)

The name of Marie Curie is as well known to the general public as that of Charles Darwin, Albert Einstein or Louis Pasteur. She was born in Warsaw on November 7, 1867 and christened Maria. Her parents came from the numerous class of minor Polish landowners. Her father Władisław Skłodowski, a kindly, erudite man, who had studied at the University of St Petersburg, was a teacher of physics; her mother Bronisława (née Boguska) conducted a private school for daughters of upper-middle-class families.

Their five children were Sofia, Józef, Bronisława, Helena, and finally Maria Salomea, the subject of this profile. By the time she was born her father was professor of mathematics and physics at a high school for boys. However, Poland was under Russian oppression and, under a policy of replacement of Polish officials by Russians, he lost his position and the apartment which went with it. Not without difficulty, he found another apartment where he could lodge boys of school age and give them tuition.

Her unfortunate father had made the mistake of investing his life savings in a business owned by his brother-in-law, which went bankrupt. From then on the family lived in a state of considerable poverty, but there were other troubles. Her eldest sister Sofia died of typhus. Her mother had developed symptoms of tuberculosis after her last pregnancy. She resigned her headship and spent a year 'taking the cure', first in the Austrian alps, then in the south of France, but the disease progressed and she died two years later. In the aftermath of the loss of her mother, Maria's health began to suffer; and she experienced some kind of nervous breakdown. At the age of fifteen, she was sent off for a year to some country relatives to recover, being forbidden to study, except that she was allowed to learn French.

Under Russian subjugation, Poland was intellectually isolated. Conventional universities were not open to women. She joined a self-improvement society called the floating university. Maria's brother Józef was studying medicine; her two sisters were planning to go into teaching, while Maria herself started work as a governess. Her first such post was a failure but the second was less so. As well as educating the ten-year-old daughter of her employer, a wealthy lawyer, she started a school for peasant children and continued her own education by reading, with an inclination towards science. She had an affair with the son of the house; marriage was out of the question because of the difference in their stations in life. She returned to Warsaw and the floating university at the age of twenty. After a year in the capital working as a governess, she returned home to her father, who had reluctantly become director of a reformatory near Warsaw. Meanwhile her elder sister Bronisława, now married, urged Maria to join her in Paris. In 1891, after some delay, she set off to study at the Sorbonne at the age of twenty-four.

Marie, as she now called herself, spent the first few months of her new life staying with her sister and brother-in-law, who had set up a medical practice in the outer suburbs of Paris. Although she was too hard up to become involved in the gay social life of *fin-de-siècle* Paris, she then rented a garret apartment near the university like many other students. She attended lectures in the physical sciences, but at first she had difficulty with the language, also she lacked the basic mathematics. However, she persevered and graduated with high honours first in physics in 1893 and then in mathematics the next year. She spent the intervening long vacation back in Warsaw, where she was awarded a scholarship for outstanding students who wished to work abroad; characteristically she repaid the scholarship money as soon as she could.

In the spring of 1894 she met Pierre Curie at the home of a Polish physicist who was staying in Paris. Her future husband was responsible for the laboratory of the Ecole Municipale de Physique Industrielle et de Chimie, a new foundation where the lectures were combined with substantial experimental work. He was tall with auburn hair and sported a small pointed beard. 'He seemed to me very young, though he was at that time thirty-five years old', she recalled, 'I was struck by the open expression on his face and by the slight suggestion of detachment in his whole attitude. His speech, rather slow and deliberate, his simplicity and his smile, at once grave and youthful, inspired confidence.' The son of a homeopathic physician, he was as shy and introverted as she was. After their first meeting she noted that 'he expressed the desire to see me again and to continue our conversation of that evening on scientific and social subjects in which he and I were both interested and on which we seemed to have similar opinions'. He was dedicated, she soon learned, to a life entirely devoted to science and the rewards its purity has to offer. He presented her with a copy of an important paper he had written 'On Symmetry in Physical Phenomena: Symmetry of an Electric Field and of a Magnetic Field'. Pierre Curie was a physicist of the first rank, a pioneer in the investigation of the magnetic properties of various substances at various temperatures. He discovered the piezo-electric effect and showed that ferromagnetism reverts to paramagnetism above a certain temperature (the Curie point). Even in his lifetime, his discoveries had widespread application, and the fame of this underpaid, overworked scientist spread far beyond the borders of France, particularly to Britain.

Quite soon after their first meeting, Pierre broached to Marie the question of marriage. He suggested they might start living together, but she had been too strictly brought up to consider that. Then he found an apartment in the Latin Quarter that could be subdivided into two parts; she said no to that too. However, at least she agreed to his suggestion of a visit to his parents. He took her to their home in the attractive township of Sceaux, whose inhabitants had once served a fine Louis XIV château and its magnificent park, and which by this time was one of the southern suburbs of Paris. Dr Eugène Curie's serene, plant-covered cottage stood in the rue de Sablons, later to be renamed after his famous son. His wife Sophie-Claire (née Depouilly) bore two sons, Jacques in 1855 and Pierre in 1859. The radical politics and anticlericalism of the doctor, a former Communard, appealed to her. Later, when she desperately needed it, the understanding which developed between them was to be of great importance.

Marie and Pierre Curie were married at a civil ceremony in Sceaux town hall in 1895. Since Pierre was a freethinker, like his father, and Marie had given up her faith, they dispensed with a religious ceremony. Her married elder sister was already in Paris; her father and younger sister came over from Warsaw for the occasion. Among the wedding presents the couple received were two bicycles; in the years to come they took cycling holidays in Auvergne, the Cevennes and along the coast of Brittany.

The young couple had few distractions from their scientific work; a bicycle outing or a rare visit to the theatre were their only recreations. His parents and her elder sister and brother-in-law seemed to be their only social contacts. She became pregnant, with accompanying sickness, and in due course gave birth to a daughter, Irène. She felt lonely and homesick for Poland. After Pierre's mother had died of breast cancer in 1897, his father helped the young couple look after the baby. Marie was studying for the *agrégation*, the certificate which would permit her to teach in a secondary school for girls. She also helped Pierre prepare his teaching courses while filling in gaps in her scientific education. The Ecole de Physique et de Chimie agreed that she could start research alongside her husband, on how the magnetic properties of various tempered steels varied with their chemical composition.

After Röntgen had discovered X-rays in 1895, many scientists began to investigate their properties. When Marie needed a thesis topic in 1897, it would have been natural to have looked in that area. Instead she chose to work on the phenomenon of radioactivity, which had been discovered by Becquerel the previous year and had attracted much less general attention. So far uranium was the only radioactive element known; she set out to discover whether there were others. Quite soon she found that thorium had similar properties, unaware that the German physicist Erhard Schmidt had made the same discovery and had already published it. However, in the laboratory the Curies were encountering substances that were much more strongly radioactive than thorium; one they called polonium was 350 times as powerful. It was in their paper describing this discovery that the term radioactive was used for the first time.

The Curies were working with a natural ore called pitchblende, which consists mainly of uranium oxide. When they had removed all known radioactive substances from this material there was still something else left, and they decided to call it radium. They were determined to isolate it and find out whether it was a new chemical element. In this quest another scientist, named Gustave Bemont, was involved; the Curies acknowledged this but just what he contributed is not clear. Others helped in different ways, for example by lending them equipment. One of them remarked that it was the amiable and self-deprecating Pierre who was the ingenious one, while Marie provided the determination which kept the research going. She was more the chemist, he the physicist. They soon concluded that, rather than just the small quantities they had been using, they needed industrial quantities of pitchblende. The chief European source of this expensive ore was the St Joachimstal mine in Bohemia, then part of Austria-Hungary. The Curies realized that, once the recoverable uranium had been extracted at the mine, the massive first stage in their work would already have been carried out. With the help of a scientist at the University of Vienna they obtained samples, which confirmed that the unwanted residue contained what they needed, and they discovered where it was being dumped. The Austrian government was helpful and in due course a four-ton load of the material arrived in sacks in the yard of the school of physics.

To refine this further they needed a much greater working space than before. The best they could obtain was an abandoned shed once used by the school of medicine as a dissecting room: 'its glass roof did not afford complete shelter from the rain; the heat was suffocating in summer, and the bitter cold of winter was only a little lessened by the iron stove, except in its immediate vicinity. We had to use the adjoining yard for those of our chemical operations involving irritating gases; even then the gases often filled our shed.' However, the first stage involved heavy labour, as well as irritating and even dangerous gases, while the later stages were extremely delicate operations; the material they prepared was already showing visible signs of radioactivity.

At the turn of the century Pierre and Marie had been married for four years. Pierre's father Eugène had moved to be near them and help look after their daughter Irène, now two years old. Not having attended one of the Grandes Ecoles, Pierre was at a disadvantage when it came to appointments; moreover, he was unduly modest about his very considerable research achievements. After being passed over for the chairs of physical chemistry and of mineralogy at the Sorbonne, he was offered an attractive post at the University of Geneva, including a physics laboratory designed to his own specifications, and an official position there for Marie. At first they were tempted to accept. However, moving to Switzerland would have seriously dislocated their research, setting it back by months, if not years, at a time when they were becoming increasingly aware of competition to isolate radium and establish that it was a new element. Some commercial firms, with far greater facilities, were manufacturing radioactive material, so that impure radium became readily available. This increased the chances that someone else would beat the Curies to the object of their quest.

The question was resolved when a vacant chair was found at the Sorbonne as a possible counter to the Swiss offer. The mathematician Henri Poincaré had been impressed by the Curies' work and used his influence to ensure that Pierre was appointed. At the same time Marie was offered a part-time post teaching physics at an advanced ladies' college in Sèvres, so they decided to remain in Paris. These appointments eased their financial situation while committing them to additional teaching and other duties. However, although others enjoyed better research facilities than they did, no-one had greater determination. By March 28, 1902 they had refined just one tenth of a gram of radium chloride. Radium proved to be a million times as radioactive as uranium; its atomic weight came out at 225.93. As soon as the news spread around, the Curies found themselves famous, in Britain even more than in France. There was great excitement when Pierre lectured about their joint work at the Royal Institution and soon afterwards the Royal Society awarded him the Davy medal.

After Marie had completed her doctoral thesis, the oral examination was an emotional occasion. The crowd in the room where it was held included family and friends, some girls from the school where she taught and many supporters who burst into applause when, after she had disposed of a few questions, the presiding examiner announced that she was now doctor of physical science in the University of Paris and added the distinction *très honorable*. By chance Rutherford was in Paris at the time; he missed the examination but met the Curies for the first time at the celebration afterwards. As we shall see when we come to his profile, he resolved to provide a definite theory of radioactive phenomena, something they had not attempted to develop.

Unfortunately the Curies, who had been handling radioactive substances for years, without any precautions, were now experiencing health problems. Those around them were concerned at how ill they looked. Pierre's fingers were so painful that he could hardly write. Her hands were painful also. She became pregnant again but the outcome was a miscarriage. International recognition culminated in their being awarded the Nobel prize for physics in 1903 for their joint researches on the radiation phenomena discovered by Becquerel, with whom they shared the prize. Marie was the first woman to become a Nobel laureate in the sciences and remained the only one until her daughter Irène was so honoured in 1935. The Curies gave much of the prize money to good causes. Of course the floodgates of publicity now swung wide open, much to their dismay. The lofty idealism of her husband forbade him to court popularity, and the honours which now were offered him were either declined or accepted with some reluctance. In 1905, after an earlier attempt had been unsuccessful, Pierre was elected to the Paris Academy, at the age of forty-six. Ironically the state of his health was such that he could no longer undertake experimental work. In the same year Marie gave birth to their second daughter Eve (or Eva) Denise. Pierre gave his Nobel lecture in Stockholm on their joint work.

A few months later, on a rainy April day, a heavy wagon drawn by two carthorses was moving down the rue Dauphine near the Sorbonne when suddenly a man holding an umbrella who was crossing the wet street appeared to slip and fall under the wheels of the wagon. It was Pierre Curie, and he was fatally injured. Marie was distraught when the news was broken to her. Her sister Bronisława came from Warsaw to comfort her. Later Marie published his collected works, but first she was determined to complete the research on which they had been engaged. Rutherford and others were not convinced that polonium was an element, and there was even some doubt in the case of radium. More dogged effort was required in order to settle these questions. Rutherford wrote to his mother to tell her how 'wan and tired Marie looked and much older than her age. She works much too hard for her health. Altogether she was a very pathetic figure.' He sat next to her at the opera one evening and could see that she was far from well; halfway through the performance she left on his arm, completely worn out.

Within a month of the death of Pierre she was back in the laboratory, having been made assistant professor, the first woman in France to reach professorial rank. Within two years she had been appointed to the chair at the Sorbonne her husband had held. On a site near the Sorbonne, the Radium Institute was established, with one part devoted to pure research into the chemistry and physics of radioactivity, under her direction, and the other part to research on the application of radioactivity to the treatment of disease. She moved with her two daughters to the Curie family home at Sceaux, where the old doctor presided over the household and, assisted by a succession of governesses from Poland, provided companionship for the girls when their mother was out at work. The wealthy philanthropist Andrew Carnegie met her and was deeply impressed, especially by her attitude as a scientist on an equal footing with men. The result was the endowment of the Curies Foundation, which was available to fund her research and provide scholarships. She could now afford assistants in her scientific work. Several of these were women, who developed a fierce loyalty to her.

For some years Marie Curie had taken her daughters for holidays to a little place called l'Arcouest on the north coast of Brittany, sometimes called Sorbonne-plage, because a small group of Paris academics, with their families and friends, used to gather there regularly each summer. Eventually it became like a second home for the Curie family. Irène and Eve found happiness there that was denied them in Paris, since their mother was not so preoccupied with her work. Later on, however, the children were usually sent to stay with relatives. In 1911, for example, they went to Poland for the first time to stay with their Aunt Bronisława. In 1913 they went hiking in the Engadine in a party that included the Einsteins. According to Einstein, Marie was like a herring, meaning that she had little capacity for either joy or pain and that the main way she expressed her feelings was by grumbling.

Meanwhile Marie Curie was proposed for election to the Paris Academy. The permanent secretary, Gaston Darboux, wrote to the press explaining why he supported her candidature and specifying the practical advantages membership of the academy would bring her. She might have been the first woman to be elected to the Académie des Sciences, although one had just been elected to the literary Académie Française. However, there was a respectable rival candidate and, after a particularly close and tense election, he was successful. Deeply hurt, Marie never allowed her name to be put forward again. Much later she was elected to the Académie de Médicine, another section of the Institut de France, due to the success of radiotherapy in the treatment of tumours, although she was never directly involved in the medical applications herself.

When she was widowed she was thirty-eight years old, strikingly beautiful, some said it was the beauty of suffering. Her closest friends, outside the family, were the physicists Paul Langevin and Jean Perrin, the mathematician Emile Borel and their wives Henrietta Perrin and Marguerite Borel. Old Dr Curie died early in 1910, after being bedridden for a year. Following the death of her husband, Marie moved from central Paris to the suburb of Fontenay-aux-Roses, where there was a colony of scientists, including the Langevin family, who had moved there to get away from the bustle of central Paris. Paul Langevin was one of those who played a part in an educational cooperative that Marie organized, somewhat influenced by the thought of Rousseau. In 1905 he had succeeded the deceased Pierre Curie as professor at the Ecole de Physique et de Chimie. Four years later he became director of studies at the school and was also made full professor at the Collège de France, having previously held junior positions there.

Langevin had recently left his wife and taken an apartment in the city, closer to the Ecole de Physique et de Chimie, where he frequently had to work late at night. Marie Curie often visited him there. After having kept to sombre clothing following her husband's death, she now began to make herself look attractive again. Her relationship with him was the only one she was to have with a man who was not many years older than herself (in fact Langevin was five years her junior). Besides offering much to enrich her middle years, including a keen interest in politics, a love of literature and music, and other common interests, he was able to provide an intellectual bridge to the emerging new physics.

Marie Curie often naively misinterpreted what she believed to be other people's reactions to her actions. That she believed that she could have a love affair with Langevin in which only a few friends and colleagues would take any interest was a disastrous miscalculation. In the summer of 1910 at l'Arcouest, encouraged by the Perrins with whom she was sharing a house, Marie started to urge Langevin to seek a divorce. Soon Langevin's wife was openly threatening to murder her.

She had been invited to attend the first of a series of small and select conferences at which leading physicists met in Brussels to survey and discuss, under luxurious conditions, the status of some important field of physics. These were called Solvay conferences, after the wealthy Belgian industrial chemist Ernest Solvay who sponsored them. Marie Curie arrived in Brussels looking better in health, but very worried. The editor of a small magazine called Le journal happened to be the brother-in-law of Langevin's wife. The headline 'The Story of Love, Mme Curie and Professor Langevin' appeared in the issue of Le journal of November 14, 1911; by the next day every Parisian newspaper had the story and it was on its way to the tabloids in other countries. Although the story that they had eloped was pure invention, the affair rolled on day after day for the rest of the month. Back in Paris she threatened to sue; there was a partial apology, but the damage to her peace of mind was already severe.

Some letters Marie sent Langevin, locked up in his desk, had been purloined and were now in the possession of his estranged wife, who was seeking a legal separation. Adultery with Marie Curie was going to be alleged when the Langevin separation case came to court. The suspense was broken on November 23 when a magazine published long extracts from the letters. Gustave Téry, the journalist responsible, had been a contemporary of Langevin's at the Ecole Normale. He accused Langevin of being a cad and a scoundrel. The existence of the letters could not be denied, but they had been edited in such a way as to make them seem more sensational than they really were. The relationship was of long standing; had Pierre Curie been driven to suicide when he became aware of it?

Among her most loyal supporters at this difficult time were the Borels. Emile Borel was at this time scientific director at the Ecole Normale Supérieure; it was his insouciante wife the writer Marguerite who was most involved. When hostile crowds started to gather outside the Curie home in Sceaux, she arrived to take Marie Curie and the children off to the Ecole Normale Supérieure, where the situation could be discussed within the safety of the official apartment. The Minister of Public Instruction warned

Borel that he should not be sheltering her in his official residence. Borel's father-in-law, Paul Appell, dean of the faculty of science, hitherto one of Marie's staunch supporters, was also furious that the Borels had become involved; he and others thought that Marie should return to the land of her birth, as did some members of her own family.

Libel actions were very expensive and unlikely to succeed; duelling was both cheaper and quicker. At this late date duelling was a ritual performance, seldom resulting in serious injury let alone death. Several duels were fought over the Curie–Langevin affair, and Langevin himself decided that he must challenge the journalist. Paul Painlevé, the future Prime Minister, acted as one of his seconds. Langevin and Téry confronted each other with loaded pistols early on the morning of November 25, went through the ritual and then withdrew.

Only a few days later, a telegram arrived from Stockholm, to say that she had been awarded the Nobel prize in chemistry for the discovery of the elements radium and polonium. She was the first person ever to be awarded a Nobel prize in science twice, and this time the prize was not shared with anyone else. However, the immediate effect was to revive the debate in the press over the Curie–Langevin affair. There were those in the Sorbonne who thought her culpable, as did many of the general public. A senior member of the Nobel selection committee wrote to advise her not to come to Sweden to accept the prize. Determined to receive the Nobel medal in her own hands, she replied that she saw no connection between her scientific work and her private life.

In fact the turmoil was beginning to subside. The separation case was settled out of court without Marie Curie's name being mentioned. However, the strain of the elaborate Nobel ceremonies proved to be the last straw. On her return to Paris she became gravely ill. She was also deeply depressed; she later told her daughters that during this period she began to consider suicide. When she had recovered she moved house and started to live under her maiden name. Then she had a relapse and spent a month in an alpine sanatorium. Her whereabouts were kept secret as far as possible.

Marie Curie was already in contact with leading members of the suffragist movement in England, one of whom invited her over to stay. She spoke English quite well and, travelling incognito, she had a blessed relief from the unwelcome attention that pursued her in France. When she returned to Paris she felt strong enough to pick up the threads of her life again, although it was no longer possible that this could be shared with Langevin. She started to use her married name once more, to the great relief of her daughters. The new laboratory which had been promised her had just been finished. For six months she was on sick leave, after medical treatment. She left the house at Sceaux and moved into an apartment on the Ile St Louis. Since her health was no better, she also spent time at various alpine spas. Whether at this stage any or all of her suffering was due to radiation sickness is not certain; there was also a suspicion of tuberculosis.

In 1914, when the First World War began, the French government moved to Bordeaux for safety. Marie Curie took her precious store of radium, a large portion of the world's supply, to a bank there and then returned to Paris. On the eastern front her native land, as so often before, was being fought over by opposing armies. In the west the Germans had crossed the Belgian border, and already casualties were mounting. Soon she received a formal request from the Minister of War to equip operators for radiographic work. She organized more than 200 mobile X-ray units and, with her elder daughter, operated one herself. Most of her scientific friends were also involved in war work.

With some reluctance she agreed to write her autobiography and to visit America for the first time, to raise funds for her research. She took her daughters with her. Rather as she had feared, she was lionized incessantly and harassed by journalists, but thanks to American generosity her tour was a huge success financially. Soon radium treatment was being given at thousands of locations, although, among the medical staff concerned, many were reporting sickness. Radiation affects healthy cells as well as cancerous cells, but at first it was not realized how serious the consequences could be. Marie Curie herself was particularly slow to recognize the dangers. She suffered from cataracts, which can be an early symptom of radiation sickness; she had them removed by surgery, but went to great lengths to keep the operation secret. Those who met her for the first time were intimidated by the glacial expression, caused by the treatment, in such contrast to her gentle voice.

Despite her declining health, Marie Curie travelled widely, often in the company of her younger daughter Eve. Irène had been the favourite daughter until her marriage, but now Eve began to take her place. Marie had already built a holiday cottage in the north at l'Arcouest, for use in the summers, which she put in the name of Irène; now she built another at Cavalaire on the Mediterranean, for use in the winters, which she put in Eve's name. Early in 1932 she had a fall and recovered from a fractured wrist unusually slowly. At the end of the following year she was taken ill again, but made a good recovery and, not without difficulty, continued to attend scientific conferences. She made a will and sent for her sister Bronisława. She felt exhausted; tuberculosis was diagnosed; the doctors again recommended an alpine sanatorium. Once she was there the diagnosis was changed to pernicious anaemia, then usually fatal. She died in her sixty-sixth year on July 4, 1934 and was buried next to Pierre, in the cemetery at Sceaux. The cause of death was given as leukaemia, caused by prolonged exposure to high-energy radiation. The first of many biographies, written by her younger daughter Eve, was particularly influential in establishing her as a legendary figure, but it does not tell the whole story. For example, the degree of financial support she received in the early stages of her research is understated, and the Curie–Langevin affair is glossed over.

7   From Millikan to Einstein

Our next five remarkable physicists were born in the twelve years from 1868 to 1879. Two came from Germany and one from each of Austria, New Zealand and America.

ROBERT MILLIKAN (1868–1953)

In America, individual physicists, such as Joseph Henry and Willard Gibbs, seem to have chosen to work in isolation, rather than develop a research school. Towards the end of the nineteenth century the situation began to change. The leading universities established something in the nature of a department of physics, with appropriate staff, and there was usually no division between experimental and theoretical physics. Our next subject was one of the key figures in this process. He wrote an autobiography consisting mainly of extracts from addresses he had given at various times, which was published two years before he died. What follows is largely based on that account. After his death a life-long friend and colleague made the following assessment: 'I do not think Millikan is a great physicist in the sense that we look upon Newton, Kelvin, Helmholtz or J.J. Thomson, that is as a man who has produced or will produce revolutionary ideas. His place is rather that of a great consolidator and experimenter, a man who is capable of gleaning by critical analysis from the suggestions of others those hypotheses which are most nearly correct, subjecting them to properly devised and carried out experimental verification and transforming them from the realm of hypothesis to the realm of exact proved fact.'

Robert Andrew Millikan's paternal grandfather was a typical American pioneer of Scottish descent who moved west from New England first to Ohio in 1825 and then on to Illinois in 1839, where he settled on the banks of the Rock river, not far from the present town of Sterling. He was one of those who helped runaway slaves to escape into Canada by the 'underground railroad'. Later he took his family yet further west, to the Upper Mississippi valley, near Prairie du Chien, Iowa. The future physicist was born on March 22, 1868, before this last move, but grew up in fertile prairie country by the great river. His parents were Silas Franklin Millikan, a Congregational

preacher, who married Mary Jane Andrews, formerly dean of women at a small college in Michigan.

The school education the boy received in rural Iowa was adequate except that there was next to no science teaching. Outside school he learned how to live in the practical self-sufficient way of the American pioneers. After leaving school he entered the first stage of the famous liberal arts college of Oberlin, which had been founded in part by a distant relative and was the *alma mater* of both his parents. The college reinforced the Christian values and goals he had acquired at home, which guided his whole life. Before he had been there long, he was asked to help teach elementary physics to other students. He was so good at this that he was appointed tutor in physics as soon as he graduated. At the same time he was employed as an instructor in the college gymnasium, having always been keen on sport.

Next Millikan moved to New York City as the sole graduate student in physics at Columbia University. At this time the study of physics in America was just beginning to develop, but, when Rutherford visited Columbia a few years later, he described the status of physical science there as 'miserable'. Millikan was taken in hand by an able young teacher from Serbia named Michael Pupin. After emigrating to America, Pupin had

gone back to Europe to take a doctorate under Helmholtz in Berlin and he
had also worked under Kirchhoff in Heidelberg. Pupin was now working
on an invention to improve long-distance telephony, which later brought
him fame and fortune. He encouraged Millikan, who by then had obtained
a Ph.D. for a thesis on optics, to go to Germany for post-doctoral work and
lent him the money he needed to do so. Millikan began at Jena, where he
became proficient in the German language. Then he made a bicycle tour,
covering 3500 miles in all. This took him to Paris, where he heard Poincaré
lecture, and to several places in Germany. He decided to base himself in
Berlin, where he could work with Planck, Warburg, Neesen and Schwartz.
He also spent a semester at Göttingen, where he took a course in thermodynamics from Voigt and another in geometry from Klein. He also saw something of Arnold Sommerfeld, then acting as Klein's assistant, but what he
found most stimulating was the seminar led by the physical chemist Walter
Nernst.

As he was about to leave the Georgia Augusta, Millikan received a
cable from the physicist Albert Michelson of the new University of Chicago
offering him an assistantship in his department. Millikan took the next train
to London, then travelled on to New York and Chicago, arriving in time
for the fall semester of 1896. He turned down the offer of a better-paid position from Oberlin because, as he told President William Rainey Harper, he
wanted the opportunity to do research. The university was a new institution, strategically located both in time and in space to exercise a profound
influence on the evolution of the American university. President Harper
was a dynamic personality who had a definite programme for transforming the American collegiate system into a university system influenced in
many respects by the German model. However, the British model was too
strongly entrenched to be easily displaced.

To start with, at the University of Chicago about half of Millikan's
time was occupied in writing badly needed textbooks and organizing
courses. In 1900 he was sent to Paris by Harper to manage the university's
exhibit at the World's Fair, which won a Grand prize. Harper was delighted
by this success, which attracted much useful publicity. Two years later
Millikan married Greta Irvin Blanchard, the daughter of a successful
Chicago manufacturer; they spent most of their seven months' honeymoon
in London and Paris, ending with a rapid tour of western Europe. While
in England he was given a warm welcome by British physicists, notably
J.J. Thomson. In Chicago he was promoted to assistant professor. The salary
was not especially generous, but he was also receiving useful royalties from

his highly successful textbooks, which was fortunate since he already had two sons to bring up and a third was on the way.

Throughout the twenty-five years they were together Michelson, a Polish *émigré*, always treated Millikan with the utmost courtesy and consideration. While Michelson pursued the experimental research which would make him the first American national to receive a Nobel prize, Millikan was left to look after the department's graduate students. With his family Millikan made another European tour, visiting Rutherford at Manchester on the eve of some of his greatest discoveries, and then went on to Berlin to revisit Planck and his research group.

In 1914 Millikan was elected to the National Academy of Sciences of Washington, at the age of forty-six, having been awarded the Academy's Comstock medal the year before. Four years earlier he had been promoted to full professor at the University of Chicago. Millikan was involved in the successful effort to develop a telephone repeater, which soon made it possible to communicate by telephone across the USA. This involved him in a major lawsuit over patent rights, but he said afterwards that he learned a great deal from the painstaking way the patent lawyers dealt with the case.

In the summer of 1916 Millikan made his first visit to the Pacific Coast, to lecture at the Berkeley campus of the University of California, and then went on to repeat his lectures at what was then the Throop College of Technology in Pasadena, later to become world-famous as the California Institute of Technology. Also in 1916 he was elected president of the American Physical Society. By this time the USA had entered the First World War and Millikan decided that he must go to Washington and devote himself full-time to the war effort. Submarine detection was the first priority; it was essential that the resources of American industry should be harnessed to the project, and for once competing firms could be persuaded to cooperate in the common cause. The new applications of physics were shared with the Allies and played an important part in the conduct of the war, especially in combatting the menace of the German submarines.

In its relation to the American state, the situation of American science has tended to be the result of war-time requirements. Thus President Lincoln first gave official recognition to science as an institution through the creation of the National Academy of Sciences to provide technological advice during the Civil War. The Academy was a semigovernmental body, the official adviser of the American government in scientific matters, but it was still a small body, had no permanent home and did not include engineering in its scope. The Royal Society of London was its model, rather than the scientific academies of continental Europe, although the complementary role played by the Royal Institution was altogether missing.

In the First World War President Wilson, finding that an honorary academy of this type was ill-suited for providing the technical advice that modern war required, added a new operational arm, the National Research Council, which could draw on the up-to-date knowledge of the active core of the American scientific community. However, the need to plan for the future, when the war was over, was not lost sight of. The outcome of much discussion was a programme of National Research Fellowships in physics and chemistry, but initially not in mathematics and biology, open to Americans and Canadians under thirty-five years of age. Each such fellow was to be attached to some American university where he or she would undertake an approved research project, the necessary funding to come from the Rockefeller Foundation. A few years later the programme was extended to include the mathematical and biological sciences, but there was no intention of extending it to include non-American universities. Meanwhile efforts were being made to establish the National Research Council itself on a permanent basis, and it was decided to combine with the National Academy over a Washington headquarters. The Carnegie Corporation agreed to pay five million dollars towards this, provided that matching funds could be raised from other sources. Millikan was heavily involved in all the negotiations and fund-raising, which took him away from his department at the University of Chicago.

Millikan had already been to Pasadena once, as we know, and now he began making regular visits. The California Institute of Technology – known informally as Caltech – was still at an early stage of development, with only three permanent buildings of its own. However, it had the advantage that the Huntington Library was nearby, as was the Mount Wilson observatory; moreover, the Pasadena area was the most rapidly growing in the whole country. The institute had the status of a university, its graduate school one of the best in the USA, but it also had an undergraduate programme. In a great university like Chicago it is practically impossible to persuade the administration to break step and push one department out in front of the others, no matter how much the general interests might demand such discrimination. However, the powers-that-be at the institute were able and willing to do just that in the case of physics, so Millikan was persuaded to move there permanently and build it up. He said afterwards that his leaving the university which had given him his chance was the greatest service he could render it, since there was a need for new blood in Chicago.

Millikan had agreed with the trustees of the California Institute of Technology that he was to be chief executive officer and would concentrate on building the best department of physics he could. Before long the privilege of doing research at Caltech was so highly prized that no less than fifteen of the new National Research Fellows were working there, as well as other gifted young scientists, notably Robert Oppenheimer. Famous visitors came to lecture from all over the world, including Max Born, Albert Einstein, Werner Heisenberg and Erwin Schrödinger. In 1923 he was awarded the Nobel prize in physics 'for his work on the elementary charge of electricity and on the photo-electric effect'. After determining the first accurate value of the charge and hence of Planck's constant, Millikan turned his attention to the mysterious cosmic rays. He measured their intensities at great heights and great depths and observed their seasonal variations.

By this time Millikan was probably the best-known American scientist of his day. He wrote magazine articles popularizing the latest achievements in science and technology. In the face of growing attacks on modern science by religious fundamentalists, he organized a group of prominent businessmen, academics and religious leaders to sign a joint statement on the complementary role they believed science and religion played in the progress of humanity. The honours he received included membership of the Paris Academy and of the Royal Society of London. In politics he was a staunch Republican, strongly opposed to the New Deal and something of a racist. After being manoeuvred into retirement in 1946, he continued to give lectures, mainly on science and religion. He died in Pasadena on December 19, 1953 at the age of eighty-five.

There is no denying that Millikan had stormy days in his varied activities. His many vigorous debates on the nature and origin of cosmic rays were major events in the world of science in the 1930s. His determination to elevate the prestige of Caltech as rapidly as possible led to him being dubbed one of the great publicity agents in the field of education. He often differed vigorously from his colleagues on matters of politics, philosophy and religion. His administrative methods were hardly conventional and often confusing, yet his strong personal leadership always pulled things back into shape. A long-time associate of his said that 'the secret of his success lay to a large extent in the simple virtues instilled in his upbringing. He had a single-minded devotion to all that he was doing, and he put his work above his personal desires and aspirations. His combination of native good sense and intellectual honesty led him far both in science and in public life. In spite of his success and high public position, he always remained a simple man of true humility.'

ERNEST RUTHERFORD (LORD RUTHERFORD) (1871–1937)
Einstein described Rutherford as one of the greatest experimental physicists of all time. In New Zealand he is rightly regarded as a national hero, held up as an example to all the aspiring young. Both his parents emigrated there as youngsters in the middle of the nineteenth century. His father, James Rutherford, had been a wheelwright working in the Scottish city of Perth. When he reached the South Island he tried his hand at farming and processing flax, cutting railway sleepers and constructing bridges. Although he was moderately successful in each of these enterprises, his family of a dozen children learned the value of hard work and thrift. The future physicist's mother, Martha Thompson, accompanied her own widowed mother to New Zealand from England and a few years later took over her mother's teaching post when she remarried. Martha was the dominant parent, who endeavoured to instil in her children an interest in literature and learning. She died in 1935 at the age of ninety-two. In their different ways both parents contributed to their son's characteristic traits of simplicity, directness, economy, energy, enthusiasm and respect for education.

Ernest Rutherford was born on August 30, 1871 at Brightwater, a settlement near the town of Nelson; when he was seven the family moved to Havelock, also near Nelson. Success in the local primary school won him a scholarship to Nelson College, which was rather like an English grammar

school. In his spare time the boy enjoyed tinkering with clocks and making
models of the waterwheels his father used in his mills. By the age of ten
he had read a scientific textbook, but otherwise there was not yet any sign
of a special interest in science; he was expecting to become a farmer when
he grew up. At Nelson College he excelled in nearly every subject, particularly mathematics, as well as becoming school captain or head boy. Another
scholarship took him on to Canterbury College, in the city of Christchurch,
which later became the University of Canterbury. At the conclusion of the
standard three-year course he was awarded a mathematical scholarship that
enabled him to remain for an extra year. He gained an M.A. in 1893, with
double first-class honours in mathematics and mathematical physics and in
physical science, and was encouraged to stay on at the college for yet another
year to gain a B.Sc. He taught briefly at the local high school, during which
time he became engaged to a fellow student at Canterbury College named
Mary Newton, eldest daughter of the landlady of the house in which he was
lodging. In a tiny basement workshop Rutherford began investigating the
radio waves discovered by Hertz not long before. He devised a magnetometer that could detect radio signals over short distances and might be useful
in lighthouse-to-shore communication. However, unknown to Rutherford,
the American scientist Joseph Henry had already thought of this.

In 1895 Rutherford, as we know, was awarded a scholarship by the
Commissioners of the Great Exhibition of 1851 and chose the Cavendish
Laboratory because J.J. Thomson was the leading authority on electromagnetic phenomena. He brought his magnetometer with him to England and
soon was able to show that he could receive radio signals from sources up
to half a mile away. This demonstration impressed a number of Cambridge
dons, Thomson included. Early in 1896, following the discovery of X-rays,
Thomson asked Rutherford to join him in studying the effect of this radiation on electrical conduction in gases. Although Rutherford was anxious
to earn enough to marry his fiancée in New Zealand, he could hardly turn
down this opportunity.

While this work was in progress Rutherford was seriously considering his future prospects. Although the Thomsons went out of their way
to be helpful, and Rutherford had been an academic and social success, he
was conscious of the prevailing Cambridge snobbery towards those who
had been undergraduates elsewhere, especially in the colonies. Seeing little
chance of a Cambridge post, he started to look further afield. Without
much hope of success, Rutherford applied for the professorship of physics
at McGill University in Canada, where the authorities were looking for

someone to direct work in their well-equipped laboratory. Older men with far greater experience had also applied, but Thomson's emphatic testimonial 'I have never had a student with more enthusiasm or ability than Ernest Rutherford' persuaded them to appoint him.

Arriving in Montréal to a warm welcome in September 1898, Rutherford found perhaps the best laboratory in the western hemisphere (it was financed by a tobacco millionaire who considered smoking a disgusting habit) and a self-denying department chairman who soon voluntarily assumed some of Rutherford's administrative and teaching duties when he recognized his genius. In research Rutherford embarked on the field which was to occupy him for the next forty years, namely the study of radioactivity. After two years he felt sufficiently well established in Montréal to return to New Zealand in order to marry his fiancée Mary Newton after their five-year-long engagement and take her back with him to Canada. Their daughter and only child Eileen was born the following year; she was to die suddenly following childbirth at the age of twenty-nine.

Rutherford's nine years at McGill, dominated by the research which made him famous, were no less busy in other ways. He was in great demand as a speaker and travelled frequently to distant places to give a lecture or a course. Much of his time was consumed in writing *Radioactivity*, the first textbook on the subject and recognized as a classic as soon as it appeared in 1904. 'Rutherford's book has no rival as an authoritative exposition of what is known of the properties of radio-active bodies', wrote Lord Raleigh in a review, 'A very large share of that knowledge is due to the author himself. His amazing activity in that field has excited universal admiration. Scarcely a month had passed for several years without some important contribution from the pupils he has inspired, on this branch of science; and what is more wonderful still, there has been in all this vast mass of work scarcely a single conclusion which has since been shown to be ill-founded . . .' So rapidly was physics advancing, however, that, when Rutherford prepared a second edition the following year, it proved to be half as long again. No sooner was this finished than he embarked on the task of writing another book arising from the Silliman lectures he had delivered at Yale University. Friends urged him to limit his outside engagements, but he was often in London to deliver some prestige lecture and to keep in touch with what was happening.

Rutherford thoroughly enjoyed this recognition for, while not vain, he was fully aware of his own worth. The Royal Society elected him to the fellowship in 1903, at the early age of thirty-two, and awarded him the Rumford medal the following year. Various universities kept trying to tempt him away from McGill, and the time arrived when Rutherford began looking for an opportunity to return to England, where he would be closer to the world's leading scientific centres. In 1907 Arthur Schuster offered to relinquish the Langworthy chair of physics at the University of Manchester on condition that Rutherford was invited to succeed him. This was agreed and Rutherford accepted. The following year he was awarded the Nobel prize, for his work on the decay of radioactive elements. Curiously it was a prize for chemistry rather than physics.

If the Cavendish, under Thomson, was the premier physics laboratory in England, then Manchester, under Rutherford, was easily the second. Schuster had built a fine structure less than a decade earlier and bequeathed to his successor a strong research department, including his invaluable young research assistant Hans Geiger. Rutherford's great and growing fame attracted to Manchester (and later to Cambridge) an extraordinarily talented research team who made profound contributions to physics and chemistry. One of them was the exceptionally able H.G.J. Moseley, whose death in action at Gallipoli in 1915 at the age of twenty-seven was such a great loss to science. It was Moseley, 'a born experimenter' according to Rutherford, who demonstrated the fundamental importance of the atomic number. Another Manchester colleague was Chaim Weizmann, better known for his promotion of the Zionist cause, who described Rutherford as being 'youthful, energetic, boisterous. He suggested anything but a scientist. He talked readily and vigorously on any subject under the sun, often without knowing anything about it . . . He was quite devoid of any political knowledge or feelings, being entirely taken up with his epoch-making scientific work. He was a kindly person but did not suffer fools gladly.'

In Rutherford's gift, as Langworthy Professor, there was a personally endowed readership in mathematical physics. He used this to bring Niels Bohr to Manchester for a period including the early years of the First World War. Although their personalities were very different, Bohr and Rutherford became close friends. In 1926 Bohr looked back on his Manchester days and described how he felt at the time:

> This effect [the large-angle scattering of alpha particles] though to all intents insignificant was disturbing to Rutherford, and he felt it difficult to reconcile with the general idea of atomic structure then favoured by the physicists. Indeed it was not the first, nor has it been the last, time that Rutherford's critical judgement and intuitive power have called forth a revolution in science by inducing him to throw

himself with his unique energy into the study of a phenomenon, the importance of which would probably escape other investigators on account of the smallness and apparently spurious nature of the effect. This confidence in his judgement and our admiration for his powerful personality was the basis for the inspiration felt by all in his laboratory, and made us all try our best to deserve the kind and untiring interest he took in the work of everyone. However modest the result might be, an approving word from him was the greatest encouragement for which any of us could wish.

Rutherford remained at Manchester for fourteen fruitful years. When J.J. Thomson relinquished the Cavendish Professorship on his appointment to the Mastership of Trinity in 1921, no-one was surprised when Rutherford was elected as his successor. Increasingly beset by outside calls on his time, Rutherford had less and less opportunity for his own research and for keeping abreast of his students' work. Yet, with the tradition of enthusiasm for research that he had established earlier, his still frequent rounds to 'ginger up' his 'boys', the Cavendish Laboratory's output remained far more than just respectable. Usually half his students came from outside the United Kingdom, and after working in Cambridge they helped to spread his teaching throughout the world.

Rutherford was outspoken, outgoing and direct; swearing, he used to say, will make an experiment work better. He liked to keep his physics and his experiments simple and described his work in straightforward and precise language. He had a loud booming voice and it is said that, in the days when counting circuits tended to be sensitive to noise, his collaborators' equipment went wrong whenever he came near. He made decisions early and firmly, and once a matter was decided did not give it any further thought. He could be rude and even unreasonable on occasion, but, when he had cooled off, would put matters right with a handsome apology. One of his remarks, made at the British Association in September 1933, 'anyone who looked for a source of power in the transformation of atoms was talking mere moonshine', is often quoted; in private, however, he said he thought there might be something in it.

During most of his career, especially the Cambridge part, his wife Mary acted as his private secretary. As well as their Cambridge house, they had a rural retreat, first a cottage in Snowdonia and later an isolated country cottage near Chute in Hampshire. He was an enthusiastic motorist, one of the pioneers, and was also a keen golfer. The eminent physicist Sir James

Chadwick, who knew him well, has given some personal impressions of Rutherford at this stage in his life:

> The Rutherfords lived at Newnham Cottage, a low house in Queen's Road with a fine old garden, which belonged to Caius College. It was encircled by a wall of dirty *Cambridge brick.* One entered through a heavy door, and walked along a covered tiled way to the house itself. There was a very fine garden, in which Lady Rutherford took great pride. Rutherford's study was on the left, immediately after entering the house. Like the desk the room was littered with books and papers. For some years a niece of Lady Rutherford's lived with them and acted as his secretary and her companion.
>
> The Rutherfords occupied separate bedrooms, and there were no overt acts of affection between them. Yet they were devoted to one another. Lady Rutherford understood little or nothing of her husband's work, but she was very proud of the honours which showered upon him and reacted violently to any criticism. She treated him in ordinary matters as a child, still attempting to correct his faults when eating, for instance. 'Ern, you've dropped marmalade on your jacket.' When, rarely, Rutherford caught a cold or influenza she nursed him with loving care.'

From 1921, when he succeeded Thomson, until his death, Rutherford also held the chair of natural philosophy at the Royal Institution. In 1922 he received the Copley medal from the Royal Society. There were also numerous other public lectures to which great honour was attached, such as the presidential lecture to the British Association in 1923. Between 1925 and 1930 he was president of the Royal Society and subsequently he became chairman of the important advisory council which had been set up to allocate public money for the support of scientific and industrial research in the United Kingdom. That involved many public appearances, such as opening conferences and new laboratories, in addition to administrative and policy-making chores.

From 1933 Rutherford was president of the Academic Assistance Council and Chairman of its Executive Committee. This body, which sought to obtain positions and financial assistance for the displaced scholars, particularly Jews from Nazi Germany, raised a fund to provide maintenance for displaced university scholars, from whatever country. The Council also acted as a centre of information, putting individuals in touch with institutions that could best help them. The British government was sympathetic, regarding it as in the public interest to 'try and secure for [the United Kingdom] prominent Jews who were being expelled from Germany and who had achieved distinction whether in pure science, applied science . . . music or art'. Many of the refugees brought intellectual riches to Britain. Of the 2600 rescued by the Academic Assistance Council and its successor body in the period before the war, twenty became Nobel laureates, fifty-four were elected Fellows of the Royal Society, thirty-four became Fellows of the British Academy and ten received knighthoods. Of the scientists, some were already famous but most were young and unknown when they arrived. Unlike Lindemann in Oxford, Rutherford made no special effort to encourage the physicists to come to Cambridge.

To continue with Chadwick's reminiscences:

In appearance Rutherford was more like a successful business man or Dominion farmer than a scholar . . . when I knew him he was of massive build, had thinning hair, a moustache and a ruddy complexion. He wore loose, rather baggy clothes, except on formal occasions. A little under six feet in height, he was noticeable but by no means impressive . . . it seemed impossible for Rutherford to speak softly. His whisper could be heard all over the room, and in any company he dominated through the sheer volume and nature of his voice, which remained tinged with an antipodean flavour despite his many years in Canada and England. His laughter was equally formidable.

He appeared to possess no fountain pen. He wrote slowly and laboriously with an old steel-nibbed pen, or more often with a short pencil. However with such a pencil he did arithmetic with surprising accuracy. Mumbling to himself he would use what appeared to be gross approximations to reduce multiplication or division to simple addition or subtraction, remembering to correct the final result by the necessary percentage. The answer he obtained was invariably within the overall accuracy of the experiment.

Rutherford smoked interminably, usually a pipe and only occasionally a cigar or cigarette. His pipe tobacco was reduced to tinder dryness by being spread on a piece of newspaper in front of the fire or on a sheet of paper placed on top of a hot water radiator in his office or laboratory. When he lit his pipe it produced sparks and even flames, like a volcano. A result of this was that his waistcoat was peppered with small holes, and he often had to brush red hot grains of tobacco from the papers before him on his desk.

On October 15, 1937 Rutherford was suddenly taken ill at his Cambridge home and was operated on the next day for a strangulated hernia. Although at first the treatment he received seemed to be successful, the internal organs did not recover their functions and he died peacefully four days later, at the age of sixty-six. After his body had been cremated the ashes were buried in Westminster Abbey, just west of the tomb of Newton. In work that may be characterized as radioactivity at McGill, atomic physics at Manchester and nuclear physics at Cambridge, Rutherford, more than any other, formed the views later held concerning the nature of matter. As was to be expected, numerous scientific and other honours came to him. Dozens of universities awarded him honorary degrees and dozens of scientific societies conferred on him honorary memberships and other distinctions. In 1914 he was knighted; later he was awarded the Order of Merit and raised to the peerage, as Baron Rutherford of Nelson, the place in New Zealand from which he came. Lady Rutherford returned to live there after her husband's death, uncomfortable socially and demanding to the end, but the one love of his life and his most devoted admirer.

LISE MEITNER (1878–1968)

Lise Meitner's name has become widely known for her part in the discovery of nuclear fission, which made nuclear power possible, as well as the atomic bomb. Among physicists she is particularly noteworthy as one of the early

pioneers in the study of radioactivity. Einstein described her as 'the German Madame Curie'; but, although most of her scientific work was done in Berlin, she came from Austria and retained her nationality, even after she became a Swedish citizen about eight years before her death. Lise Meitner was born on November 7, 1878 in Vienna, where she spent the first third of her life; she remained very much attached to the imperial city, never more splendid than in those last autumn days of its glory. Another third of her life was spent in Germany. When Austria was taken over by the Nazis she found refuge in Sweden, where she lived for twenty-two years. It was only at the age of eighty-one that she gave up scientific research and retired to England to live out the rest of her days in Cambridge.

Lise Meitner's father Philipp was a respected lawyer and keen chess player. His ancestors came from Moravia, now part of the Czech Republic; the family of her mother Hedwig (née Skovran) came from Russia. Being the third of eight children, she was used both to being ruled by her two older sisters and ruling over the four younger children. Although her parents were Jewish, her father was a freethinker and the Jewish religion played no part in her education. Indeed, all the children were baptized, and Lise Meitner grew up as a Protestant; in later years her views were very tolerant, although she would not accept atheism.

Lise Meitner said that she became a physicist because of a burning desire to understand the working of nature, a desire that appears to go back to her childhood. At the end of her school career she first had to pass the state examination in French, so that, if necessary, she would be able to support herself as a teacher; only then did she obtain permission to sit for the *Matura*, the school-leaving examination, equivalent to the German *Abitur*, that qualified her to enter the University of Vienna in 1901. For two years she worked intensively to prepare herself to pass this hurdle, coached by Arthur Szarvasy, later professor of physics in Brno. Her sisters used to tease her; she had only to walk across the room for them to predict that she would fail because she had interrupted her studies to do so. She was one of the four women who passed that year, out of fourteen candidates.

In her university days she encountered occasional rudeness on the part of the students (a female student was regarded as a freak) but also much encouragement from her teachers. In later years she often spoke particularly highly of the lectures of Boltzmann; it was probably he who inspired her with his vision of physics as a search for the ultimate truth, a vision she never lost. When she obtained her doctorate in physics in 1905 she was only the second woman in Vienna to have done so. She remained at the university for another year or so, clearing up a question raised by Rayleigh. Thus encouraged to aim for a career in theoretical physics, she obtained her father's permission and promise of financial support for post-doctoral study in Berlin.

She had already made her first contact with the new subject of radioactivity, which was to become the chief topic of her life's work. In 1905 it was not known for sure whether certain rays were deflected in passing through matter; she designed and performed one of the first experiments in which some degree of deflection could be observed. She had met Planck briefly when he had visited Vienna in response to the invitation to be Boltzmann's successor and wanted to study under him. On arriving in Berlin in 1907, she arranged to attend his lectures. After he had invited her to his home, she recalled that 'even with my first visit I was very impressed by the refined modesty of the house and entire family. In Planck's lectures, however, I fought a certain feeling of disappointment at first . . . Boltzmann had been full of enthusiasm . . . he did not refrain from expressing his enthusiasm in a very personal way . . . Planck's lectures, with their extraordinary clarity, seemed at first somewhat impersonal, almost dry.' Although Planck, like so many of his colleagues, did not believe that women generally should be permitted to study at universities, he was prepared to make exceptions and Lise Meitner received his whole-hearted support. In her early days in Berlin she found cheerful, informal company and good music in the Grünewald house. She was often to seek Planck's advice in the years to come.

At first she had some difficulty in finding anywhere to carry out experimental work. Then she met the young chemist Otto Hahn, whose profile comes next. He was a frank and informal man of her own age from whom she felt that she could learn a great deal. He was looking for a physicist to help him with his own research into radioactivity. There was the difficulty that Hahn was to work at the chemical institute under the Nobel laureate Emil Fischer, who banned women from his laboratory (they might set fire to their hair); in any case, women were not allowed access to laboratories used by male students. However, an old carpenter's workshop was equipped for doing radiation measurements, and this was where Lise Meitner was permitted to work. Soon she and Hahn were collaborating in research on radioactive substances, Lise Meitner taking responsibility for the more physical aspects and Hahn being more concerned with the chemistry. In the years leading up to the First World War, they published a large number of papers on radioactivity, most of which are no longer of interest. To Hahn, the chemist, the discovery of new elements and the study of their chemical properties was the most exciting part of the work; Lise Meitner was more interested in understanding their radiations.

In 1908 Rutherford, as we know, was awarded the Nobel prize and, on the way back from Stockholm, the Rutherfords spent a few days in Berlin. While Rutherford had a valuable discussion with Hahn, Meitner was sent off to help Rutherford's wife Mary with her shopping. Although she and Hahn became close friends as well as colleagues, they never had a meal together except on formal occasions. She was very reserved, even shy, and had been brought up strictly.

In 1914 the outbreak of war caused their programme of research to be interrupted; Hahn was called up and Lise Meitner volunteered to serve as an X-ray nurse with the Austrian army. It was a harrowing time for her, working up to twenty hours a day with inadequate equipment and coping with large numbers of Polish soldiers with every kind of injury, without knowing their language. In the study of radioactive substances, measurements at fairly long intervals may be needed in order to let activities build up or allow unwanted ones to decay. Periodically she went back to Berlin on leave to make such measurements, and Hahn sometimes succeeded in synchronizing his leave with hers.

By then they were no longer working in the carpentry shed. The ban which kept her out of the chemical laboratories had been lifted in 1909, when women were at last admitted to academic studies in Germany. Hahn was offered a small independent department in the Kaiser Wilhelm Institute for chemistry, which was opened in 1912 on a (then) rural site in Berlin-Dahlem. Lise Meitner worked there with him for twenty-five years, first as a 'guest' and from 1918 as head of a department of physics. In her laboratory she maintained strict discipline, so that in a quarter of a century it never became contaminated with radioactivity, despite the large amounts of radio-elements that were handled in the same building. Although her students feared her strictness, they came to her with their personal problems even so, and later her warm, practical humanity was remembered with fondness.

Lise Meitner regularly attended the weekly colloquium at the University of Berlin, where new papers were discussed before an impressive bench of Berlin scientists. She was assistant to Max Planck from 1912 to 1915 and in 1922 received the *venia legendi*; the subject, cosmic physics, of her inaugural lecture was reported as cosmetic physics in the press. In 1926 she was made titular professor, but never gave any courses of lectures.

The discoveries of the neutron in 1932, the positron in 1933 and artificial radioactivity in 1934 caused a turmoil in the world of nuclear physics, reflected in a number of short papers in which Lise Meitner and her collaborators tried to keep pace with the rapid new developments. After Hitler came to power, when 'non-Aryan' scientists lost their university posts, the scientists in the institutes of the Kaiser Wilhelm Gesellschaft were less vulnerable, being partly controlled by industrialists. Even so, the Nazis tried to enforce party loyalty by various forms of infiltration, and Hahn and Meitner had to be increasingly cautious in order to avoid open conflict and to avoid losing those of their staff who were partly Jewish or who refused to join the Nazi party. Although her *venia legendi* was rescinded in 1933 and she lost her external professorship at the university, this did not affect her position at the institute. People in her position were no longer allowed to give reports at scientific meetings; she stopped attending them. Her name was dropped from citations of papers of which she was a co-author.

The *Anschluss* (annexation) of Austria by Germany in March 1938 created serious problems for Lise Meitner: she was no longer a foreigner protected by her Austrian nationality and consequently became subject to the racial laws of Nazi Germany. As a person who was 'over 50 per cent non-Aryan' she could expect to be dismissed from her post and to suffer other penalties. Her honesty did not permit her to conceal her Jewish descent (as some people did), and her dismissal could only be a question of time. Her position looked even worse when her friend and colleague Max von Laue said that he had heard of an order, issued by Heinrich Himmler, head of the Gestapo, that no university teachers – whether Jewish or not – should in future be allowed to leave Germany without special permission. An attempt by the president of the Kaiser Wilhelm Gesellschaft to obtain a permit for her was unsuccessful. There appeared to be a very real risk that she might not only lose her position in Germany but also be prevented from seeking a new one abroad. She decided that she must leave without delay and arrangements were made by sympathizers in the Netherlands for her to escape to their country. On the day of her departure she had just an hour and a half in which to pack her most necessary belongings. In the laboratory no-one but Hahn knew that she was leaving Germany for good; he gave her a diamond ring to sell in case of need.

The Netherlands has a distinguished tradition in physics but at that time lacked good facilities for nuclear research, so Lise Meitner quickly moved on to Denmark, where for some weeks she enjoyed the hospitality of the Bohrs. The facilities for nuclear research in Copenhagen were excellent, and there were some young and active physicists at work. It was probably her wish not to compete with those younger people that led to her decision not to remain there but instead to accept an invitation to join Manne Siegbahn, head of the new Nobel Institute for Physics in Stockholm. In 1924 Siegbahn had been awarded the Nobel prize in physics for his research work in the field of X-ray spectroscopy; he and his pupils had created a Swedish tradition of precision physics.

It was shortly after her arrival in Sweden in 1938 that she made her most spectacular contribution to science. After she left Germany, Hahn and his assistant Strassmann had continued the research, as before, until they started discovering barium in the products of uranium bombardment. Hahn wrote to consult Lise Meitner for an explanation of what was happening in physical terms. His letter reached her before the discovery had been published, and thus she became the first scientist outside Germany to learn of this extraordinary phenomenon. It arrived during the Christmas vacation when she was visiting friends in a small Swedish village. In the party there was another refugee physicist, her nephew Otto Robert Frisch. Lise talked to him about Hahn's letter, but at first the young man did not believe that uranium atoms could split into two almost equal parts. He thought that Hahn and Strassmann must have made a mistake.

In order to talk the matter over at leisure, aunt and nephew took a long walk in the snow. Physical exercise, they thought, might clear their minds. Lise Meitner did most of the talking, urgently, convincingly. At last she persuaded Frisch that Hahn and Strassmann had made no mistake, that uranium atoms underwent fission, and that the energy released in the process was probably very great. Once they felt quite sure they hastened to Copenhagen to break the news to Niels Bohr. He listened eagerly to what they told him, exclaiming 'Oh, what idiots we have been. We could have foreseen it all. This is just as it must be.' He suggested an experiment whereby they might measure the energy released when uranium atoms split. Bohr was so engrossed in this extraordinary new phenomenon that he almost missed the ship which was about to take him to New York for a meeting of the American Physical Society. When he told the American physicists the news of the sensational discovery of nuclear fission, they said that his gaze, troubled and insecure, moved from person to person but stopped on no-one. In a state of great excitement, the experimental work was repeated by several research groups and confirmed. In Berlin Hahn was deeply worried: 'God cannot have intended this', he is reputed to have said.

The 1945 Nobel prize for physics went to the theoretician Wolfgang Pauli, although there were many who thought that Lise Meitner deserved it more. Bohr nominated Meitner and Frisch for the Nobel prize for physics in

1946 and for chemistry later, but to no avail. However, her work did not go unrecognized, since later she shared with Hahn the Max Planck medal of the Society of Physics, and the institute of nuclear research in Berlin was named the Hahn–Meitner Institute. Her relationship with Otto Hahn remained close, but he became increasingly inclined to undervalue her contribution to their great discovery. Yet she had been the physicist member of the team for thirty years and Hahn had always deferred to her judgement when it was a question of physics. Later Strassmann commented that 'she was the intellectual leader of our team, and therefore she belonged to us – even if she was not present for the discovery of fission'. This close working relationship between the head of a scientific institute in Nazi Germany and an *émigrée* Jewish scientist was a tribute to Hahn's stability of character and personal loyalty. However, it was widely felt that, by encouraging her to leave Berlin when she did, he might have saved her life, but he had effectively blighted her scientific career.

Lise Meitner remained in Sweden for twenty-two years, during which a cyclotron was constructed in Siegbahn's institute, the first to be built on the mainland of Europe. The experience of Lise Meitner was invaluable for making the best use of this new atom-splitting machine and for training students in the required ancillary techniques. She acquired a good command of the Swedish language, built up a small research group of her own and published a number of short papers, mostly describing the properties of some new radioactive species formed with the help of the cyclotron. Inevitably she felt cut off at the Nobel Institute since Sweden, as a neutral country, was isolated during the war; she had few students and lacked the stimulus that Hahn had given her during her years in Berlin. She also felt that Siegbahn was more interested in his precision physics than in the comparatively crude measurements that were possible in the study of radioactive isotopes. Initially there seemed to be a chance that she could go to Britain, which she would have preferred, but at Oxford Lindemann had the reputation of being unsympathetic to women, while from Cambridge there was such a lukewarm response that she did not think it worth following up. In July 1939, just before the war began, she nevertheless accepted Lawrence Bragg's invitation to visit Cambridge and was pressed to remain on a three-year research contract, but she hesitated and, to her lasting regret, returned to Sweden. During the war she contributed to the Allied war effort by helping to provide information about nuclear research in Germany, but she was adamant that she would have nothing to do with the atomic bomb.

In 1946 she spent half the year in the USA as visiting professor at the Catholic University in Washington and was nominated as 'woman of the year' by the American press. In 1947 she retired from the Nobel Institute and accepted an offer from the Swedish atomic-energy committee to set up a small laboratory for her at the Royal Institute for Technology. Later she moved to a laboratory of the Royal Academy of Engineering Sciences, where an experimental nuclear reactor was being built deep down in a hall blasted out of the solid granite on which Stockholm stands. There she remained for the rest of her time in Sweden, first directing the work of a research assistant, later mainly engaged in reading, attending colloquia and discussing problems with other physicists. Her mind was still active when in 1960 she retired to Cambridge in order to be nearer her relatives, including her nephew Otto Robert Frisch, who by then held the Jacksonian chair of natural philosophy at the Cavendish.

In Cambridge, Lise Meitner led a quiet life, but she still travelled a good deal, to meet friends and give lectures, often about the rightful place of women and, in particular, of women scientists. In 1963 she went to Vienna to address a conference about 'fifty years of physics', a talk that was later published in English. She had always taken great pleasure in music, as did all her brothers and sisters (one sister became a concert pianist). In old age she still fondly recalled the weekly musical evenings at the home of Planck. She went to concerts as long as she could walk, and tried to follow contemporary trends in music, although loss of hearing made this difficult.

After an exhausting visit to the USA at the end of 1964, Lise Meitner suffered a heart attack, which caused her to spend some months in a nursing home, from which she returned to her flat much enfeebled. Yet her strength only failed slowly, and in 1967 she made a good recovery after a fall in which she broke her hip. After that accident she did not travel any more and gradually gave up all other activity. For the last two months she was in a nursing home in Cambridge, where her life slowly ebbed away. She died on October 27, 1968, a few days before her ninetieth birthday, having outlived all her brothers and sisters, and was buried in a country churchyard, where her youngest brother had been buried some years previously.

In spite of her close friendship with Planck and other great physicists, Lise Meitner never quite lost the shyness of her youth, but among her friends she could be lively and cheerful, and was an excellent story-teller. She was interested in almost everything; always ready to learn and ready to admit her ignorance of things outside her own field of study. Within that field, however, she moved with great assurance and was convinced of the power of the human mind to arrive at correct conclusions from the great laws of nature. The advance of knowledge was always her first concern and she felt the delight of every good scientist in an excellent piece of work whoever it was done by.

OTTO HAHN (1879–1968)

The borderline between physics and chemistry is one that should not be drawn too definitely. The chemists Dalton and Lavoisier have been mentioned incidentally and would certainly have been given profiles if this had been a book on remarkable chemists. However, Marie Curie is one exception, and Otto Hahn, who figured prominently in the preceding profile, is another. They seem to me to be chemists whose lives were so bound up with those of physicists that it is only natural to profile them here.

Otto Hahn was born on March 8, 1879 in Frankfurt. His father, Heinrich Hahn, was a glazier by trade and came from the village of Gungersheim near Worms. The family derived from Rhenish peasant stock but, while some of its members were farmers, others had pursued professional careers, becoming either teachers or doctors. Heinrich settled in Frankfurt in 1866, where he met a young widow named Charlotte Stutzmann (née Giese), who belonged to a North German family of some distinction. They

were married in 1875 and had three children, their sons Heiner, Julius and Otto, besides Charlotte's son Karl from her former marriage. One of Otto Hahn's cousins, Friedrich Thimme, was a historian who became director of the *Landesbibliotek* in Hanover, and another, Heinz von Trubzschler, was a member of the German foreign service and later became ambassador in Dublin.

The construction boom in Frankfurt after the Prussian victory in the Franco-Prussian war provided an opportunity for small tradesmen to expand their businesses and Heinrich Hahn was among those able to do so, becoming a contractor with several employees. Even so, the family lived plainly with few luxuries. Karl attended the Goethe Gymnasium and specialized in classics, while the other sons, including Otto, went to the Klinger Realschule instead. In childhood his health was not good, but after the age of fifteen this ceased to be a problem. His brother Heiner went into the family business, as did Julius at first, before opening what became a successful art gallery. Otto was sent to a technical school since his father had wanted to become an architect himself and hoped that his son would enter the profession. However, the youth soon came to the conclusion that architecture did not suit him and instead he turned to science, especially chemistry.

On the advice of a friend, he went to the University of Marburg, a town not far from Frankfurt. The science courses did not strike him as very inspiring: the chemistry lectures were scholarly but not well presented, the physics dull and the mathematics too difficult, although in later years he regretted that he had not pursued these subjects further. Following German custom, he did not remain at Marburg but migrated to the University of Munich. Until then he had managed to avoid taking part in the regular activities of the student fraternities, such as drinking beer and picking quarrels, but was obliged, in the Easter of 1899, to fight another student who had insulted him in the street by calling him a sissy. Although it was not unknown for one of the contestants in a duel to be permanently disabled, Hahn received only a few scratches. He began work on his doctorate, while living an active social life as chief officer of a student fraternity, which involved much carousing and other traditional activities. Nevertheless, the thesis he wrote was accepted.

In most of Europe a period of military service was required of able-bodied young men, two years in Germany and three years in Austria-Hungary. However, those with sufficient education and social standing could offer themselves for just one year's training as officer cadets. Hahn did so, but, although he passed out successfully at the end of the year, he did not apply for a commission. Instead he returned to Marburg to work as an assistant demonstrator, planning to go into industry afterwards. For this a knowledge of foreign languages would be an advantage, so in 1904 he went to work at University College, London, under Sir William Ramsay, discoverer of the inert gases. Ramsay was impressed by the young German and advised him to continue research in radiochemistry, rather than go into industry. Hahn took his advice and gained a place at the University of Berlin, but, before taking this up, he arranged a six-month visit to Rutherford's laboratory at McGill. Hahn was delighted with the easy, informal and yet stimulating atmosphere he found there.

Hahn returned home to write up his research and found time to holiday in the Tyrol with his brother Julius before settling down to work in Berlin in October 1906. The first step was to report to Geheimrat Fischer at the institute of chemistry. After six months he was given the status of Privatdozent, on the strength of his published research. The next year Lise Meitner, as we know, came from Vienna to study theoretical physics under Planck. She too wished to do experimental work in the chemical institute and Fischer agreed that she could work with Hahn.

Hahn was a handsome man who took pleasure in the company of women, both as friends and as colleagues. When he was on an excursion to Stettin by steamship, he met Edith Junghans, the daughter of the chairman of the Stettin city council; soon they had become engaged. Hahn's professional position became more secure with the formation of a separate department for radioactive work, enabling them to get married in March 1913. When the First World War broke out the following year, Hahn, being a reservist, was called up at once; he was involved in action on the western front at an early stage and participated in the fraternization between the British and German troops which took place at Christmas 1914.

Shortly afterwards he was ordered to attend a meeting at army headquarters in Brussels, where he was told that poison gas was to be used to achieve a breakthrough. When Hahn protested that this would contravene the Hague Convention, he was told that the French had already started using gas; moreover, if the war could be ended quickly through the use of chemical weapons, many lives might in the end be saved. So Hahn took part in their development and was horrified later when he saw the results and realized that his own work was partly responsible. A decision had been taken to attempt to break through the Italian line with the support of a gas attack. Hahn was sent there in September 1917 to make the necessary preparations and the attack the next month was completely successful. Soon he was back on the western front, taking part in chemical operations throughout the spring and early summer of 1918. Finally he went to Danzig for experiments with a new type of chemical weapon.

After the end of the war there were disturbances and strikes in Berlin for some time, and at one stage Hahn found himself acting as a stoker in a power station. His wife Edith was expecting a baby, after nine years of marriage; their first and only child, Hanno, was born in April 1922. Conditions were still difficult in post-war Germany and financial problems, which became increasingly severe during the remainder of the year, reached a climax in the early part of 1923 when the currency lost all value. Although domestic life was by no means easy under hyperinflation, scientific work at the institute continued much as before.

By 1932 Hahn had been invited to lecture on radioactivity at Cornell University. He gave a review of what he and Lise Meitner had been doing, later published as a book: *Applied Radiochemistry*. The alarming news from Germany made him cut short his stay and return to Berlin, when he went at once to Planck with a proposal for an organized protest by German scientists against the persecution of Jewish scientists. Planck advised him that this would have no chance of success; it was already too late for protest: 'if today thirty professors get up and protest against the government's actions, by tomorrow there will be 150 individuals declaring their solidarity with Hitler, simply because they are after the jobs'. Although Hahn steadfastly refused to join the Nazi party and continued to occupy his post without molestation, as the weeks went by he became increasingly aware of the political constraints. When the student fraternity to which he belonged decided to exclude non-Aryans, Hahn promptly resigned, while keeping in touch with some of the individual members. More significantly, he resigned his lectureship at the University of Berlin, although as a member of the Berlin Academy he could still lecture there whenever he wished.

In 1933 Hahn was made interim director of the Kaiser Wilhelm Institute of physical chemistry until a loyal Nazi could be found to take over. He was persistently and openly contemptuous of the Nazis, but at the end of his few months of office he reported that 'he had done his best with the unpleasant and thankless task of cleansing the institute of non-party members'; one of them, as we know, was Lise Meitner. At this point the able Fritz Strassmann was brought into the research project. He had originally joined the institute in 1929 and by 1935 was already an experienced radiochemist. Since he wished to enter academic life, Hahn had urged him to apply to the university for admission to the academic staff, but when he was informed that he must first join one of the Nazi organizations, he refused to proceed with the application. His attitude towards the new political masters of Germany was therefore much the same as Hahn's; they were to remain close associates and collaborators for the next ten years.

Five years later, as we also know, the experimental work reached a climax with the discovery of what proved to be nuclear fission. Once it had been established that huge amounts of energy were released by the fission process, the excitement grew intense. After celebrating his sixtieth birthday, Hahn went to the Scandinavian centres and to England to lecture on the discoveries. However, the impending conflict was sharpening official interest and obstacles began to be placed in the way of free publication. An office for nuclear research was established by the German war department as early as the summer of 1939, and leading nuclear scientists, including Hahn, were drawn into the national discussions.

Hahn's position during the war was in many ways typical. He had always been a loyal German. Although he felt that he owed no loyalty to the Nazis, who could be held responsible for starting the war, he nevertheless recognized that the whole German nation was involved. His presence at high-level meetings and the cooperation of his institute were bound to be regarded as essential. The most useful thing they could do was to characterize the fission products and to study relative yields, a programme that was in fact a continuation of Hahn and Strassmann's personal research and therefore quite acceptable to them.

Five years later the war in Europe was coming to an end. The bombing offensive against Berlin intensified. Hahn's institute was destroyed and the staff were moved to the small town of Tailfingen in southwest Germany. The Hahns' son Hanno joined his parents to convalesce after losing an arm on the eastern front. The details of what happened after the area was occupied by the Allied forces will be told in the profile of Heisenberg, who was more closely involved with the German nuclear project. Suffice it to say here that Hahn was taken to England in 1945, with nine other leading German nuclear scientists, and interned.

Soon afterwards the first atomic bomb was exploded over Hiroshima, and then a second dropped on Nagasaki, killing altogether more than a hundred thousand people. Hahn felt partly responsible for the Japanese casualties and contemplated suicide. Then a letter arrived from Planck saying that he wished to retire as president of the Kaiser Wilhelm Gesellschaft and that Hahn was being proposed to succeed him. Hahn was duly appointed; soon afterwards he was awarded the 1944 Nobel prize in chemistry for the discovery of nuclear fission. Lise Meitner was not included, but Hahn gave ten per cent of the prize money to Strassmann.

The internees returned to Germany in January 1946 and were resettled in Göttingen, where the Kaiser Wilhelm Gesellschaft was restarted, under Hahn's presidency, and renamed the Max Planck Gesellschaft. At first it mainly operated in the British zone of occupation, one of four. Göttingen was close to the border with the Russian zone; Hahn went to bed each night wondering whether he might be kidnapped and wake up in Russia. Rumours kept circulating that the Germans had also constructed atomic bombs; Heisenberg and Hahn vehemently denied this, saying 'at no time did Germany have any atomic bombs or installations for the manufacture of atomic bombs'.

Hahn was now almost sixty-seven, a respected figure in the new Germany. This rather unassuming scientist, who had spent most of his life in the laboratory, spent the next few years helping to rebuild German science after the war. He wrote two autobiographical books, giving a somewhat biased account of the Nazi period and of the post-war years when many of those who, thanks to political opportunism or conviction, had flourished during the Nazi regime still held important positions and were using their influence to the detriment of their anti-Nazi colleagues. With other German scientists, he presented to the West German government a statement that they would not cooperate in the development of nuclear weapons.

Hahn was awarded various honorary degrees, lectured in Germany and England, travelled in many other parts of Europe, but his declining years were clouded by misfortune. One evening as he opened the door of his house in Göttingen he was shot by a disgruntled inventor, who wished to draw attention to the neglect of his ideas by established scientists. He had hardly recovered from this incident when his wife Edith had a nervous breakdown. Then, in 1952, he was injured in a car accident; the following year he had a heart attack. His son and daughter-in-law were killed in a motor accident. Edith never recovered from the shock and remained an invalid for the rest of her life. Hahn himself had a fall getting out of a car, became progressively weaker and died in Göttingen on July 28, 1968, at the age of eighty-nine. His wife survived him by no more than a fortnight.

Albert Einstein (1879–1955)

Einstein realized that each of the separate fields of physics could devour a short working life without having satisfied the hunger for deeper knowledge, but he had an unmatched ability to scent out the paths that led to the depths

and to disregard everything else, all the many things that clutter the mind and divert it from the essential. This ability to grasp precisely the particular simple physical situation that could throw light on questions of general principle characterized much of his thinking.

Albert Einstein was born in the peaceful German town of Ulm, in the state of Württemberg, on March 14, 1879. He was the only son of Hermann and Pauline (née Koch); both sides of the family were from Swabia. His father Hermann was a kind, inoffensive but somewhat ineffectual person; his mother the more dominant parent. His father's brother, who lived with the family, was a trained electrical engineer; father and uncle together ran a business designing and manufacturing electrical apparatus, such as dynamos. Not long after Albert was born the family moved to Munich, and a year after that a sister Maria (known as Maja) was born. There were no other children.

The future physicist grew up in suburban Munich. Although the family was Jewish, he attended a Catholic primary school before proceeding to the Leopold Gymnasium, a conventional school of good repute. His scientific interests were awakened early by a small magnetic compass his father gave him when he was about four; by the algebra he learned from his uncle; and by the books he read, mostly popular science works of the day.

A textbook on Euclidean geometry that he studied at the age of twelve made a deep impression. We have an account of his childhood from his son Hans Albert:

> He was a very well-behaved child. He was shy, lonely and withdrawn from the world even then. He was even considered backward by his teachers. He told me that his teachers reported to his father that he was mentally slow, unsociable and adrift forever in his foolish dreams. Very early Einstein set himself the task of establishing himself as an entirely separate entity, influenced as little as possible by other people. In school he did not revolt, he simply ignored authority. His parents, although Jewish, were largely indifferent to religion. Einstein, while still a schoolboy, deliberately emphasized his Jewish origin and went through a period of religious fervour which he later described as his 'first attempt to liberate myself from purely personal links'. At the age of twelve he finally freed himself from conventional religious belief, although he retained a firm belief in some rather undefined 'cosmic religion', which was entirely suprapersonal. When asked about this, later on in his life, he used to say that although he did not think there was a God who was interested in people he thought, like Spinoza, that there might be one who created the universe.

As a child Einstein was echolalic, repeating to himself what he heard in order to make sure that he had heard it correctly. He did not learn to speak before he was three and did not speak fluently until he was seven. Later in life he was a confusing lecturer, giving specific examples followed by seemingly unrelated general principles. Sometimes he would lose his train of thought while writing on the blackboard. A few minutes later he would emerge as from a trance and go on to something different. He explained that 'thoughts do not come in any verbal formulation. I rarely think in words at all. A thought comes and I try to explain it in words afterwards'. He did poorly at school, where so much teaching is verbal rather than visual; he is believed to have been dyslexic. He was also a loner: 'I'm not much with people', he would say. The mature Einstein impressed everyone who met him by his gentleness and wisdom, but, as one of his biographers remarks, 'he has never really needed human contacts; he deliberately freed himself more and more from all emotional dependence in order to become entirely self-sufficient'.

When the family business failed in 1894 after an over-ambitious attempt to compete with much stronger firms, the rest of the family moved from Munich to Pavia in Lombardy, leaving the fifteen-year-old Einstein in the care of distant relatives. The intention was to allow him to continue his education, but he felt abandoned. He found the authoritarian Leopold Gymnasium, with its emphasis on classics, increasingly unbearable. He became ill and before long left, officially on medical grounds, but perhaps partly to avoid liability for military service, and joined the family in Pavia. One of his first actions was to renounce his German citizenship, thereby becoming stateless. After spending most of a year enjoying life in Italy, he resumed his education, but the family business was again failing and he could expect no financial support from his parents. An aunt in Genoa gave him a monthly allowance to see him through school and university, but after that he would need to support himself.

Einstein's aim was to enter the Polytechnikum (ETH) in Zürich, but to do so he had to pass the entrance examination. After one unsuccessful attempt, due to a poor performance in non-scientific subjects, he was advised to complete his school education first. Accordingly, he spent a year at the liberal gymnasium in the Swiss town of Aarau, where the regime was influenced by Pestalozzi. His teachers thought him lazy and were unimpressed, but he gained the certificate needed for university entrance. By the time he was admitted to the Polytechnikum his main interests centred on theoretical physics rather than mathematics. His mathematical abilities were not exceptional; he was slow and made mistakes. The professors of mathematics, Hurwitz and Minkowski, were outstanding, but Einstein had not yet fully realized the creative value of mathematics in physical research. Later he attributed his failure to learn from them as due to his lack of mathematical instinct. He avoided normal classes and spent most of his time studying the classics of physics, especially the works of Clerk Maxwell. Einstein was impressed both by the successes and by the failures of the old physics and was attracted to what he later called the 'revolutionary' ideas of Maxwell's field theory of electromagnetism. His study of the writings of the nineteenth-century masters received a new impetus from Ernst Mach's *Science of Mechanics*.

After graduation Einstein became a Swiss citizen; this again made him liable for military service, but he was rejected on medical grounds. For two years he applied for schoolteaching posts but was unable to obtain regular employment. While supporting himself by occasional tutoring and substitute teaching, he published several scientific papers. Then, in 1902, he was appointed an examiner at the Swiss patent office in Berne. The seven years Einstein spent there, examining applications for patents in electrotechnology, were the years in which he laid the foundations of large parts of twentieth-century physics. He liked the fact that his official work, which occupied only part of the day, was entirely separate from his scientific work, so that he could pursue this freely and independently, and he often recommended this arrangement to others later on.

In 1905 Einstein received his doctorate from the University of Zürich with a dissertation entitled *Eine neue Bestimmung der Moleküldimensionen*, which contained the germ of his later theory of Brownian motion. At this stage in his life he was about five and a half feet tall, with regular features, warm brown eyes, a mass of jet-black hair and a slightly raffish moustache. In 1903, against strong opposition from his mother (his father had died the previous year), Einstein married Mileva Marić, a Serbian woman, a fellow physics student from the Polytechnikum, more or less his contemporary academically although five years senior in age. Their two sons were born in Switzerland, Hans Albert in 1904 and Eduard in 1910. A previous child, Lieserl, was born at the home of Mileva's parents and given for adoption; there is no trace of her afterwards. Hans Albert emigrated to the USA before the Second World War and became professor of hydraulic engineering at the University of California; for various reasons he felt bitter towards his father. Eduard was a gifted child; as a young man his resemblance to his father was said to be 'almost frightening'. He suffered from paranoid schizophrenia, and after he had been institutionalized his father had nothing more to do with him.

Einstein never publicly acknowledged any contributions by his wife to his work; neither did she make any such claim. Yet some of his letters at the beginning of the twentieth century refer to her role in the development of 'our papers'; for example, in one of them he wrote 'How happy and proud I will be when both of us together will have brought our work on relative motion to a successful end'. Whatever this may mean, the consensus among his biographers seems to be that she gave up scientific work to devote herself to raising a family, while he did the opposite.

All biographers agree that he had a passion for music, as a way of experiencing and expressing emotion that is impersonal. Einstein was an enthusiastic violinist; Mozart, Bach and Schubert were his favourite composers. When he was world-famous as a physicist he is reported to have said that music was as important to him as physics: 'it is a way for me to be independent of people'; on another occasion he described it as the most important thing in his life. Photographs of him playing with Born, Ehrenfest, Hadamard, Hurwitz or Planck show a different Einstein from the more familiar images.

Einstein started his scientific work at the beginning of the twentieth century. The closing decades of the nineteenth century were the period when the long-established goal of physical theory – the explanation of all natural phenomena in terms of mechanics – came under serious scrutiny and was directly challenged. Mechanistic explanation had achieved many successes, particularly in the theory of heat and in various aspects of optics and electromagnetism; but even the successful mechanistic theory of heat had its serious failures and unresolved paradoxes, and physicists had not been able to provide a really satisfactory mechanical foundation for electromagnetic theory. It was a time of startling experimental discoveries, but the problems which drew Einstein's attention and forced him to produce the boldly original ideas of a new physics had developed gradually and involved the very foundations of the subject.

In 1905, one marvellous year, Einstein produced three masterly papers on three different subjects, which revolutionized the way scientists regarded space, time and matter. These papers dealt with the nature of Brownian motion, the quantum nature of electromagnetic radiation and the special theory of relativity. Einstein considered the second paper, on the quantum of light, or photon, as the most important, and it was for this that he was to be awarded the Nobel prize, but it was relativity that caught the popular imagination. It was not a new idea. Poincaré had already considered relativity from a mathematical point of view, in which the speed of light was regarded as an absolute constant. Of course, the idea of treating time as a fourth dimension has a long history; it had already been suggested by d'Alembert in the eighteenth century. The structure for the space-time continuum known as Minkowski space was developed by Minkowski only after he had moved to Göttingen; at the Polytechnikum he lectured on analytical mechanics. While Einstein was by no means alone in thinking about relativity, notably Langevin had been thinking on similar lines, it was only he who understood its revolutionary implications and worked out its consequences. Significantly, his original paper on the subject contained no references and very little mathematics.

Although Planck strongly influenced Einstein, it was more Henrik Antoon Lorentz who was his scientific father-figure. From his student days Lorentz had been a disciple of Clerk Maxwell, one of the few people in the Netherlands who understood the theory of electromagnetism. He was also an admirer of the work of Fresnel and of Hertz. He loved mathematics and was invited by the University of Utrecht to become professor in the subject, but preferred physics. Before long a new chair was created at Leiden, his *alma mater*, and at the beginning of 1878 he returned there as professor of theoretical physics, although he was not yet twenty-five. Three years later, when he was twenty-seven years old, Lorentz was elected to the Royal Academy of Sciences at Amsterdam. Amongst many other honours he received the Nobel prize for physics in 1902. When Lorentz died in 1928, at the age of seventy-four, Einstein represented the Berlin Academy at his funeral and said 'I stand at the grave of the greatest and noblest man of our times. His genius was the torch which lighted the way from the teachings of Clerk Maxwell to the achievements of contemporary physics . . . his life was ordered like a work of art down to the smallest detail. His never failing kindness and magnanimity and his sense of justice, coupled with an intuitive understanding of people and things, made him a leader in any sphere he entered. Everyone followed him gladly, for they felt that he never set out to dominate but always simply to be of use. His work and his example will live on as an inspiration and guide to future generations.'

It took a few years for Einstein's research to receive recognition. When he submitted the relativity paper to support his application to become Privatdozent at the University of Berne, it was rejected, although he was invited to give some lectures. His academic career did not really get started until three years later, when he was appointed associate professor at Zürich University; two years after that he became full professor at the German University in Prague and then returned to Zürich as full professor at the ETH, as the former Polytechnikum had become, the following year. Finally, in the spring of 1914, Einstein was persuaded to move to Berlin as a professor of the Berlin Academy, free to lecture at the university or not as he chose, and as first director of the Kaiser Wilhelm Institute of physics. He had mixed feelings about the move, partly because he disliked the Prussian life-style and partly because in physics he felt that he would be expected to produce one brilliant idea after another. As it turned out, however, he found the atmosphere in the German capital quite stimulating, although he missed the freedom of Switzerland, and he greatly appreciated having Planck, Nernst and, later, Schrödinger and von Laue as his colleagues. In the First World War Einstein refused to join in the widespread support of the German cause by German intellectuals and did what he could to preserve a rational international outlook and to urge an immediate end to the war.

However, while Einstein's scientific work was flourishing, his private life was not. His marriage had been under strain for some years. He was physically attractive to women and had a number of affairs. His wife Mileva and their two sons followed him to Berlin, but before long they returned to Zürich, which remained Mileva's home for the rest of her life. Legal separation and finally divorce followed soon after the end of the war. Earlier, when Einstein became ill and was bedridden for some months, he was nursed back to health by his cousin and childhood friend Elsa Löwenthal, a widow with two daughters; when the divorce came through, in which violence towards Mileva and adultery with Elsa were cited, they married. She was three years older than he was, more maternal and protective towards him than Mileva had been, and totally ignorant of science. The film actor Chaplin, who knew her later, described her as 'a square-framed woman with abundant vitality; she frankly enjoyed being the wife of the great man and made no effort to hide the fact; her enthusiasm was endearing'. Einstein gradually lost interest in her.

Meanwhile Einstein had been developing the general theory of relativity, the kind of theory which the mathematician Riemann may have been seeking in vain, wherein gravitational forces arise from the geometry of space-time. For the necessary mathematics Einstein turned for assistance to Tullio Levi-Cività, an Italian geometer of the old school; later the theory was recast in a more modern fashion by Hermann Weyl. A new scientific theory needs to be tested by experiment, and an opportunity for this came in 1919 when the deviation of light passing near the sun, as predicted by the general theory of relativity, was observed during the solar eclipse. Already famous among scientists, Einstein now became a celebrity to the general public. The publicity, even notoriety, which ensued changed the pattern of his life. He crossed the Atlantic for the first time and was lionized to an embarrassing extent. He spent three months in Leiden and then went on a grand tour of China, Japan, Israel and Spain. In 1921 he was awarded the Nobel prize in physics, for his 1905 paper on light quanta; the prize money he received went to his former wife Mileva, as part of the divorce settlement.

As usual the Nobel prize, to which he attached little importance, was followed by a whole cornucopia of other honours. He was now able to put the prestige of his name behind the causes he believed in and he did this, frequently naively, always bravely, but trying not to misuse the status his scientific reputation had given him. The two movements he backed most vigorously in the 1920s were pacifism and Zionism, particularly the creation of the Hebrew University in Jerusalem. He also took an active part for a few years in the work of the Committee on Intellectual Cooperation of the League of Nations.

Einstein's chief outdoor recreation was sailing a dinghy on the numerous lakes formed by the river Havel around Berlin. He was very skilful at manipulating the little boat, enjoying the gliding motion and the quiet mindsoothing scenery. He could be seen almost every day out sailing, but he had no mooring for his boat. As the date of his fiftieth birthday approached, the municipality conceived the plan of giving its most distinguished citizen a birthday present: a house beside the lakes, which would give him perfect quiet and direct access to the water. Unfortunately the project became so entangled in politics that Einstein rejected the idea and simply built a lakeside house for himself. It was there that he began meditating on the final goal of his scientific life, the discovery of one unifying theory that would bring together the hitherto separate phenomena of gravitation and the electromagnetic field. With characteristic concentration and obstinacy, he advanced numerous ideas on this subject during the rest of his life, but none of them commanded widespread acceptance.

After the end of the war it had been the policy of the organizing committee of the Solvay conferences to exclude Germans, but this restriction was lifted for the fifth conference in 1927, the last time that Lorentz was in the chair. Much of Einstein's scientific work after the end of the war had been concerned with quantum mechanics, so he took the opportunity to present an extended critique of Bohr's ideas, with which he did not agree. There was a vigorous discussion after which most of the participants departed believing that the positivist Copenhagen view had prevailed, but Einstein and his followers were not convinced. The controversy continued for years; even today there are problems with quantum theory, although it works so remarkably well in practice.

In Germany Einstein became increasingly the target of the anti-Semitic extreme right. He was viciously attacked in speeches and articles, even his life was threatened. Despite this treatment, he remained based in Berlin, declining many offers of positions elsewhere. He still went regularly to the Netherlands to see Lorentz. In 1931 he spent some time in Oxford as Rhodes Memorial Lecturer, staying at Christ Church, where he accepted a five-year research fellowship, with no duties attached, the only condition being one of residence. Although in the end he did not take this up, he returned to Oxford in 1933 as Herbert Spencer Lecturer. He also visited the California Institute of Technology for three successive winters.

By this time the Nazis had seized power in Germany and the attacks on him were intensifying. His papers on relativity were publicly burned before the Berlin State Opera House and all his property was confiscated. When resigning from the Bavarian Academy he wrote 'To the best of my knowledge the learned societies of Germany have stood by passively and silently while substantial numbers of scholars, students and academically trained professionals have been deprived of employment and livelihood. I do not want to belong to a society which behaves in such a manner, even if it does so under compulsion.' He had been considering an arrangement that would have enabled him to divide his time between the Berlin Academy and the Institute for Advanced Study in Princeton. Such a compromise was clearly no longer viable and so he simply resigned from the Berlin Academy and moved to Princeton. With Einstein on the faculty, the institute was in a strong position to attract other leading scientists. Fortunately most of Einstein's scientific correspondence had been saved and brought to America by diplomatic bag. Having automatically become a German citizen again when he was appointed to the Berlin Academy, he relinquished this just before the Nazis could deprive him of it. Many other Jewish scientists were also leaving Germany; some stayed behind and, ironically, a few of them tended to blame Einstein for what befell them.

So Princeton became Einstein's home for the remaining twenty-two years of his life. He described it as 'a wonderful piece of earth and at the same time an exceedingly amusing ceremonial backwater of tiny spindle-shanked demigods'. His *Heldenzeit* lasted a good twenty years, but well before he left Europe his great days were over. He was noticeably aged; scientifically the Princeton years were much less fruitful than what came before. Although he thought the chances of success were small, he continued to seek a unified field theory and became increasingly isolated from the mainstream of physical research. Princetonians respected his desire for solitude. On one occasion he said that really his only friend in Princeton was Kurt Gödel, the mathematical logician, who used to call for him every morning at 11 o'clock so that, whatever the weather, they could walk together to Fuld Hall. He continued 'I do not socialize because social encounters would distract me from my work and I really only live for that, and it would shorten even further my very limited lifespan. I do not have any close friends here as I had in my youth or later in Berlin with whom I could talk and unburden myself. That may be due to my age. I often have the feeling that God has forgotten me here. Also my standard of decent behaviour has risen as I grew older: I cannot be sociable with people whose fame has gone to their heads.'

During the 1930s Einstein renounced his former pacifist stand, since he became convinced that the menace to civilization embodied in Hitler's regime could only be put down by force. He never returned to Europe, in fact the only time he left America was in 1935 as part of the process of becoming an American citizen. He did not participate in the American efforts that eventually produced the nuclear reactor and the atomic bomb. After the bomb had been used and the war ended, he devoted his energies to the attempt to achieve a world government and to abolish war once and for all. He also spoke out against repression, arguing that intellectuals must be prepared to risk everything to preserve freedom of expression. Nonetheless, despite his concern for world problems and his willingness to use whatever influence he had towards alleviating them, his ultimate loyalty was to science. As he said once with a sigh to an assistant during a discussion of political activities: 'yes, time has to be divided this way between politics and our equations. But our equations are much more important to me because politics is for the present but an equation like that is something for eternity.'

After 1936, when his second wife Elsa died, Einstein was looked after by his sister Maja, his stepdaughter Margot and his secretary-housekeeper Helen Dukas. Maja had come to live with her brother in 1939; she suffered a stroke in 1946, after which she was bedridden, and died in 1951. Einstein had retired from the Institute for Advanced Study in 1945 and was living almost like a recluse, trying to avoid the endless stream of people who wanted to see him about something, or just to see him. He suffered much harassment by press photographers; no other scientist has become so well known to the public in appearance. He was generally quite a merry person, with a strong sense of humour and a loud laugh; on one occasion he put out his tongue to express his annoyance and that photograph has been reproduced endlessly. Around Princeton he could often be seen at the local cinema – he was particularly fond of Western films. Although he kept a wardrobe of seven identical suits to wear on formal occasions, his ordinary dress was casual, he favoured sweatshirts, leather jackets and sandals. He never learnt to drive a car, but used to sail a dinghy on Lake Carnegie. Otherwise he stayed peacefully at home in number 112 Mercer Street, a colonial-style house no different from others in the neighbourhood.

For many years Einstein experienced recurrent health problems, including anaemia and digestive attacks, and he also suffered from an enlarged heart. He was drafting a speech on the tensions between Israel and Egypt when he became ill and died a few days later, on April 18, 1955, in his seventy-sixth year; the immediate cause was a haemorrhage after a large aneurysm of the abdominal aorta burst. One of his last acts was to sign a plea, initiated by Bertrand Russell, for the renunciation of nuclear weapons and the abolition of war. He left his brain for use in research, his body for cremation, which was carried out privately, and all his scientific and other papers to the Weizmann Institute in Jerusalem; this did not prevent historians having difficulty in obtaining access to study this material and permission to publish in scholarly works.

Einstein never identified with any particular country, living and working in many different places, and, although he had quite a few individual collaborators, he never set out to create a research school in any sense. In his own words: 'I have never belonged wholeheartedly to any country or state, to my circle of friends, or even to my own family. These ties have always been accompanied by a vague aloofness, and the wish to withdraw into myself increases with the years. Such isolation is sometimes bitter, but I do not regret being cut off from the understanding and sympathy of other men. I lose something by it, to be sure, but I am compensated for it by being rendered independent of the customs, opinions and prejudices of others, and am not tempted to rest my peace of mind upon such shifting foundations.'

8   From Ehrenfest to Schrödinger

Our next five remarkable physicists were born in the eight years from 1880 to 1887. Two came from Austria and one from each of Denmark, England and Germany.

PAUL EHRENFEST (1880–1933)

In Austria-Hungary the laws controlling and restricting Jewish life had been greatly relaxed by the eighteen-seventies, after which Jews began first to enter and then to dominate Viennese intellectual and cultural life. Their position in commerce and finance also grew to one of great strength. Unfortunately anti-Semitism grew stronger at the same time and boys who were visibly Jewish suffered from it constantly. In Austria, as in Germany, the universities were citadels of anti-Semitism and it was difficult for even the most distinguished Jewish scholars to obtain professorships.

The physicist Paul Ehrenfest was born in Vienna on January 18, 1880. His parents had moved to the imperial capital about twenty years earlier from Loschwitz, a small Jewish village in Moravia. His father Sigmund worked in a textile mill until he married Johanna Jellinek, the daughter of a merchant in the same village, and set up a grocery business. The business thrived and, by the time their son Paul had been born, the family was reasonably well off. They had four older sons, Arthur (1862), Emil (1865), Hugo (1870) and Otto (1872); a daughter was lost at birth. As the youngest by eight years, born when his father was forty-two and his mother thirty-eight, Paul was the baby of the family, very much his father's favourite. When he was ten years old his mother died of breast cancer; her place in his upbringing was taken by his widowed maternal grandmother. Paul's older brothers played a major role in his early life. Arthur was completing his studies at the Technische Hochschule, Emil was his father's right-hand man in the business, while Hugo and Otto were going through secondary school. In due course Paul too passed through the school system, entering the *akademisches Gymnasium* in 1890 but transferring in 1897 to the Kaiser Franz Josef Gymnasium to join his friend Gustav Herglotz for the last two years of school.

In October 1899 Paul Ehrenfest enrolled at the Technische Hochschule, listing chemistry as his major field, but also taking courses in

a wide range of other scientific subjects, including mathematics. Herglotz, already recognized as a promising mathematician, was at the University of Vienna. However, they were able to maintain their friendship and to become acquainted with other young mathematicians, particularly Hans Hahn and Heinrich Tietze. The four of them often took walks together in the hills of the Wienerwald. University students in the German-speaking world, as we know, were not obliged to limit themselves to a single institution, but could attend the lectures of particular professors wherever they might be found. Thus Hahn studied at Strasbourg, Munich and Göttingen, while Herglotz went off to Munich to study astronomy, and Tietze joined him there after military service.

By that time Ehrenfest was certain that he wanted to be a theoretical physicist, due to the influence of Boltzmann, whose course on the mechanical theory of heat he had been taking. It was Boltzmann who initiated Ehrenfest into both the substance and the spirit of theoretical physics, as he did for so many others, and the Boltzmann influence was to shape Ehrenfest's own teaching and research in the years to come. After Boltzmann left Vienna for Leipzig in 1900, Ehrenfest stayed on for another year before migrating to

Göttingen, where he found a much richer scientific life than he had known in Vienna. He signed up for no fewer than fifteen courses in the first year, gradually reducing the load as the year went on. While most of these were on physics, it was the mathematical lectures which he found most exciting, particularly Hilbert on potential theory and Klein on mechanics.

Among the physics students in Göttingen was a young Russian woman, Tatyana Alexeyevna Afanassjewa, who was accompanied by her aunt Sonya. Her father, the chief engineer of the Imperial Russian Railways, had died when she was still a child, and she went to live with a childless uncle, who was a professor at the Technische Hochschule in St Petersburg. During the period leading up to the First World War, St Petersburg offered special university-level institutions for women, which to some extent shadowed the imperial university (then still reserved for men). Tatyana attended first the women's paedagogical school and then the institution which offered women courses in arts, sciences and law. She shone in mathematics and had gone on to study physics at the Georgia Augusta, where she met Paul Ehrenfest. Before long the two young physicists had decided on marriage. She was just a few years older than he was and imposed various conditions: among them that he must read the novels of Tolstoy as soon as possible and was never to smoke tobacco. There was also the problem that Paul was Jewish and Tatyana was Russian Orthodox. The laws of the Austro-Hungarian empire did not permit the marriage of a Jew to a Christian. Such a marriage could take place only if the couple officially declared themselves 'unchurched', and foreswore all religious affiliations.

First, however, Ehrenfest needed to complete his Ph.D. thesis. He decided to study under Lorentz in Leiden, while working his way through the novels of Tolstoy. After a quick trip back to Vienna to see Boltzmann, who had now returned from Leipzig, Ehrenfest spent some time with his fiancée in Göttingen. He then fitted in an Italian tour, spent a year working on his thesis in Vienna, completed it in Dubrovnik and had it accepted in June 1904. That summer, with his doctorate in hand, Ehrenfest was ready to marry and Tatyana, who had remained in Göttingen until then, joined him in Vienna. They proceeded to comply with the formalities of renouncing their respective religions and were married in December that year. He promised her that before long they would move to Russia and settle there.

Fortunately both Paul and Tatyana had inherited small incomes, which made it possible for them to maintain a modest life-style and continue with scientific work. To start with they stayed on in Vienna, where their life was centred around the university. Although Paul held no position there, he continued to participate actively in the Boltzmann seminar with Lise Meitner and others, and continued to devour books at a huge rate. Tatyana went through a serious illness early that summer – mumps, with high fever and delirium. In 1905 their first child was born, a daughter named after her mother. The following spring they left Vienna, never to live there again. After a summer in Switzerland they returned to Göttingen, where Paul was invited by Klein to talk at his seminar about some joint work on statistical thermodynamics that he and Tatyana had just completed.

One of the consequences of her renunciation was that it would be difficult for Tatyana to return to her homeland, since Czarist Russia was reluctant to allow entry to non-believers, but fortunately she was able to persuade the Russian consul in Vienna to grant the necessary visas. In the autumn of 1907 they moved to St Petersburg, where Pavel Sigismondovich Ehrenfest, as he was known there, was given a warm welcome. During the five years they lived in Russia the Ehrenfests spent their summers at Kanuka, a tiny Estonian village on the Gulf of Finland some 90 or 100 miles west of St Petersburg. Boltzmann had been writing a review article on statistical mechanics for the *Encyclopaedia of the Mathematical Sciences* which Klein was organizing, but after Boltzmann's suicide Klein turned to Ehrenfest to replace him. In addition to this, Ehrenfest wasted a lot of time trying to qualify for the *Magister*, the essential prerequisite for a faculty position in Russia; his research doctorate was irrelevant. Paul could speak Russian tolerably well and gave some exemplary lectures on the differential equations of mathematical physics at the Polytechnic Institute.

Tatyana published her first paper in theoretical physics in 1905 and went on publishing during these Russian years. In 1910 she gave birth to a second child, another daughter, named Anna. The Ehrenfests began to feel short of money and so Paul tried again to get his foot on the bottom rung of the academic ladder by becoming a Privatdozent, first at Leipzig and then at Munich, but without success. Early in 1912 he set out to visit other people who might be able to help, including Planck in Berlin, Herglotz in Leipzig and Sommerfeld in Munich, but in vain. He returned to Vienna to consult his brothers Arthur, Emil and Otto; the fourth brother Hugo had emigrated to America, and Paul considered following his example. He went to Brno to see his old friend Tietze and to Prague to stay with the Einsteins. Long afterwards Einstein recalled that 'within a few hours we were true friends – as though our dreams and aspirations were meant for each other'. They played the Brahms sonatas for violin and piano. He also was a success with their little son, the seven-year-old Hans Albert.

Einstein was a full professor at Prague but had already decided to move back to Zürich, where he was going to be professor at his *alma mater*, the ETH; who could be a better successor in Prague than Ehrenfest? However, Prague, being in the Austro-Hungarian empire, required a formal religious affiliation. Although Einstein assured him that this was just a formality, which no-one took seriously, Ehrenfest's conscience would not allow him to conceal the renunciation he had made in order to marry Tatyana. When he saw Ehrenfest off at the railway station, Einstein said that he would try to find an opening for him in Zürich. Meanwhile every effort was being made to persuade Ehrenfest to overcome his scruples and accept the position in Prague.

At this point Ehrenfest received a letter from Lorentz, whom he had not seen for nine years, congratulating him on his Encyclopaedia article: 'highly interesting', 'beautiful and profound'. Lorentz went on to say that he had decided to give up his position at the University of Leiden, after thirty years, and retire to nearby Haarlem. He had been hoping to persuade Einstein to succeed him, but Einstein was already committed to the Zürich post. Lorentz then made enquiries about the suitability of Ehrenfest for Leiden. Sommerfeld wrote that 'he lectures like a master. I have hardly ever heard a man speak with such fascination and brilliance. Significant phrases, witty points and dialectic are all at his disposal in an extraordinary manner. His way of handling the blackboard is characteristic. The whole disposition of his lecture is noted down on the board for the audience in the most transparent possible way. He knows how to make the most difficult things concrete and intuitively clear. Mathematical arguments are translated by him into easily comprehensible pictures.' Lorentz wanted to know whether Ehrenfest would be interested in coming to Leiden.

Meanwhile the Ehrenfests had set off on a cruise down the river Volga; they were away two weeks and on their return found a letter from Sommerfeld offering Ehrenfest the position of Privatdozent in Munich, followed by the one from Lorentz. Deeply moved, both by the nature of Lorentz' enquiry and by the delicate way in which it was put, he wrote a long and intimate letter in reply, saying that only a chair in Switzerland would appeal to him more. Because the appointment had to be made by the government, there was some delay; then near the end of September it was confirmed, and the Ehrenfests could prepare to move to Leiden and settle down. They arrived with two daughters; a third child was born, in May 1915, a boy named Paul after his father, and then in August 1918 another, Vassily, who was afflicted with Down's syndrome.

When they arrived in October 1912 they had been given a warm reception and Lorentz could not have been more helpful. Once the formalities were over, one of the first things Ehrenfest did was to attend a special meeting of the Berlin Academy held in Göttingen, where he had the opportunity to see both the older generation of German physicists and some of the new people, such as Courant and Weyl, also the exotic Lindemann. Meanwhile Tatyana was planning their new house, on an open plan quite different from the traditional design. When it was built it proved to be more costly to maintain than they could afford. The hospitable Ehrenfests liked to entertain rather informally; they were vegetarians, offered no alcoholic drinks and did not permit smoking.

Just before Einstein moved to Berlin he spent a week in Leiden, lecturing on his latest ideas and going to see his old friend and mentor Lorentz. During the First World War, in which the Netherlands was neutral, Ehrenfest kept trying to persuade Einstein to visit Leiden again. Eventually the bureaucratic difficulties were overcome and Ehrenfest was able to bring Einstein and Lorentz together once more:

> in his usual way, Lorentz first saw to it at dinner that Einstein felt
> himself enveloped in a warm and cheerful atmosphere of human
> sympathy. Later, without any hurry, we went up to Lorentz' cozy and
> simple study. The best easy chair was carefully placed in position next
> to the large work-table for the esteemed guest. Calmly, and to forestall
> any impatience, a cigar was provided for the guest, and only then did
> Lorentz quietly begin to formulate a finely honed question concerning
> Einstein's theory of the bending of light in a gravitation field. Einstein
> listened to the exposition, sitting comfortably in the easy chair and
> smoking, nodding happily, taking pleasure in the masterly way
> Lorentz had rediscovered, by studying his works, all the enormous
> difficulties that Einstein had needed to overcome before he could lead
> his readers to his destination, as in his papers, by a more direct and less
> troublesome route. But as Lorentz went on and on, Einstein began to
> puff less frequently on his cigar. He sat up straighter and more intently
> in his armchair. When Lorentz had finished, Einstein sat bent over the
> slip of paper on which Lorentz had written mathematical formulae to
> accompany his words as he spoke. The cigar was out, and Einstein
> pensively twisted his finger in a lock of hair over his right ear. Lorentz,
> however, sat smiling at an Einstein completely lost in meditation,
> exactly the way a father looks at a particularly loved son – full of

confidence that the youngster will crack the nut he has given him, but eager to see how. It took quite a while, but suddenly Einstein's head shot up joyfully; he had it. Still a bit of give and take, interrupting one another, a partial disagreement, very quick clarification and a complete mutual understanding, and both men with beaming eyes skimming over the shining riches of the new theory.

Even more than usual this particular meeting with Einstein left Ehrenfest inspired and invigorated. They always played violin and piano sonatas when they met. Ehrenfest's favourite composer had been Beethoven. He also played and enjoyed music from Haydn to Brahms, but he had never particularly liked the music of Bach. Einstein opened the world of Bach's music to his friend in Leiden, and Ehrenfest was completely captivated. Once the war was over, Ehrenfest made a determined effort to persuade Einstein to move to Leiden from Berlin. Einstein said no to this, but agreed to a regular visiting professorship, which brought him to Leiden for three or four weeks annually from 1920 onwards. Already the anti-Einstein faction was gathering strength in Berlin; the visits he made to Leiden came as a welcome relief from their attacks.

However, all was not well with Ehrenfest. He was partially estranged from Tatyana, who had gone back to St Petersburg. He took very personally the growing threat posed to his fellow scientists by the rise of the Nazi party in Germany. At the same time he seems to have felt overwhelmed and inadequate to deal with the continuing change taking place in physics during the early 1930s. Much as his students loved him, he lost his self-confidence more and more. Niels Bohr wrote that 'he is a very clear-sighted man, fertile in ideas, but his temperament is so troubled I have never encountered anything like it'. Increasingly prone to depression and bizarre behaviour, the end came in Amsterdam on September 25, 1933, when he committed suicide at the age of fifty-three, after shooting and blinding the mentally handicapped Vassily, then aged fourteen.

Afterwards Einstein wrote of Ehrenfest that 'he was not only the best teacher in our profession I have ever known; he was also passionately preoccupied with the development and destiny of men, especially his students. Unfortunately the accolades of his students and colleagues were not enough to overcome his deep-rooted sense of inferiority and insecurity'. Tatyana, 'whom he loved', said Einstein, 'with a passion the likes of which I have not often witnessed in my life', was in St Petersburg when the tragic events took place. She returned to Leiden and spent the rest of her life there. Although she never completed a doctorate or held a regular university teaching post, her writings substantially enriched physics in the Netherlands. She wrote two major monographs in the 1950s, enlarging her readership by publishing in English as well as Dutch. Her last work, published in 1960 when she was eighty-four, was a treatise on the teaching of mathematics. The surviving son Paul was killed in an avalanche while skiing in the French alps in 1939; Tatyana lived until 1964.

MAX BORN (1882–1970)

In these pages, we follow the fortunes of several of the physicists who had to leave Germany after the Nazis came to power. We have already seen how Lise Meitner went to Sweden and Albert Einstein to the USA. Later we will describe how Erwin Schrödinger settled in Ireland. Of course there were many others. No two cases were exactly alike, but since Max Born was one of the founders of quantum mechanics and he wrote an informative autobiography, I have chosen him as the subject of the next profile.

The son Max of Gustav and Margarethe Born was born on December 11, 1882, in Breslau, the chief city of what was then the Prussian province of Silesia. Gustav Born was a well-known embryologist who occupied a chair at the University of Breslau, and whose contributions to embryology

anticipated some modern developments in our knowledge of sex hormones. His wife Margarethe (née Kauffmann) came from a wealthy Silesian family in the textile business; she died from gallstones when Max was only four years old. It was probably from her that her son inherited his life-long love of music; one of his most treasured possessions was an album that had belonged to his mother, containing autographs of Johannes Brahms, Clara Schumann, Xaver Scharwenka, Pablo Sarasate and many other celebrated musicians. For the four years after the loss of their mother, Max and his younger sister Käthe were placed under the care of governesses. In 1890 their father married again, and, although his second wife Bertha (née Lipstein) proved an admirable stepmother, she never quite replaced Margarethe in the affections of her two stepchildren.

The family home, with its atmosphere of scientific and general culture, provided a stimulating environment during Max's formative years. His father's circle of friends at this time included Paul Ehrlich, the pioneer of chemotherapy, and the bacteriologist Albert Neisser. The young Born's schooling at König Wilhelms Gymnasium in Breslau was of the usual humanistic type, with Latin, Greek and German as the principal objects of study, together with some mathematics, physics, history and modern languages. Although Born does not appear to have been a particularly outstanding scholar, the enthusiasm of his mathematics teacher, the geometer Heinrich Maschke, who also taught a little physics, communicated itself to him. At this time Marconi's experiments on wireless telegraphy were becoming known, and Maschke repeated them with some of his pupils. He succeeded in transmitting a signal to an adjoining room. Born used to recall his feeling of chagrin when his headmaster, summoned from his humane studies to behold the latest miracle of technology, was not in the least impressed.

Shortly before Born graduated from the gymnasium in 1901, his father died. Following paternal advice not to specialize too early at the university but to sample lectures on a variety of subjects before coming to any decision on his future career, Born started by attending courses on chemistry, zoology, philosophy, logic, mathematics and astronomy. Those he found most interesting were the last two, and he had thoughts of becoming an astronomer. Following the custom of students in Germany at that period he did not spend all his time at Breslau, but migrated for the summer semesters to the University of Heidelberg in 1902 and the University of Zürich in 1903, enjoying to the full the amenities and cultural opportunities of these cities. It was at Heidelberg that he met James Franck, who was to become a life-long friend and later colleague at Göttingen; at Zürich that he received his introduction to advanced mathematics from Hurwitz' lectures on elliptic functions. By returning to the university of his native city for the winter semesters and for most vacations, he was able, at the home of his late father's friends the Neissers, to meet many writers and musicians, including such celebrities as Gerhard Hauptmann, Ferruccio Busoni, Artur Schnabel, Edwin Fischer and Carl Flesch.

Among Born's fellow students at Breslau were Otto Toeplitz and Ernst Hellinger, both of whom were destined to become mathematicians of distinction. It was they who told him of the three prophets of Göttingen – Klein, Hilbert and Minkowski – and inspired him to make the pilgrimage thither. At the Georgia Augusta he began by attending lectures by Hilbert and Minkowski. Before long Hilbert offered him the unpaid post of *Privat-assistent*, his primary duty being to prepare a fair copy of the professor's lecture notes. The main attraction was the privilege of close contact with Hilbert and Minkowski, accompanying them on their rambles in the woods around Göttingen, during which they would discuss not only mathematics but also philosophical, social and political problems. A seminar they conducted on the electrodynamics of moving bodies directed Born's attention to the problems of what was to become known as the special theory of relativity. Later on he wrote a useful elementary introduction to the theory, but at that time Einstein's first paper on relativity had only just appeared. Minkowski was developing his own four-dimensional formulation of electrodynamics.

Another seminar that Born attended was one on elasticity, which Klein conducted jointly with the applied mathematician Runge. Born, who had offended Klein by irregular attendance at his lecture course, was called on to give an account of a problem in elastic stability at very short notice due to a fellow student falling sick. Lacking sufficient time to study the literature, he treated the subject *ab initio*; Klein was so impressed by his performance that, with Born in mind as competitor, he set the problem for the annual university prize competition. At first Born refused to enter his name, thus giving fresh offence to 'the great Felix', but eventually he capitulated, submitted his entry and carried off the prize.

When it came to the oral examination for the doctorate, Born deemed it inadvisable to risk having Klein question him on geometry, which he had intended offering as one of his subjects, so he offered astronomy instead. Having already started studying the subject at Breslau, at Göttingen he was accepted into the astrophysical seminar of Karl Schwarzschild. Born's relations with Schwarzchild were happier than they had been with the formidable Klein, and the doctoral examination in January 1907 passed off successfully with Schwarzchild as examiner in place of Klein. Born's thesis was based on his prize dissertation on elastic stability; for the rest of his life he retained an affection for his first scientific offspring, through which he first tasted the joy of independent investigation of a problem and the satisfaction of finding the predicted results in harmony with experiment.

At Göttingen Born came into contact with a remarkable group of younger mathematicians and physicists, not only Hellinger and Toeplitz, but also Richard Courant, Erhard Schmidt and Constantin Carathéodory. Among the subjects he studied were the kinetic theory of gases, electrodynamics and the aberrations of optical instruments. After taking his doctorate, Born would normally have had to undergo a year's compulsory military service, but a tendency to asthma enabled him to shorten the period. The experience, he tells us in his autobiography, confirmed his antipathy to all things military. A visit of six months' duration to Cambridge followed. As an 'advanced student' at Caius College he attended lectures by Larmor and Thomson. Larmor's lectures he found much inferior in content to Minkowski's, quite apart from problems with the Irish accent, but Thomson he found most stimulating.

On returning to Breslau, Born engaged in some experimental work, but soon turned to theory again. By combining Einstein's special theory of relativity with Minkowski's mathematical foundation, he found a new and more rigorous way of calculating the electromagnetic mass of the electron, and sent the manuscript to Minkowski in Göttingen. As a result, Minkowski invited him to return to the Georgia Augusta to assist him with his work on relativity. Sadly, the possibility of this promising collaboration was cut short by Minkowski's untimely death. In his autobiography Born relates how downcast he felt by the ruin of all his hopes and how he again fell foul of Klein, but managed through the good offices of Runge to convince Hilbert of the soundness of his ideas. He also attended theoretical and experimental courses under Woldemar Voigt, who offered Born a position as Privatdozent. Among his colleagues on the teaching staff at the Georgia Augusta were Richard Courant, Hermann Weyl and von Kármán. With the last of these he developed the Born–Kármán theory of the specific heats of solids. This was the beginning of an ambitious programme of research that was to occupy Born and his pupils for many years, namely the explanation of the physical properties of solids – in particular crystals – on the basis of their lattice structures.

Although he was still no more than a Privatdozent, an invitation from Albert Michelson to visit Chicago took Born to the USA for the first time in 1912. The next year Born married Hedwig (née Ehrenburg), the daughter of the professor of jurisprudence at Göttingen; her forebears included Martin Luther. Between them, she and her husband could claim an extended family of great intellectual distinction. They had three spirited and sometimes turbulent children, Irene, who became a well-known singer, Margaret and Gustav, who became a prominent biologist.

The outbreak of war in August 1914 coincided with the offer of an associate professorship at the University of Berlin, where Born would have Planck as a colleague. Upon arriving in Berlin in the spring of 1915, Born was soon drawn into the war effort, but not before he had completed the manuscript of a book on the dynamics of crystal lattices. After a short time as a radio operator in the German air force, he was seconded to the artillery with commissioned rank for research on acoustical range-finding. Characteristically, he conceived it to be his duty to have as many as possible of his former colleagues and students recalled from the front-line to work in his section. After the war was over, although conditions in Germany were extremely difficult, Born was able to appreciate the life of Berlin, particularly the scientific life. Among the physicists he particularly enjoyed the friendship of Einstein, with whom he had long been in correspondence but now could know as a colleague and neighbour. Born was an accomplished pianist and they often played violin sonatas together.

In 1919 von Laue, at that time professor in Frankfurt, proposed an exchange of chairs with Born in Berlin in order that he could more easily work with his beloved teacher Max Planck. The exchange was arranged with the agreement of both universities, and as a result Born moved to Frankfurt in April that year as professor and director of the institute of theoretical physics. Two years later, he moved on to a similar post at Göttingen. He was joined by his friend James Franck, who took charge of the experimental side of physics at the Georgia Augusta. During the early years of this, his third period in Göttingen, Born and his students carried on the work on lattice dynamics. He also wrote a long survey article for Klein's Encyclopaedia, later published as a separate book, and edited the collected works of Gauss.

Before long Born's chief research interest shifted towards quantum theory, where he was particularly fortunate in having as his junior colleagues Pauli and Heisenberg. During the winter of 1925/6 Born was in America again for a lecture tour, including a course on 'problems of atomic dynamics' at the Massachusetts Institute of Technology. This was written up and became the first book to be published on quantum mechanics. Three years later he visited Russia with a party of European scientists. He had been feeling the strain of directing the institute in Göttingen, which had become a place of pilgrimage for large numbers of young theoretical physicists from all over the world, and, although the Russian tour might have provided a respite, instead it triggered a nervous breakdown, which forced him to inter-rupt his teaching and research for the following year. He maintained that afterwards he never fully recovered his earlier capacity for intensive work. Nevertheless, the publication in 1933 of his classic textbook, *Principles of Optics*, demonstrated his ability to complete a major undertaking on top of his other commitments, even in a field that was not central to his interests.

In May of that same year Born, being Jewish, was effectively deprived of his Göttingen chair and left Germany. After a short rest in the Italian Tyrol, where he was consulted by Lindemann, he accepted an invitation to go to Cambridge and work with Infeld on a non-linear modification of Maxwell's electromagnetic theory, which they thought might remove the difficulty of the infinite self-energy of the electron, but the results could not be reconciled with quantum theory. For Born, accustomed to directing a large department, there was a significant problem of adjustment, but he came to feel, like many other refugees, that it was a rejuvenating experience. He wrote the well-known textbook *Atomic Physics*, which went through numerous editions, and a popular work, *The Restless Universe*. After Cambridge he made a visit to the Indian Institute of Sciences in Bangalore, where it was hoped to recruit some of the displaced German scientists, and then seriously considered the offer of a more permanent post in Moscow, but just at this time the Tait Chair of Natural Philosophy at the University of Edinburgh fell vacant, and in 1936 Born was elected.

In Edinburgh, Born rapidly established a research school on the con-tinental model, although he was not altogether successful in grafting this onto the undergraduate teaching in his department. A member of the Born school around 1940 described what it was like:

> The theoretical research was essentially carried out in one large room in the basement of the physics building (the old infirmary) in Drummond Street. There was Born's writing desk and a number of long tables at which usually about a half-a-dozen research workers (including myself) sat. When Born arrived in the morning he first used to make the round of his research students, asking them whether they had any progress to report and giving them advice, sometimes

presenting them with sheets of elaborate calculations concerning their problems which he had done himself the day before. The apparent ease with which he could switch from one subject to another during this inspection tour was truly amazing. Being such an incredibly fast worker himself he could on occasion become quite impatient when he found that a student had not managed to complete the calculations which had been suggested to him only the day before. The rest of the morning was spent by Born in delivering his lectures to undergraduate honours students, attending to departmental business, and doing research work on his own. Most of the latter, however, he used to carry out at home in the afternoons and evenings.

His seventeen-year tenure of the Edinburgh chair afforded Born many opportunities for visits, sometimes extended, to conferences and universities at home and abroad, including congresses in Paris, Bordeaux and the Soviet Union; he also spent a term in Egypt and gave the Wayneflete Lectures in Oxford. The war years brought little disturbance to his routine of research, apart from a temporary decline in the number of his research students. Although he was a master of atomic physics, he did not contribute to the efforts that went into the development of the atomic bomb. After the war Born's research school continued to be active, but, at the end of the session 1952/3, having reached the retirement age of seventy, Born returned to Germany and settled at a small and secluded spa within easy reach of Göttingen. Some amends were made to him for his treatment by the Nazis by the restoration of his confiscated property and pension rights, and by the conferment on him (along with Courant and Franck) of honorary citizenship of Göttingen, whose university he had served with such distinction. He revised several of his books for new editions and wrote his autobiography, but his scientific career was essentially over. The award of a shared Nobel prize in physics in 1954, for his research in quantum mechanics, was one of those cases in which the prize was in recognition of work done long before.

In his later years Born became increasingly active in the cause of the social responsibility of scientists, and wrote and lectured indefatigably on what he saw as the appalling dangers inherent in the technological explosion and the concurrent collapse in ethical standards. In 1957, when the nuclear policy of the Federal Republic of Germany was a matter of active debate, Born was one of the leaders of the 'Göttingen eighteen' who made public their belief that nuclear armament was a suicidal policy and declared that they would refuse categorically to collaborate in any scientific work associated with nuclear weapons; although Born himself was not religious, he was a staunch pacifist as was his Quaker wife Hedwig. He was a kindly, even-tempered but rather formal person; he could sometimes become surprisingly inflexible in matters of scientific controversy. He died in hospital on January 5, 1970, at the age of eighty-seven, and Hedwig survived him by only two years.

NIELS BOHR (1885–1962)

The first son, Niels Henrik David, of Christian and Ellen Bohr was born in Copenhagen on October 7, 1885. His older sister Jenny was born two years earlier and his younger brother Harald two years later. Their mother, a generous, intelligent and liberal woman, came from the wealthy Jewish Adler family, which was prominent in Danish banking and parliamentary circles. Their father was a university professor, a famous physiologist and lover of science, also the founder of the university football club; although he too was of Jewish extraction he had been converted to the Lutheran faith. The three Bohr children were brought up as Christians in a patrician home of culture where they were exposed to a world of ideas in animated debates in which conflicting views were examined rationally and in good humour,

and they developed a respect for all those who seek deeper knowledge and understanding. It was a close-knit family; the two brothers were inseparable in childhood and remained close throughout life.

The school career of Niels, the elder brother, was academically successful without being outstanding. Later, looking back, he stated: 'my interest in the study of physics was awakened while I was still at school, largely owing to the influence of my father'. As well as a growing interest in physics and mathematics, he early showed an ability to inspire affection in others, forming friendships at school that were to last throughout his life. For a glimpse of him as a schoolboy we have some reminiscences by classmates: 'in those days Niels was tall, rather coarse of limb, and strong as a bear. He was not afraid to use his strength when it came to blows during the break between classes. He seemed to be quite an ordinary boy, gifted but not smug, a promising honours student, but otherwise a young man like the rest of us.'

It was at the University of Copenhagen that Niels Bohr's potential as a scientist was first recognized. In 1907, at the age of twenty, he was awarded the Gold Medal of the Royal Danish Academy of Sciences and Letters for a prize exercise on the measurement of surface tension by the study of vibrating fluid jets. This careful and complete piece of research, both experimental and theoretical, drew upon and extended the work of Rayleigh, and served later on to give Bohr particular insight in his liquid-drop model of the atomic nucleus. Although Bohr worked very hard, he also played hard, so no account of his life would be complete without mention of his enthusiasm for sports, especially association football, although he was not quite up to the standard of his younger brother Harald, who captained the Danish team at the 1908 Olympic Games. The brothers became nationally famous for their achievements on the football field; their father had helped to make soccer the Danish national game.

Both for the master's degree and for his doctorate, Bohr submitted studies of the application of electron theory to the explanation of the physical properties of metals. Although these studies were not without some success, it was through this work that he began to be aware of the difficulties and limitations of classical physical theory in the description of electron behaviour and of the need for some radically different mode of description of atomic processes. He began to recognize the limitations of ordinary language in the description of phenomena and the need to accommodate apparently conflicting aspects in order to form complete descriptions. He was convinced that 'there is no point in trying to remove such ambiguities; we must rather recognize their existence and try to live with them'. Such ideas reappeared later in his principal contribution to physics and epistemology: the complementarity argument.

Harald Bohr obtained his doctorate two years earlier than his older brother Niels and went on to become a distinguished mathematician. Their sister Jenny was also talented; she studied first history at the University of Copenhagen and then English at the University of Oxford before embarking on a career in schoolteaching, but later became mentally ill and died at the age of fifty. Niels was an assiduous correspondent; he left a legacy of interesting and apparently spontaneous letters: in fact they were far from spontaneous but, like his scientific papers, only reached their final form after multiple drafts and painstaking revisions. As early as 1911 he began enlisting the aid of an amanuensis – at first his mother and, later on, his sons, his colleagues or his wife.

In 1910 Niels Bohr met his future wife Margrethe Nørlund, the daughter of a pharmacist, and they soon became engaged. The next year, in the period before marriage, he made the first of many visits to Britain. In Cambridge Bohr failed to interest Thomson in the work he had been doing on the theory of metals, instead Thomson gave Bohr a rather routine research project to work on. Bohr spent his first few months in England feeling fairly frustrated until he met and impressed Rutherford, who had recently devised his new model of the atom. The model was like a miniature solar system, consisting of a dense, positively charged nucleus with a family of negatively charged electrons in orbit around it.

In August 1912 Niels and Margrethe were married in the town hall of Slagelse, where she had been brought up. They took their honeymoon in Scotland, calling at Cambridge and Manchester en route. Then Bohr, taking the Rutherford atomic model as a starting-point, began to work out the ideas on atomic stability that were to lead to the quantum description of atomic structure. In 1913 he published these ideas in three articles in the *Philosophical Magazine*, where he set out his bold attempt to combine aspects of classical physics with Planck's concept of the quantum of action, the potential of which remained largely unexploited. This new theory, in which the electrons were restricted with regard to the orbits they could use, yielded impressive quantitative agreement with measurements of atomic spectra. These three articles, known as the *Trilogy*, formed the foundation of Bohr's early reputation. Although it was not immediately accepted by everyone, his concept intrigued his contemporaries and made them aware of the need for a new way of describing events at the atomic level. The

Bohr atomic model, although it has been superseded scientifically, persists today in the minds of many people as a vivid image of what atoms look like.

The association between the exuberant experimenter from New Zealand and the thoughtful young Danish theoretician developed and deepened into a kind of father–son relationship during a quarter-century of friendship and collaboration. In Manchester Bohr found a stimulating atmosphere and congenial colleagues. It was there, in the spring and early summer of 1912, during a period of almost continuous research, that he made several important contributions to atomic physics: he helped to clarify the nature of radioactive transformation, was probably the first person to recognize the basis of nuclear isotopy, and developed a theory of the energy loss of alpha particles as they passed through matter – a topic that interested him all his life.

During the years leading up to the First World War Bohr held some fairly junior appointments at the University of Copenhagen. It was during this period that Courant first met him; he described Bohr as a somewhat introverted, saintly, extremely friendly yet shy young man. Throughout his life Bohr spoke with a quiet voice, hardly above a whisper, and listeners had trouble understanding him in any language. In what he said or wrote he was always conscious of the many limitations and conditions that restrict the validity of any statement: as he liked to express it: 'truth and clarity are complementary'. He was always concerned about hurting any person's feelings. Also he had great difficulty in making definite plans, only too often changing them almost as soon as they had been arranged.

Bohr spent the early years of the First World War in Manchester, working with Rutherford, while hoping to obtain a more senior post in Copenhagen. He and Margrethe enjoyed the carefree life of Manchester very much: 'we have met with so much kindness and feel so much at ease'. Bohr and Rutherford had much in common scientifically; their collaboration was one of the most brilliant, fertile and fortunate in the history of science. Both were capable of enormous enthusiasm for a promising idea in physics. Both refused to be deflected by unimportant details, although they could give painstaking attention to detail when it mattered. Both regarded mathematics as an important tool in formulating and applying the laws of physics, but never as an end in itself. Rutherford was fond of making disparaging remarks about theoreticians who were too attached to formal mathematics, so much so that he is sometimes believed to have been opposed to theory altogether. Bohr was too polite for such remarks, but restricted himself to the minimum of mathematics in his own work. Both were untidy lecturers but could fascinate and stimulate an audience.

In 1916 Bohr heard that a new chair in theoretical physics was being created at his *alma mater* and that he was expected to become the first holder. When Rutherford wrote a letter of recommendation for him it was in no uncertain terms: 'in my opinion Dr Bohr is one of the most promising and able of the younger mathematical physicists in Europe today. I think any university would be fortunate who is able to acquire the services of such an original and fruitful investigator.' So in 1916 Bohr returned to his homeland as the first professor of theoretical physics at the University of Copenhagen. As soon as the war was over, young physicists began flocking to the Danish capital, where research could be pursued in an atmosphere free of politics. Not everyone fell under his spell, however; Einstein, for one, was too independent-minded and too involved in the theory of relativity.

In 1916 also the Bohrs' first child, Christian Alfred, was born. Five more sons followed: Hans Henrik in 1918, Erik in 1919, Aage Niels in 1922, Ernest David in 1924 and Harald in 1928. Hans, in later life, gave a picture of the family milieu in which he grew up: 'father always took an interest in us and from the beginning tried to teach us something about the things he himself liked best and thought important . . . the dinner table was generally a meeting-place at which father was eager to hear what each of us had done, and to relate what he had been doing himself . . . he was no doubt not a teacher in the accepted sense of the word, but if you were patient and listened, a wide and rich perspective opened up . . . he was nearly always occupied with one problem or another.'

Although he was still only thirty-one years old when he returned to Copenhagen, Bohr's subsequent influence was exerted not so much through his own original research as through the way he inspired others, for whom he provided an ideal environment for scientific work. He began to delve into the logical and philosophical foundations of physics. Even his first papers from Copenhagen are often essays in search of verbal understanding rather than mathematical analyses of crucial problems. His lecturing style was a discursive mumble, but with small groups and especially in one-to-one discussions he was unequalled in his enthusiasm, his empathy and his contagious love for the subject. He wrote many scientific and other papers in the course of his life, but never a full-scale book.

Bohr found an invaluable assistant in Hendrik Antonie Kramers, a young physicist who had previously studied with Lorentz and Ehrenfest in Leiden and first came to Copenhagen as a place where science could be pursued in an atmosphere free of politics. The two men worked together, collaborating not only on scientific papers but also on the planning and administration of the new university institute of theoretical physics, later to be named the Niels Bohr Institute. Despite the original name, it was to undertake experimental as well as theoretical research. Within a very few years of its opening in 1921, this unpretentious building in the capital city of a small country was to become one of the best-known centres of physics in the world. The list of visitors over the next two decades – some of whom came for a few weeks, some for months or even years – reads like a roll-call of the founders of quantum mechanics. This intensely interactive enterprise, constantly changing in membership but always led and shaped by the mind and personality of Niels Bohr, came to be known as the Copenhagen school of physics. In his forties, the jovial Bohr became a father-figure to scores of young physicists from all over the world, who liked to call him the Great Dane.

In 1918 Bohr had published his paper 'On the Quantum Theory of Line Spectra', which presented a detailed elaboration of the correspondence principle he had introduced five years earlier. In the skilled hands of Bohr and Kramers, the principle proved a powerful tool for elucidating the fine structure of spectra and for predicting spectral intensities, transition probabilities and selection rules with considerable accuracy; however, it was never fully understood or much exploited by other physicists. Bohr made his next major contribution to physics in 1922 through several papers on the theory of atomic structure and the periodic system of the elements. In the same year he was awarded the Nobel prize in physics 'for his services in the investigation of the structure of atoms and of the radiation emanating from them'. In his acceptance speech, Bohr surveyed the state of quantum theory and the progress that had been made in applying it to the problems of atomic structure, but he took pains to point out the limitations and weaknesses of the theory. He was more acutely aware of these and more perturbed by them than were others who had accepted his ideas less critically.

After the Nobel prize, Bohr received honours from academies, universities and other institutions too numerous to mention. In 1921 Planck wrote to him about the possibility of moving to Berlin, and Rutherford sounded him out about possibilities in England. However, Bohr's attachment to his homeland was too great; all he was prepared to do was offer to make visits. In 1923 he made his first visit to North America, and, although he concluded that he would not like to live there, he was happy to make further visits to the USA and later on take advantage of American philanthropy. Notably the

Rockefeller-financed International Education Board was particularly helpful when he needed more accommodation for visitors to the institute. One
of these visitors was Erwin Schrödinger; according to Werner Heisenberg,

> the discussions between Bohr and Schrödinger began already at the
> railway station in Copenhagen and were continued each day from
> early morning until late at night. Schrödinger stayed at Bohr's house
> and so for this reason alone there could hardly be an interruption in
> the conversations. And although Bohr was otherwise most considerate
> and amiable in his dealings with people, he now appeared to me
> almost as an unrelenting fanatic, who was not prepared to make a
> single concession to his partner in discussion or to tolerate the
> slightest obscurity. It is hardly possible to convey the intensity of
> passion with which the discussions were conducted on both sides, or
> the deep-rooted convictions which one could perceive equally in Bohr
> and in Schrödinger every spoken sentence . . . so the discussion
> continued for many hours throughout the day and night without any
> consensus being reached. After a couple of days Schrödinger fell ill,
> perhaps as the result of the enormous strain. He had to stay in bed
> with a feverish cold. Bohr's wife Margrethe nursed him and brought
> tea and cakes, but Niels Bohr sat by the bedside and spoke earnestly to
> Schrödinger; 'but surely you must realize . . .'

'No real understanding could be expected', said Heisenberg, 'since neither
side was able to offer a complete and coherent interpretation of quantum
mechanics.'

In the years that followed, Bohr continued to publish work in atomic
physics, including one highly controversial paper with Kramers and John
Slater. Although Bohr himself had formulated the first quantization rules for
the emission process of radiation, this paper revealed his continuing disin-
clination to accept the concept of the photon. With the arrival of Heisenberg
and Pauli in Copenhagen, Bohr's role became less that of an initiator in the
progress of quantum physics and more that of a supporter, a mentor and a
penetrating critic of those who were leading the way. His mind returned to
some of the preoccupations of his early years: the physical interpretation
of the mathematical formulations of quantum mechanics, the importance
of seeking to define the boundary between a measuring apparatus and a
measured object, and the place of language in making explicit the outcome
of such measurements. His mature views on these questions were brought
together in the paper 'The Quantum Postulate and the Recent Development

of Atomic Theory', presented at the conference to commemorate the centenary of Volta's death held at Como in September 1927. In his presentation,
Bohr spoke of the epistemological problems of quantum mechanics, and
again set out his complementarity argument – the principle which he was
to continue to develop and extend until the end of his life and which he
came to believe was his main contribution to our understanding of nature.
These ideas were soon put to stringent test in the debate with Einstein
at the Solvay conference of 1927, which took place shortly after the Volta
Commemoration. Compared with Bohr, Einstein's spectrum of scientific
interests was much broader; Bohr concentrated almost entirely on quantum theory and its ramifications. There was another important difference
between them in that Bohr identified very strongly with his native Denmark
and created a major research school in Copenhagen. Although he never
supervised research students, he was stimulated by the endless stream of
visitors and research workers at the institute.

In 1930 Bohr's mother Ellen died; he had lost his father long before,
in 1911. Further distress was caused in 1934 when the Bohrs' first-born son
Christian was drowned when sailing out at sea with his father and some
friends; he was just seventeen years old, not uninterested in science but more
interested in the arts, especially poetry. At Carlsberg, near Copenhagen,
the Carlsberg Foundation owned a beautiful mansion that was given to be
used by 'an outstanding citizen of Denmark, most prominent in science
or literature or the arts'. In 1931 it was offered to Bohr for the rest of his
life, and the Bohrs moved there the next year from their quarters in the
institute, where they had for many years offered hospitality to physicists
from all over the world. The mansion, which was ideal for entertaining,
became not only a family home for the Bohr's children and grandchildren,
but also a haven for the many colleagues and visitors who stayed there and
a convivial meeting place for scientists, artists and politicians from all over
the world. The Rutherfords were the first of many guests who enjoyed their
hospitality.

For the next few years Bohr continued to develop his ideas, both in
physics and in epistemology. He published works dealing with the problem
of measurement in quantum electrodynamics; an essay entitled 'Light and
Life', explaining how the complementarity principle could be applied to
biology, and a paper entitled 'Can the Quantum-mechanical Description
of Reality Be Considered Complete?', a response to the well-known paper
by Einstein, Podolsky and Rosen of the same title. After Bohr and Einstein
first met in 1920, they initiated a series of discussions, which, conducted

with mutual pleasure and respect, continued for many years and became part dialogue, part duel and part crusade. They disagreed on many things, including causality, the meaning of relativity and the incompleteness of quantum-mechanical descriptions. For nearly twenty-five years each tried to convince the other, by ingenious argument and subtle logic. Eventually Bohr's interpretation of quantum mechanics became the orthodox view.

During this period Niels and his brother Harald, who was at that time director of the mathematics institute of the University of Copenhagen (which had been built alongside the institute for theoretical physics) became deeply involved with a Danish group that had been formed to offer support to scientists and other intellectuals who were being forced to flee their homelands by the racial policies of the Nazi government in Germany. These two institutes acted as a temporary refuge for many of them, and the brothers dedicated themselves to finding new posts for those such as Lise Meitner who had suddenly become stateless and unemployed.

In 1934 Bohr made his first visit to the Soviet Union, where he was impressed by what he was shown, and in 1937 he made a world tour including China and Japan as well as Russia. Although they were reluctant to leave their younger sons, especially after the loss of Christian, his wife Margrethe generally accompanied him on these longer journeys. Beginning about 1934, Bohr made his next major contribution through his work in nuclear physics, his theory of the compound nucleus and his elaboration of the liquid-drop model. This work reached its climax in 1939 with a paper, written jointly with John Wheeler, on the theory of nuclear fission, although Bohr continued to work and publish on these topics well into the 1940s. In 1940 Germany occupied Denmark; at first there was a fiction of self-rule, but this did not last. The institute continued to function, but Bohr's own role changed. As a public figure and a focus of national admiration and pride, he felt a responsibility to help maintain Danish science and culture under the prevailing conditions. It was during this period that he was asked to write the introduction to the book *Danish Culture in the Year 1940* – a task on which he lavished the same care and attention to detail and language as that which he devoted to his scientific writings.

By August 1943 the position of the Danish Jews had become perilous, and the following month Hitler ordered that they were all to be rounded up and deported in two freighters that had docked in Copenhagen. Some were caught, but the great majority escaped to safety in Sweden in the greatest mass-rescue operation of the war. Bohr himself left Denmark for Sweden in this way and spent some time there making sure that the refugees would be well treated. However, crossing the Kattegat was for him only the first stage in a journey to Britain, arranged by the British government. His son Aage, also a physicist, followed a few days later and went on with him to America. In London Bohr was briefed on what was then known as the Directorate of Tube Alloys, the precursor of the Anglo-American Manhattan project to produce atomic bombs, which came as a revelation to him. He agreed to work on it, although he remarked later that 'they did not need my help in making the atom bomb'.

Once in the USA he ran into security problems because of his firm belief that the Soviet Union should be made aware of what was going on at Los Alamos. Although he was scientifically interested in the progress towards the production of the fission bomb, Bohr turned his attention almost immediately to the political significance of the project and to the need for early and clear recognition of the threat it would pose to post-war stability. According to Oppenheimer, 'Bohr at Los Alamos was marvellous. He took a very lively technical interest. But his real function, I think for all of us, was not a technical one. He made the enterprise seem hopeful, when many were not free of misgiving . . . His own high hope that the outcome would be good, that the objectivity, the cooperation, of the scientists would play a helpful part, we all wanted to believe.'

In the spring of 1944 Bohr returned to London, in the hope of conveying his concerns to Winston Churchill. With some misgiving Lindemann, Churchill's scientific adviser, arranged for Bohr to meet the Prime Minister, but Bohr never succeeded in getting his message across. He was more successful later when, on his return to America, he had an audience with President Roosevelt, who was impressed by him at the time but afterwards had second thoughts. A few months later, after Churchill and Roosevelt had met privately at the latter's Hyde Park estate in the Hudson Valley, Churchill sent a note to Lindemann: 'the President and I are much worried about Professor Bohr. How did he come into this business? He is a great advocate of publicity. He says that he is in close correspondence with a Russian professor [Kapitza], an old friend of his in Russia, to whom he had written about the matter and may be writing still. The Russian professor has urged him to go to Russia in order to discuss matters. What is all this about? It seems to me that Bohr ought to be confined or at any rate made to see that he is very near the edge of mortal crimes . . . I do not like it at all.'

In August 1945, a few days after the bombing of Hiroshima, the editorial page of *The Times* newspaper carried an influential article by Bohr entitled 'Energy from the Atom', including the following passage:

The formidable power of destruction which has come within reach of
man may become a mortal menace unless human society can adjust
itself to the exigencies of the situation. Civilisation is presented with
a challenge more serious perhaps than ever before . . . we have reached
the stage where the degree of security offered to the citizens of a
nation by collective defence measures is entirely insufficient . . . no
control can be effective without free access to full scientific
information and the granting of the opportunity of international
supervision of all undertakings which, unless regulated, might become
a source of disaster . . . the contribution which an agreement about
this vital matter would make . . . can hardly be exaggerated.

In the autumn of 1945 Bohr returned to Copenhagen and once again
took up his role as honoured teacher and stimulating friend of a new generation of physicists. He devoted much time and thought to promoting a sane
and realistic policy for nuclear armaments and, in 1950, in an open letter to
the United Nations, he made a heartfelt plea for world cooperation. He and
Kramers, who had succeeded Ehrenfest at Leiden and was then chairman of
a United Nations committee on nuclear policy, both worked for peace, but
their efforts were overtaken by events and Bohr's proposal for openness and
free exchange of information was never tried out.

In Denmark Bohr had become a much respected elder statesman and
was called upon to guide the government's policy on atomic energy. For
the next ten years much of his time was given to the detailed planning
and completion of the research establishment of the Danish Atomic Energy
commission at Risø. He lent his support to the plan for the establishment
of a centre for European cooperation in science. In the early and delicate
days of the organization, the division of theoretical physics was housed
and grew strong in Copenhagen before moving to Geneva and becoming
CERN. In 1955 he reached the mandatory retirement age for a university
professor and was succeeded in his chair by his Nobel laureate son Aage;
but he continued as director of the institute. Two years before he had lost
his brother and closest confidante, the mathematician Harald.

At the beginning of the 1960s, Bohr and his colleagues began to plan for
a meeting in 1963 to celebrate the fiftieth anniversary of the publication of
the original research on atomic structure. They hoped to renew the intimate
atmosphere of the interwar Copenhagen meetings by inviting back members
of the institute from those years and giving them the opportunity to review
the past half-century of progress and to speculate about what the future

might hold. However, before this could happen Niels Bohr died suddenly in his seventy-seventh year at his home in Carlsberg, on the afternoon of November 18, 1962; the cause was given as heart failure. Nevertheless, the planned reunion was held, and many of the surviving members of the Copenhagen School returned to exchange their latest opinions and ideas, as he would have wished. Afterwards, at the invitation of Margrethe Bohr and her sons, they gathered once more in the beautiful mansion which they had come to know so well. She survived him by twenty-two years.

FREDERICK LINDEMANN (LORD CHERWELL) (1886–1957)

The arrival in the world of Frederick Alexander Lindemann on April 5, 1886 at the German spa of Baden-Baden provided him with a life-long grievance against his parents, since a German birth-place caused much annoyance to someone who became in so many ways a member of the British establishment. His mother Olga, a forceful and handsome American, was born in New London, Connecticut, in 1851; her British-born father was a successful civil engineer. She was the widow of a wealthy banker and had three children, two girls and a boy, when she married Adolphus Frederick Lindemann, by whom she had four more children, three boys and a girl. The Lindemanns were Catholics from Langenberg in Alsace. He was a clever man who made

a fortune as an entrepreneur concerned with the making and laying of the early Atlantic cables and with the construction of waterworks in Bavaria. After the Franco-Prussian war, he settled in England, became naturalized and remained there for the rest of his life.

All three Lindemann sons were gifted. Charles, the eldest, was a physicist whose early promise was comparable to that of his brother Frederick, but who gave up science for a military career; later he led a life of wealth and ease in the upper ranks of society. The youngest son, James Septimus, turned into a devil-may-care spendthrift who resided in a villa on the French Riviera, overlooking the harbour at Villefranche. He was an excellent linguist of irresistible charm, an *habitué* of the Paris nightclubs and a sore trial to his brothers; Frederick, who thoroughly disapproved of him, was amazed when he heard that his wastrel brother had been a British agent in occupied France during the Second World War. Their sister married twice. Her first husband was killed in the First World War; her second was so disapproved of by Frederick that he had little further to do with her.

The family lived in 'Sidholme', a large Victorian mansion in the town of Sidmouth on the Devonshire coast; this belonged to their mother, having been bought by her first husband, and was sold when she died in 1927; there was considerable ill-feeling about this at the time. To the house, built in 1826 by the Earl of Buckingham, had been added a well-equipped scientific laboratory and an observatory, for their father had a serious interest in astronomy. There was music and tennis and much else to amuse and instruct the children, whose early education was entrusted to tutors. Frederick first went to school at the age of thirteen. The school, of no great distinction, was run by a friend of his father's. It specialized in preparing boys for a career in the army; since it was situated in the Scottish Highlands, Frederick was expected to wear a kilt, which he hated. He learned to play golf, but later gave it up because he believed that it was bad for his tennis, the game at which he excelled. He already showed striking mathematical ability and an extraordinary gift for lightning calculation. Throughout his life he could recite huge quantities of figures, entirely without interest, just for amusement. From an early age he had an aversion for black people amounting to phobia; he attributed this to a particular doll he had been given in his infancy.

After three years in Scotland, his parents decided that Frederick should continue his education in Germany, specifically in Darmstadt, the seat of the Grand Duke Ernst Ludwig, a grandson of Queen Victoria. He began at the admirable Lyceum and then went on, at the age of eighteen, to the

Technische Hochschule, where he was joined by his brother Charles. The Grand Duke took an interest in the young Lindemanns, encouraging them in the game of tennis, of which Frederick became an exceptionally fine player. For some reason the brothers challenged another man to a duel, not just the ritual exercise popular among young Germans but a serious contest with loaded pistols: fortunately their opponent apologised when he realized that they were in earnest.

After Darmstadt the brothers moved to Berlin, where Frederick took his Ph.D. under Nernst in 1910. It was an exciting time, when Planck and Einstein were creating the new physics. Observing the shy genius of Einstein at close quarters, Lindemann concluded that the towering intellect was combined with a pathetic naivety in the ordinary affairs of life. When the brothers first came to Berlin they brought with them a cook-housekeeper to run their bachelor establishment, where young ladies were often entertained and spent the night, but later they moved to the luxurious Adlon Hotel. One of Lindemann's fellow students recalled, perhaps a little inaccurately:

> I worked in Professor Nernst's laboratory in the Bunsenstrasse when Lindemann was there. He was recognized then as the most distinguished of the researchers working for the Ph.D. He had good ideas in physics; he was a good mathematician as well as a good experimenter. He also had a flair for designing instruments. We all started work like good little boys at 9 a.m. Lindemann rarely appeared before noon, and rarely stayed after four but he was known to be a great night worker. He was always well-dressed, often in morning coat. No-one ever saw him with his coat off working, so to speak at the bench; he directed others in his research. He was then very thin, a tall maypole of a man with dark shiny hair. He spoke German softly in a thin voice with an English accent. Most of us lived in cheap 'digs' in Charlottenberg, ten miles away. Lindemann lived in one or other of the hotels in Unter den Linden, less than a quarter-mile away. His staying at the Adlon hotel there impressed the German element in the laboratory enormously. He kept aloof from the English and American group in the laboratory as well as from the German but without giving offence. He was always pleasant and helpful when approached but never joined in any extra-laboratory activity. Then as later he was self-indulgent over food, patronising the best restaurants in the Unter den Linden where the staff knew that although he was extremely particular about his meals he liked to have exactly the same thing

served, day after day. He had acquired these peculiarities as a small child from his mother; during wartime when food supplies became scarce these self-imposed restrictions may have damaged his health.

On reading this, his brother Charles remarked that Frederick never wore a morning coat except on formal occasions. As for the Adlon hotel, he had been brought up in an atmosphere of wealth and comfortable living, and throughout his life rarely got to work before lunch-time. Nernst gave formal dinner parties: on one occasion the Germans present claimed that they knew Shakespeare better than did the English, to which Lindemann retorted that the English knew Goethe better than did the Germans, reciting a few lines of verse that he had just made up to illustrate the point. Another colleague was struck by his skill in glass-blowing, which in Germany was left to technicians.

The research which went into his thesis was published in a joint paper with Nernst. It provided an effective test of the theory which Einstein had developed for describing how the amount of energy stored in a solid would increase with increasing temperature. The Lindemann–Nernst theory to account for this was soon superseded, but at this time quantum theory was still in a state of flux. In 1911 a signal honour was paid to him when he was invited with his friend the duc de Broglie (Maurice, not Louis) to act as co-secretary to the Solvay conference in Brussels. Throughout the conference he impressed the eminent participants, the leading physicists of the time, with his ability. Then in 1913 he was invited by Millikan to give a graduate course at the University of Chicago. He accepted and lectured on kinetic theories.

After Lindemann had returned from America, just before the beginning of the First World War, he joined the Royal Aircraft Factory at Farnborough. This unique institution had evolved from an establishment concerned with the military use of balloons but by this time was studying the new science of aerodynamics. Those who worked there were remarkable people. By a process of improvization that has since become legendary, they were to design and construct aeroplanes capable of remaining in the air for hours at a time and to devise the instruments by means of which they were navigated and controlled. Lindemann believed that the scientists at Farnborough ought to know how to fly. Despite his poor eyesight, he gained a pilot's certificate, after minimal training, and was able to put his ideas about flight to the test. He was accustomed to enter and leave his plane with bowler hat and umbrella. Because of his birth-place the ground staff firmly believed that he was German and would not allow him enough fuel to fly across the English Channel.

In the early days of aviation, an aircraft that went into a tailspin was generally doomed. Lindemann worked out a theoretical solution to the problem and was determined to try it out in practice. He did so successfully, although there is some doubt as to whether he was the first to do so. In his own words:

> in 1916 many pilots were killed flying our recently designed planes by spinning into the ground. Although various people had succeeded in getting out of a spin, nobody quite knew how, nor indeed how or why aircraft spun at all. Anyone watching a spinning plane could see that the rate of turn did not increase on the way down. I concluded therefore that the lift on both wings must be equal; and this could only be true – since the outer wing is beating against the air whereas the inner is not – if its effective angle of incidence was on the high side of the angle of maximum lift, whereas for the inner wing it was the other way round. This being so, if the speed were increased the aeroplane would no longer spin. Experiments proved this idea was correct . . . therefore pilots were taught to push the stick forward – the very opposite of the instinctive reaction of pulling back in order to get the nose up – and to straighten out the rudder and then pull out of the dive in the ordinary way.

This was by no means his only contribution to the war effort. For example, he also designed a bomb sight and, characteristically, demonstrated it himself. At the end of the war he could look back on his work at Farnborough with satisfaction. His foreign education ended, the love of England which had been nourished in boyhood by the hills and lanes of Devonshire, and which was to burn with so fierce a flame for the rest of his life, now possessed him, becoming the focus of his loyalty and devotion.

The study of physics in Oxford, or experimental philosophy as it was then called, began in 1749; by 1860 there was a professorship in the subject. Lindemann was elected to this in 1919 and as a result became head of the Clarendon Laboratory, then in something of a decline. The chair was attached to Wadham College, where he became a fellow and member of the governing body, but the set of rooms he was offered there was unmodernized, without even a bathroom and lavatory, and his offer to install these at his own expense was turned down. Before long he migrated to Christ Church, the college which had endowed his professorship, where he became a

Student (i.e. Fellow) and was provided with a more comfortable set of rooms. He was not a member of the Christ Church governing body, but regularly attended the one at Wadham, which he helped to turn into a college noted for science.

Lindemann was already generally known as 'the Prof', although his earlier nickname of 'Peach' lingered on. As a lecturer he was not a success, because he spoke in such a low voice that he was difficult to hear. Already he had many enemies, who thought of him as a once-great physicist who had deserted the profession in which he had been trained and for which he was so eminently gifted. Yet, despite being a wealthy man, he was determined to continue his scientific career; if he had not secured the Oxford post he intended to try for one in America. He was elected a fellow of the Royal Society in 1920, at the early age of thirty-three.

At Oxford a great deal of his time and energy was expended on building up the Clarendon Laboratory to one comparable with the Cavendish at Cambridge. The peculiar history of the Clarendon was as follows. The first earl of Clarendon was a great Restoration statesman and historian, the author of a history of the Civil War. In 1751 his grandson, Lord Cornbury, left some of the first earl's papers to trustees, with the direction that the money derived from their sale or publication should be used for the benefit of the University of Oxford. After other proposals had been considered and rejected, including one for a riding school and another for a swimming pool, the trustees had been persuaded that it was high time that there was a laboratory for physical science.

The Clarendon Laboratory was built from the funds available and opened its doors in 1872. It was the first purpose-built physics laboratory at an English university and its design gave it a curiously ecclesiastical appearance. In 1900 a new professorship, called the Wykeham because it was partly funded by New College, was established in physics; and ten years later an electrical laboratory was built. The arrangement was that the professor of experimental philosophy, at the Clarendon, would take responsibility for mechanics, heat, light and sound, while the professor of physics, at the new laboratory, would take responsibility for electricity and magnetism. The separation of responsibility for experimental physics between two laboratories and heads of department led to bitter rivalries, which lasted until the laboratories were merged in 1945.

When Lindemann was appointed he was unaccustomed to the peculiarities of Oxford. His predecessor had been entirely opposed to research; there was no staff at the Clarendon and hardly any apparatus for carrying out experiments, except for an extravagant provision of the best obtainable optical apparatus. The laboratory was badly in need of modernization; the water supply was inadequate, for example, and there was no electricity supply at all. When Lindemann took over, his first task was to change this deplorable state of affairs. It says much for his perseverance, his skill in choosing men and his political ability, and perhaps not a little for the broad-mindedness of Oxford, that, long before he retired, the New Clarendon Laboratory which he had persuaded the university to build was comparable in importance to any department of physics in England. However, he did not achieve this in the normal way by setting up a school based on his own particular line of work and achieving pre-eminence in it. Being wealthy himself, he was good at raising money for his laboratory; and he had his own way of finding talented staff, as we shall see. His friend Max Born has given us some insight into his character:

> I have often pondered what feature of his character prohibited this versatile mind from producing a great new concept of physics, comparable with the quantum of action. Planck stuck to classical concepts as long as possible, but when he had convinced himself that the facts of observation could not be explained in the frame of classical theory he was just as decided to develop new and strange ideas. He was a revolutionary not by character but through his willingness to acknowledge the power of evidence. Lindemann, although conservative in many respects, had a natural revolutionary strain which found its outlet in physical theory. He had little respect for traditional thinking and as soon as some new facts appeared not to fit in within current theory he jumped to conclusions about fundamental assumptions without analysing the evidence in detail. His occupation with other things, the administration of his laboratory, university affairs and politics, allowed him little time for following up his ideas and for looking in the physical literature for confirmation.

Lindemann, who was not himself Jewish, appeared mildly anti-Semitic at times, but he was a key figure in the efforts made to rescue Jewish scientists from the Nazis. Although he had no love for the Jews as such, he deplored the way they were being treated in Germany. The kindness Lindemann showed towards Jewish *émigrés*, and the fact that he became a close personal friend of some of them, shows that his superficially hostile attitude towards Jews was never a deep one. He was one of the first to realize that, owing to the folly and brutality of Hitler, these brilliant and highly trained men and women were now available to work in Britain. He wasted no time, going at once to Germany to visit them with a view to inviting some of them to continue their research work in his laboratory; there was, he thought, room for at least six of them to continue their work at Oxford and to assist in the development of the Clarendon. He persuaded Imperial Chemical Industries to provide thirteen two-year grants for Jewish refugees. Other firms began to follow this lead, so that there were funds to pay their salaries.

To return to Max Born's reminiscences:

> In the spring of 1933 I was compelled to leave Germany as a consequence of Hitler's accession to power. We spent the summer in Selva, Val Gardena, South Tyrol, and here Lindemann appeared with his car and chauffeur, to discuss the political situation with me, in particular the fate of the numerous scientists who lost their position in Germany. He explained to me his plan to improve the situation of science in Oxford by inviting refugee physicists to the Clarendon Laboratory. I was not available for the project because Cambridge had already offered me a post. But I was deeply impressed by Lindemann's idea which was not only generous to the homeless scientists but clever and far-sighted in regard to the future of Oxford.

'He did not shrink from travelling far and wide through Europe to achieve his aim, and thus he found me in Selva', Born continued, 'it is known how he made the Clarendon one of the centres of physics in Great Britain by recruiting Simon and his group, Kurti, Mendelssohn, Kuhn and others, providing them with decent positions, excellent working conditions and – last but not least – encouraging them through a deep understanding of their projects – although he practically gave up physical research he was always amazingly up-to-date, not only about the work done in the Clarendon, but elsewhere in the world.'

Lindemann's initiative in securing the German scientists was sometimes wrongly attributed to a selfish desire to increase the prestige of Oxford science and his own Clarendon Laboratory in particular. In fact, he did excellent work in placing distinguished German Jewish scientists in other English universities besides Oxford and in obtaining funding where necessary. He insisted that the object was not philanthropy but the promotion of scientific research in Britain.

Sir Cyril Hinshelwood, sometime president of the Royal Society, summed up Lindemann's scientific work in the interwar years as follows:

Before he came to Oxford Lindemann's most important contribution
had on the whole been theoretical rather than experimental, he was
one of the most brilliant analytical minds I have ever known, and he
continued throughout life to take a deep interest in the fundamentals
of science. His views on all matters of theory were always worth
hearing. He was always interested in the work of others, and although
his criticism at times tended to be destructive, he was always fertile in
suggestions about interpretations which at their best could be
intensely illuminating.

He had been a brilliant young man moving in the most
distinguished European scientific circles, and he came, still young,
into a university where science was still looked down upon by the
dominant schools. He found himself playing two very important and
stimulating roles. On the one hand he was something of an oracle in
scientific circles, and, on the other hand, he began the rehabilitation of
science among those people who were forced to respect his brilliance
of mind and to recognize in this rather glamorous continental figure
someone very different from the image they liked to make of the man
of science. Add to this Lindemann's charm and mondain predeliction,
and it is hard not to see him devoting a great deal of his energy to the
playing of a kind of Socratic role.

As far as physics (and chemistry) went this was by no means a
dereliction because his analysis often was of the greatest help to other
people, and I would say that he made a great and real contribution to
the modern rise of scientific studies in Oxford. He continued to think
deeply about the major problems of his subject. He did not solve them;
they have not been solved yet. And we must bear in mind that
[Lindemann] was someone who would have scorned to publish
anything in the nature of a pot-boiler, even to use that word in its least
pejorative sense. He stimulated many things for which he never
claimed credit . . . Some people have said that he preferred social
success to science . . . he preferred it to anything but the finest vintage
of science, and because it was not given to him to solve the deepest
problems he was not much interested in the lesser things. Perhaps to
all this one must add a certain element of indolence.

At Christ Church it was a long time before Lindemann gained general
acceptance in the Senior Common Room. He assailed the philosophers with
instinctive relish and loved to tease the historians with questions about

events of which he had expert knowledge but they did not. In the last twelve years of his life he had become a character, the most popular member of the Room, whose appearances were eagerly awaited. As one of his friends wrote:

> If we had some interesting guest dining, our eyes strayed over to the door at about 9.15 hoping that he would appear and add new life to our party. We were usually not disappointed. The door opened slowly and his well-known figure appeared. He walked with measured tread, for he was already ailing. He took off the heavy greatcoat, and placed it methodically on the table and the bowler hat on top of it. Then he came forward to join us, ready and anxious to be interested in the affairs of whoever was there, attentive, quite unassuming, and full of jests and anecdotes appropriate to the person. Or if there was no-one who especially wanted to talk to him, and a bridge table was out, he liked to stand behind and watch the game. He knew all about it but very rarely interjected a comment. Over and over again we tried to persuade him to play himself, but he declined to do so on the ground that, if he made a mistake, that would cause him to lie awake all night replaying the hand.

His college rooms were on the first floor of Meadow Buildings. Outside his windows the flat tranquil meadow stretched down to the river, where the painted barges then lined the bank. His set consisted of a sitting room, dining room and bedroom, and a spare room; he had installed a bathroom and lavatory. Although he lived in great comfort, he had little taste. His personal needs were attended to by his factotum Harvey, or Harvey's assistant, and his cars were normally driven by a chauffeur. When forced to take the wheel himself, he did so reluctantly and was an execrable driver, at once timid and dangerous, and in a state of sustained tension that made him abusive to other motorists. On the continent, he travelled in patrician comfort, his progress resembling that of some English milord in the eighteenth century. He loathed the English winter and usually managed to escape to somewhere warmer.

Lindemann was a welcome guest at many of the grand houses of England and as remote from the proletariat as an aristocrat of the French *ancien régime*. He was frequently at Blenheim Palace, the seat of the duke of Marlborough and conveniently near Oxford. Blenheim was the birth-place of Winston Churchill, whom Lindemann had first met at the end of the First World War. He became better acquainted with Churchill in 1932, when he accompanied him with others to follow the route of Marlborough's famous march from the Netherlands to the Danube, and then three years later when he joined Churchill on a cruise along the east coast of Spain to Tangier. Later they became close friends.

Long before most people, Lindemann foresaw the coming war. He made a vain attempt in September 1933 to meet the dictators Hitler and Mussolini. He stood for Parliament as a university member in 1935, but was not elected. He was one of the minority pressing for Britain to rearm and particularly to strengthen its defences against air attack, but there were different views regarding what, if anything, should be done A major feud developed between Lindemann, backed by Churchill, and Tizard, a scientific czar who was more influential with the service chiefs. Both scientists had studied with Nernst and worked at Farnborough. From being a strong supporter of Lindemann, whom he thought the cleverest man he had ever known, as clever as Rutherford, Tizard turned against him. The abilities of both scientists and their former friendship made the pettiness with which they conducted their feud depressing to everyone else; there were faults on both sides, but it was one of those situations in which Lindemann's German birth was held against him. In 1937 he stood for Parliament again in a by-election but was not elected. At the outbreak of the Second World War Lindemann entered public life at a high level when Churchill, as First Lord of the Admiralty, appointed him his personal adviser; his role was primarily but not exclusively a scientific one. When Churchill became Prime Minister, he continued in the role, and in 1941 he was raised to the peerage; he chose for his title the name of Cherwell, the tributary which joins the river Thames at Oxford.

The next year he was appointed to membership of the Privy Council and given a seat in the cabinet as Paymaster-General, an office with no specific duties. After the crushing defeat of the Conservative Party in the general election of 1945, Cherwell, as we should now call him, returned to Christ Church with relief. However, Churchill asked him to continue as a member of the shadow cabinet, and in the House of Lords he often spoke about economic and scientific affairs. Like his scientific lectures at Oxford, his speeches were generally inaudible, but they could be read afterwards. At the Clarendon, which he had done so much to build up, he was still titular head but not greatly involved in its activities. In 1951, when Churchill became Prime Minister once more, the reluctant Cherwell was persuaded to take office again and to base himself next door to 10 Downing Street so as to be within call. However, like Churchill himself, he was past his prime, and his influence was waning. In 1953 he was made a Companion of Honour and three years later elevated to a viscountcy.

Cherwell, at the age of seventy, suffered from mild heart trouble and from diabetes, which was not ameliorated by the peculiar dietary regime he followed, but otherwise he enjoyed reasonably good health. However, on July 2, 1957 he had a heart attack and early on the following day he died from a coronary thrombosis in his college rooms. The funeral service was held in the cathedral; among those present was the estranged sister he had not spoken to for forty years. Being tall and saturnine, he would have made an arresting figure in any society. Owing to his sardonic manner, the first impression Cherwell made was often unfavourable. This was partly the result of a caustic tongue and too sharp an understanding of the weaknesses of others. People were black or white to him; no greys were allowed. In a sense he was a lonely man, but there was no sign that this was distasteful to him.

Erwin Schrödinger (1887–1961)

Lindemann's efforts to recruit scientists who wanted to leave Nazi Germany did not always meet with success. The case of Erwin Schrödinger is in many respects unique. The only child of Rudolf and Georgine Schrödinger, he was

born in Vienna on August 12, 1887. The families of both parents had lived in the city for three or four generations. His mother, whose maiden name was Bauer, was the daughter of an able analytical chemist, who became professor of general chemistry at the polytechnic. Erwin's father Rudolf, whose family had originally been Bavarian, had studied under the maternal grandfather before inheriting a small but profitable business manufacturing linoleum and oilcloth, which he carried on without much ability or enthusiasm, preferring to spend his time painting. After a sheltered upbringing, Erwin's cheerful and undemanding mother, sickly by nature, became rather helpless in the face of life's problems. As for his father, Erwin later described him as a man of broad culture, a friend, teacher and inexhaustible partner in conversation, to whom he was always grateful for giving him a comfortable upbringing and a good education.

The boy entered the *akademisches Gymnasium* in the autumn of 1898, having just turned eleven, and, after the usual course of humanistic studies, including a little mathematics, matriculated at the University of Vienna in the autumn of 1906, bringing with him from school the reputation of being an outstanding student. Although he did not keep himself aloof, all the other students regarded him as something special. Later opinion was divided between those who considered him to be a person of the most amazing modesty and those (the majority) who thought that he was one of the most conceited men they had ever met. The main focus of his interests was the course in theoretical physics given by Friedrich Hasenöhrl, the successor of Boltzmann, who was youthful, full of energy and a brilliant lecturer. Hasenöhrl lectured neither from notes nor from memory, but simply relied on the strong logic of the science and developed it as he went along. He often invited groups of students to his house, where his beautiful wife Ella presided. He was an able mountaineer and expert at skiing and other winter sports.

Schrödinger became a Privatdozent in 1910 and the next year, after performing the obligatory military service, he was appointed to an assistantship in experimental physics, in charge of the large practical class for freshmen. He discovered that he was not cut out to be an experimentalist and in any case the university was not properly equipped for experimental work. Nevertheless, he had no regrets about the experience. At this time there were few good academic opportunities in Austrian physics. There were more in Germany, but, as Schrödinger fully realized, applicants must already have achieved something quite special; standards were higher than in Austria.

The theoretical papers he submitted to his Habilitation committee were one on the kinetic theory of magnetism and another on the kinetics of dielectrics. Although the work was accepted, with some dissent, he was allowed to proceed to the final stage by giving a lecture on the magnetron. Although he had cleared the first hurdle, Schrödinger's prospects of an academic career were not good, so he could not afford to get married. He had a serious affair with the daughter Felicie of some friends of his parents named Kraus, but her parents were opposed to the match, perhaps because he was not of the right social standing. After this rebuff he tended to have affairs with women of a lower social class.

There was an important international scientific congress in Vienna in 1913, the eighty-fifth meeting of German scientists and physicians, with over 7000 participants, at which Einstein, already widely recognized as a great theoretical physicist, lectured on 'The present status of the problem of gravitation'. Schrödinger was deeply impressed and, like Einstein himself, became attracted by the idea of finding a unified field theory, including both electromagnetic theory and gravitation. The next year he sent to the *Annalen der Physik* the most significant of his early publications, 'On the Dynamics of Elastically Coupled Point Systems', which harks back to one of the persistent themes in the scientific work of Boltzmann. In this paper we encounter for the first time the authentic Schrödinger style, with its urbane confidence and its ability to relate the question in hand to deeper philosophical concerns of mathematical physics.

Schrödinger began his career in physics during the last peaceful years of the Danubian monarchy; the question was not whether but when it would disintegrate. At the beginning of the conflict which soon turned into the First World War he was called up for active service. He first learned of Einstein's general theory of relativity when he was stationed at the front and recognized its great importance at once. In the field he was unable to keep up with the scientific literature, but in the spring of 1917 he was transferred to Vienna and able to start scientific work again. Active service had not dulled his theoretical skills, but neither had it led to an outburst of original thinking about the deepest problems of physics.

After the armistice in 1918 and the disintegration of the empire, conditions in Vienna became appalling. However, Schrödinger was appointed assistant at the University of Jena and this enabled him to get married. His bride Annemarie Bertel was twenty-three, while he was thirty-two. Her fervent admiration of everything about her husband was one of her great attractions for him; she had little education and was no intellectual. He treated her as a sort of superior domestic servant, whose main function was to provide him with a comfortable home. There were no children to the marriage. After his interest in her as a sexual partner had disappeared, they remained friends and she even helped him to find other women, while she became interested in other men.

The Schrödingers arrived in Jena, a pleasant German city of about 70 000 inhabitants, and received a warm welcome. The university, founded in the mid-sixteenth century, had gained a reputation for scientific research, thanks largely to the presence of the large Zeiss optical works nearby. He gave his inaugural lecture on recent developments in atomic theory and made such a favourable impression that he was promoted to associate professor almost at once. However, the post was not permanent and, when the opportunity to move to the Technische Hochschule of Stuttgart as an associate professor with tenure arose, he did so. During their Stuttgart period his mother died of breast cancer; due to hyperinflation she had become almost destitute after his father died at the end of 1919.

Although Schrödinger had been passed over for full professorships in Austria, offers arrived from major German universities, and of these he chose Breslau. Thus the Schrödingers made their third move in eighteen months. A few months later they moved again, this time to the University of Zürich. At the age of thirty-four he had achieved a full professorship at a leading university, despite the fact that he had not yet accomplished a truly outstanding piece of work in any particular field. Neutral Switzerland had been spared the ravages of war, and living conditions were much better than in Austria or Germany, although there was high unemployment and general economic depression. Almost as soon as he had arrived Schrödinger was diagnosed with suspected tuberculosis and sent to an alpine sanatorium in Arosa to recover. It was there that, stimulated by Weyl's influential book *Raum–Zeit–Materie* (*Space, Time and Matter*), he wrote one of his most important papers, 'On a Remarkable Property of the Quantized Orbits of an Electron'.

When he finally took up his professorial duties, Schrödinger found that he had a heavy work-load. A student who attended one of his lectures at this period recalled that it was

> extremely stimulating and impressive. At the beginning he stated the subject and then gave a review of how one had to approach it, and then he started exposing the basis in mathematical terms and developed it in front of our eyes. Sometimes he would stop and with a shy smile

confess that he had missed a bifurcation in his mathematical development, turn back to the critical point and start all over again. This was fascinating to watch and we all learned a great deal by following his calculations, which he developed without ever looking at his own notes, except at the end, when he compared his work on the blackboard with his notes and said 'this is correct'. In summertime when it was warm enough we went to the bathing beach on the Lake of Zürich, sat with our own notes on the grass and watched this lean man in bathing trunks writing his calculations before us on an improvised blackboard which we had brought along. At that time few people came to the bathing beach in the morning and those that did watched us from a discreet distance and wondered what that man was writing on the blackboard!

While Schrödinger's physical health was returning to normal, emotionally it was a difficult time. He and Annemarie were having problems with their marriage, which was at a high point of disagreement and tension, with constant talk of divorce. However, for his work in theoretical physics, 1925 proved to be a marvellous year. Schrödinger was by now particularly concerned with fundamental problems in atomic physics and quantum theory, especially the nature of radiation and how it interacts with electrons and atoms. These new interests brought him into closer relation with work in progress in the schools of Born in Göttingen, Sommerfeld in Munich and especially Bohr in Copenhagen, all of whom had close connections with each other.

He invited 'an old girl-friend in Vienna' to join him in Arosa while his wife remained at home in Zürich. This former girl-friend might just possibly have been his first love Felicie, who had an infant daughter, although her husband had now left her. Whoever it may have been, the effect on Schrödinger's creative powers was dramatic, and he began a twelve-month period of sustained creative activity; Weyl once said that Schrödinger did his great work during a late erotic outburst in his life, a striking example of the association in some people between sexual activity and scientific discovery, something Schrödinger himself was quite open about.

When he was enthralled by an important problem, Schrödinger was able to achieve intense and absolute concentration, bringing to bear all his great mathematical powers. Soon after his return to Zürich he had found the relativistic wave equation to which his name is attached. When asked whether he had enjoyed the skiing at Arosa, he said that he had been distracted by a few calculations. If he had taken with him a copy of *Methods of Mathematical Physics* by Courant and Hilbert, of which the first volume had just been published, he might have found in it the kind of mathematics he needed. Fortunately Weyl was at hand and could be consulted in person.

The original inspiration for the new theory came from the young French physicist Louis de Broglie, whose profile occurs later. Schrödinger published it, together with applications, in a famous series of papers in the *Annalen*. When Planck received a copy of the first one, he wrote that he had read it 'like an eager child hearing the solution to a riddle that had plagued him for a long time'; and after the second, 'you can imagine with what interest and enthusiasm I plunged into the study of this epoch-making work'. He showed the papers to his colleague Einstein, who wrote 'the idea of your work springs from pure genius'. A rival theory had been developed by Bohr's disciple Heisenberg, and a protracted argument developed between them; there was a good deal of controversy, but Schrödinger had earned the professional and personal esteem of the mandarins of German physics. It was at this point that he went to visit Bohr in Copenhagen. Although deeply impressed by Bohr personally, he was not persuaded by his arguments, as we have seen. Next Schrödinger, who spoke English well, took up an invitation to cross the Atlantic and give a course at the University of Wisconsin. He went via New York and Chicago, both of which he disliked intensely, but he liked Madison very much. The lectures he gave there were well received; he turned down the offer of a permanent position on the faculty. Afterwards he went on a tour that took him to Pasadena, where Millikan and Lorentz, who happened to be visiting, sat in the front row of his lecture. On the way back he stopped in Baltimore, where he was offered a position at Johns Hopkins, but again he declined, because he had heard from Germany that he might be invited to move to Berlin in succession to Planck, who was retiring as head of physics at the Kaiser Wilhelm Institute. The testimonial drawn up for this purpose provides a useful summary of his achievements at that date:

> For some years already he has been favourably known through his versatile, vigorously powerful, and at the same time very profound style in seeking new physical problems that interested him and illuminating them through deep and original ideas, with the entire set of techniques which mathematical and physical methods at present provide. He has proved this method of working to be effective in the treatment of problems in statistical mechanics, the analysis of optical

interference, and the physiological theory of colour vision. Recently he has succeeded in an especially daring design through his ingenious idea for the solution of the former particle mechanics by means of wave mechanics in the differential equation he has set up for the wave function . . . Schrödinger himself has already been able to deduce many consequences from this fortunate discovery, and the new ideas that he has inspired with it in many fields are even more numerous . . . it may be added that in lecturing as in discussions Schrödinger has a superb style, marked by simplicity and precision, the impressiveness of which is further emphasized by the temperament of a South German.

Of the three being considered for the Berlin chair, Sommerfeld was preferred to Schrödinger, who in turn was preferred to Born. When Sommerfeld declined to leave Munich, Schrödinger was offered the post. Every effort was made to persuade him to stay in Zürich. The physics students organized a torchlight parade around the university to the courtyard of his house, where they presented him with a petition. Schrödinger was deeply moved, but in the end it was a personal appeal from Planck that persuaded him to accept the Berlin offer; as the result of doing so he automatically became a German national.

Before beginning his duties in Berlin, Schrödinger travelled to Brussels for what proved a historic occasion, the most important of the Solvay physics conferences. Although he had attended previously, this time he was invited to give one of the prestigious lectures. The topic of the meeting was electrons and photons. Schrödinger's lecture, on wave mechanics, aroused considerable debate; Born and Heisenberg attacked it quite vehemently. Schrödinger was always something of an outsider, sustained by his superior mathematical abilities and sceptical of any orthodoxy.

Among the science courses at the University of Berlin, Schrödinger's was considered the best. He introduced an informal style of lecturing that was new in a place where formality still prevailed. Many professors practically read their lectures; he spoke without notes. Professors were expected to dress formally; Schrödinger usually wore a sweater and bow tie in winter, an open short-sleeved shirt in summer. Later on he was often seen in Tyrolean costume. For Schrödinger, as for other Berlin physicists, the weekly colloquium was the great event, where new discoveries and theories were discussed; Einstein played a leading role, as with careful questionings and elucidations he sought to reach the heart of every problem presented. Schrödinger was elected to the Berlin Academy, at forty-two the youngest member of the august society. Although his marriage was still floundering, he had an active social life, including various affairs.

When the Schrödingers moved to Berlin, the German economy was recovering. The capital had the reputation of being the most licentious city in Europe. Theatrical and musical life was flourishing, against an ominous political background. As the economic recovery faltered, the Nazis saw their chance and, after the 1933 elections, seized power and, as we know, instituted the policies which deprived Jewish scientists of their positions. At first Schrödinger was one of those who thought that the 'Nazi madness' would be over in a couple of years, but soon he became convinced that they would be in power for a long time. When Lindemann came to discuss the situation, one of the people he saw was Schrödinger and he was surprised to find that, although not one of those personally affected, his disgust with the Nazis was so great that he was prepared to leave Germany if Lindemann could arrange something for him in Britain. Lindemann persuaded Magdalen College, Oxford, to offer Schrödinger a fellowship, to be supplemented by a research appointment from industry, giving him an income comparable to that of an Oxford professor. Schrödinger took study-leave from his position at Berlin, although he was not expected to return. He and Annemarie deposited some of their possessions in Zürich for safety and then went on to join Born and Weyl in the Italian Tyrol. For some time Weyl had been having an affair with Annemarie, while Schrödinger himself had been having an affair with Hildegunde March, the wife of one of his Berlin colleagues. Before long he was treating her almost as his mistress; she had a child by him, a daughter christened Ruth Georgie Erica.

En route to Oxford Schrödinger attended the seventh Solvay conference, this time on nuclear structure and nuclear reactions. His formal admission to Magdalen was accompanied by the news that he had just been awarded the Nobel prize in physics, jointly with the Cambridge physicist Paul Dirac, whose profile follows later. Although Lindemann and others did all they could to meet Schrödinger's requirements, for example by providing somewhere for the March family to live, Schrödinger was dissatisfied with his status in Oxford.

Across the Atlantic at Princeton there was a prestigious chair in mathematical physics to be filled. Although the Princeton physicists were hoping to appoint Heisenberg, they invited Schrödinger to come over on a visiting lectureship. When he did so his lectures were, as usual, models of scientific exposition, and he was offered the vacant professorship on the spot. After he had returned to Oxford he declined the offer; it seems that he might have accepted had it been an offer from the Institute for Advanced Study, where he would have had Einstein and Weyl as colleagues.

Although the funds Lindemann had obtained from British industry were almost exhausted by the end of 1935, Schrödinger was given a two-year extension of his grant. However, when he heard that a professorship in physics was soon to become vacant at Graz, he went over to Austria, where he was offered a full professorship at Graz combined with an honorary professorship at Vienna. While awaiting official confirmation of this, he received an offer of a chair at Edinburgh, but, owing to a bureaucratic delay, the necessary permission for permanent British residence had still not arrived when the formal offer came from Graz. So he went back to his homeland and, as we have seen, it was Max Born who moved to Edinburgh instead. Annemarie spent most of the time in Vienna with her mother while Hildegunde and the child Ruth were living in part of their house.

For the next few years Schrödinger's research was inspired by the cosmological theories of Arthur Eddington, who in his prime had virtually created the discipline of stellar astrophysics, although later he became a lonely and controversial figure who believed that scientific knowledge derived not from the external world but from the abstract structure of human thought. Schrödinger loved the Tyrol, but was hoping that a vacancy in Vienna would arise so that he could move to the Austrian capital. Unfortunately the political situation in his homeland was deteriorating rapidly. In 1938, after the crisis which led to the *Anschluss*, the Nazis straightaway extended to a compliant Austria the anti-Semitic policies which were in force in Germany. The new Nazi Rector of the university advised Schrödinger to make a 'repentant confession', which was then published in the press, beginning as follows: 'in the midst of the exultant joy which is pervading our country, there also stand today those who indeed partake fully of this joy but not without deep shame because until the end they had not understood the right course', with much more in the same vein. In future years Schrödinger would always regret this letter, which at first his friends thought could only have been written under duress, until they found out that the Schrödingers were enjoying a peaceful skiing holiday in the Tyrol.

Although Schrödinger had nothing but contempt for the Nazis, he never expressed any public criticism of the regime. By this time the German physicists, notably Heisenberg, had adapted themselves to the absence of their Jewish colleagues, and at the meeting in Berlin to celebrate the eightieth birthday of Planck the Schrödingers were warmly welcomed. However, on his return to Austria he found a letter of dismissal from the honorary position he held in Vienna. Despite the notorious letter quoted above, he was regarded as anti-German and within a few months he received another letter of immediate dismissal, this time from the University of Graz, for 'political unreliability'. The manner in which he had left Berlin had not been forgotten.

Realizing that there was no time to lose, the Schrödingers took the train to Rome with just a few suitcases, leaving almost everything else behind. The Italian physicist Enrico Fermi met them at the station and lent them some money. Like Fermi, Schrödinger was one of the original members of the Papal Academy of Sciences. In Rome he heard that Eamonn de Valera, the prime minister of Eire, wanted to see him in Geneva, where he was presiding over a meeting of the League of Nations. De Valera, who had an interest in mathematics, explained that he was preparing legislation for the Irish parliament to establish an Institute for Advanced Studies in Dublin and that he wanted Schrödinger to be a member of it. He warned Schrödinger that war was imminent and advised him to go to England or Ireland as soon as possible.

Schrödinger agreed to de Valera's proposal, but did not proceed directly to Dublin. Instead he spent some time at the University of Ghent writing an important research paper, the first for some years, on the expanding universe. But for the outbreak of war, the University of Ghent would have conferred on him his first honorary degree. The Schrödingers were joined by Hildegunde and Ruth from Germany, and then they all travelled as far as Oxford. Lindemann and others who had made such efforts to be helpful before were not pleased to see them again. There was a problem obtaining British transit visas, since they were now classed as enemy aliens, but Lindemann did them one last favour by using his influence, and by October, 1939 they had reached Dublin.

The Schrödinger *ménage*, including Hildegunde and Ruth, settled into a small house on the coast; they had come down in the world, but Schrödinger recognized that the new institute was being generously funded by the taxpayers of a relatively poor country with many demands on its resources. Schrödinger adjusted surprisingly well to life in Dublin, in those days by no means a prosperous city but one with a rich cultural life. Under his leadership the institute became a lively centre for advanced study in theoretical physics. There was a physics colloquium with forty-five participants, including a number from Britain, notably Dirac and Eddington (both monomaniacs, according to Schrödinger). Like Einstein in Princeton, he dedicated himself to the vain quest for a unified field theory encompassing both electromagnetism and gravitation.

Rather than discuss his fruitless efforts in that direction, it is more interesting to describe what he did in biology. This was not an entirely new field to him; at a lecture in Berlin in 1933 he had discussed the question of why living organisms contain so many atoms. He did not pursue the question at the time, but now he returned to it and reached the conclusion that the chromosome is a message written in code. The genetic code of course is one of the fundamental principles of the new science of molecular biology. A few earlier works had hinted at such an idea, but Schrödinger was the first to state the concept in clear physical terms. A book he wrote entitled *What is Life?*, which was published in 1944, had an enormous influence, although it did not go down well in Catholic Ireland because it contained a scornful debunking of western religious teaching. The book had a major influence on the microbiologist Francis Crick.

Schrödinger was nearly sixty by the time the war ended, but he had energy for two more love affairs. He was convinced from previous experience that scientific activity would be promoted and sustained by erotic excitement. The first was with the Irish actress and political activist Sheila May Green; her own marriage was childless, but she had a child by Schrödinger who was brought up by her husband after she had left him. Sheila was followed by yet another conquest; this again produced a daughter. Meanwhile Schrödinger began to attack the quest for a unified field theory with fresh enthusiasm. He announced a great breakthrough at the institute, which was trumpeted by the press, but Einstein was soon able to convince Schrödinger that he was mistaken. After this embarrassing debacle he began to devote more of his time to philosophy.

He and Annemarie now became Irish citizens, while retaining their Austrian nationality. When he attended the first post-war Solvay conference, he was left with the impression that no-one was much interested in his current research. He was invited to give a course at Harvard, on his philosophical ideas, but declined when he found out that he would have to stay on afterwards to grade the papers of the students who took the course. He wanted to return to Austria, especially to his beloved Tyrol, but the country was under four-power occupation. He was afraid to enter the Russian zone, but in 1951 spent a term at the University of Innsbrück, in the French zone, where Hildegunde's husband was on the faculty. At one point it appeared that Schrödinger might remain there on a permanent basis, but nothing came of it.

By 1955 the Russians had withdrawn and the way was clear for him to return to Vienna. He arrived there to be treated as a celebrity, festooned with honours. After he had given some lectures, he was appointed to a special professorship at the University. His final academic year was 1957/8, although he remained active as an emeritus professor. He was decorated with the prestigious German order 'Pour le Mérite' in 1957, as was Lise Meitner at the same time, but he never went back to Germany. Already Schrödinger's physical health, especially the condition of his heart and lungs, was causing concern; he was close to the end of his life. He died peacefully on January 4, 1961, at the age of seventy-three, and was laid to rest at Alpbach in the Tyrol, which he used to say was his favourite place on earth.

9 From de Broglie to Fermi

Our next five remarkable physicists were born in the ten years between 1892 and 1901. Two came from France and one from each of India, Italy and Russia.

LOUIS DE BROGLIE (1892–1987)

The de Broglie family, one might almost say dynasty, is one of the most illustrious in the history of France. Of Piedmontese origin (the name is pronounced de Broy), they came to serve the French kings in the seventeenth century. Some achieved military distinction; four became Marshals of France. The first duc de Broglie was a warrior and diplomat in the first half of the eighteenth century; his son was created a prince of the Holy Roman Empire after leading the defeat of the Prussians in the Seven Years War. The family remained prominent in French public life throughout the nineteenth century, but, as well as politicians and soldiers, it ended the century by producing two distinguished brothers, of whom the elder was a notable experimental physicist and the younger a theoretical physicist of genius.

The family background of Louis Victor Pierre Raymond de Broglie is an essential clue for the understanding of his long, prestigious and solitary career. He was born in Dieppe on August 15, 1892, the youngest of a family of five, by all accounts an adorable child, remarkably good-looking, as witnessed by early photographs, bright, gay, overflowing with good spirits and impish pranks. His elder sister gives the following description of him in her memoirs:

> this little brother had become a charming child, slender, svelte, with a small laughing face, eyes full of mischief, curled like a poodle. Admitted to the great table, he wore in the evenings a costume of blue velvet, with breeches, black stockings and shoes with buckles, which made him look like a little prince in a fairy tale. His gaiety filled the house. He talked all the time even at the dinner table where the most severe injunctions of silence could not make him hold his tongue, so irresistible were his remarks. Raised in relative loneliness he had read much and lived in the unreal . . . he had a prodigious memory and

knew by heart entire scenes from the classical theatre that he recited
with inexhaustible verve . . . he seemed to have a particular taste for
history, in particular political history . . . hearing our parents discuss
politics he improvised speeches inspired by the accounts in the
newspapers and could recite unerringly complete lists of ministers of
the Third Republic, which changed so often . . . a great future as a
statesman was predicted for Louis.

As befitted a member of the de Broglie family, the young prince was educated
at home by private tutors. In 1906, when he was aged fourteen, his father duc
Victor died. Maurice, then aged thirty-one, took a hand in the education of
his young brother. On his advice Louis was sent to the Lycée Janson de Sailly,
where he spent three years before graduating in 1909, at the age of seventeen,
with both the *baccalauréat* of philosophy and that of mathematics.

We are fortunate in having an account of Louis' formative years by
his brother duc Maurice: 'Having experienced myself the inconvenience of a
pressure exercised on the studies of a young man I refrained from imparting a
rigid direction to the studies of my brother, although at times his vacillation
gave me some concern. He was good at French, history, physics, philosophy,
indifferent in mathematics, chemistry and geography, poor in drawing and
foreign languages.' It is interesting to note that, although his sister, thanks

to an English nurse, was bilingual from an early age, the respectable clerics who nurtured the young prince did not give him a taste for foreign languages. This may help to explain his isolation from the foreign scientific community in later years. By 1910 Louis de Broglie was a student at the Sorbonne. He was not sure what he wanted to do; a military or diplomatic career did not attract him. At first he read history but was repelled by the uncritical way it was taught at that time. Next he spent a year studying law in preparation for a career in the civil service, until he found his true vocation through reading Poincaré's epistemological masterworks, *La valeur de la science* and *La science et l'hypothèse*; in the end it was physics, particularly theoretical physics, he chose to devote his life to. However, as taught at the Sorbonne, physics was not an intellectual adventure but a rigorous discipline based on standard subjects such as rational mechanics and wave optics, ignoring recent developments such as Maxwell's electromagnetic theory and statistical thermodynamics. Practically the only French source for advanced theoretical physics was Poincaré, whose lectures on electrodynamics, thermodynamics, celestial mechanics and other subjects Louis de Broglie duly attended. Fortunately he could read English and German; French translations of foreign textbooks were often of poor quality.

Louis de Broglie was deeply attached to his sister, the princess Pauline. She was twenty years older than he was and already well known in literary circles. Around his twentieth year she married the count Jean de Pange, and this seems to have triggered some kind of emotional and psychological crisis in her young brother. He experienced a change of personality, losing the gaiety and high spirits of his youth. His self-confidence was badly shaken when he failed an examination in general physics. He had begun to lose faith in himself when he started reading the report of the first Solvay conference on quantum theory. After he had studied it in depth, his confidence began to return and he became convinced that his career would definitely be in theoretical physics. He had now reached the age at which he needed to discharge his obligation for military service. However, the First World War broke out before he had been in the army for long and so he remained there for the next six years. He began by serving as a sapper in the Corps du Génie and was sent to the fort at Mont Valérien, where he was bored stiff. After his elder brother Maurice had pulled strings, Prince Louis was transferred to the radiotelegraphy section; he was based at the bottom of the Eiffel Tower, on which a radio transmitter had been installed. Later he would say that the practical experience he gained in this way was of value in his scientific work.

As duc Maurice puts it, 'he was able to serve his country while working as an electrician, taking care of machines and wireless transmissions and perfecting heterodyne amplifiers then in their infancy'. Louis did not take lightly the waste of six of the best years of his life, but of course he was fortunate not to have been sent to the trenches. As soon as he was demobilized in August 1919, with the rank of 'adjutant' at the age of twenty-seven, he attended a seminar given by Langevin on quantum theory and then a course on relativity, which impressed him by its beauty. He recalled that

> demobilized in 1919 I returned to the studies I had given up, while following closely the work pursued by my brother in his private laboratory with his young collaborators on X-ray spectra and on the photoelectric effect. Thus I made my first steps towards research by publishing a few results in the fields studied by my brother.
>
> In a first series of publications I considered the absorption of X-rays, its interpretation by the theory of Bohr, and its relation with thermodynamic equilibrium . . . Some of the reasonings I used were questionable but they led me to formulae which gave an acceptable account of the facts. At the same time I had long discussions with my brother on reinterpretation of the beautiful experiments that he pursued on the photoelectric effect and corpuscular spectra. I published, with him or separately, a series of notes on the quantum theory of these phenomena which, although classical now, was not well established then.

If Louis de Broglie had done nothing else during his long life, these notes would be enough to immortalize his name. It is best to let him describe the contents himself: 'in the first, inspired by relativistic considerations, I established the relation, well known today, between the motion of a free particle and the propagation of the wave that I proposed to associate with it, and showed how these new ideas gave a simple interpretation of the quantum stability conditions for the motion of the intra-atomic electrons. In the second I applied these ideas to photons and I sketched a theory of interference and diffraction, compatible with the existence of photons. Finally in the third, I showed how my conceptions led to Planck's law for black-body radiation and I established the now classical correspondence between the Maupertuis principle of least action in analytical mechanics and the Fermat principle, applied to the propagation of the associated wave . . .'

He expanded the three notes of 1923 the following year into a doctoral thesis, and that document is the basis for what is known as wave mechanics.

The solitary genius whose ideas kindled those of de Broglie was undoubtedly Einstein. It was by striving to understand, in the framework of relativity, the dual wave–particle nature of the photon, inherent in the theory of the photoelectric effect given by Einstein in terms of light quanta, that de Broglie was led to the dual wave–particle nature of the electron. At the same time, being in daily contact with the photo-electric effect, through his brother's laboratory, must have been a further stimulus in the search for the solution of this riddle. The whole discovery could be summed up in the following phrase: 'Because the photon, which, as everyone knows, is a wave, is also a particle, why should not the electron (or any material particle) also be a wave?' It is hard to overestimate the extraordinary daring and the far-reaching consequences of this simple idea, but at first it was not taken very seriously. Ralph Fowler, the quantum expert at Cambridge, reported on it in the British journals. Langevin gave an enthusiastic account of the thesis to Einstein, who replied 'Louis de Broglie's work has greatly impressed me. He has lifted a corner of the great veil. In my work I obtain results which seem to confirm his. If you see him please tell him how much esteem and sympathy I have for him.' Einstein reported on de Broglie's thesis to the Berlin Academy, thus ensuring its rapid acceptance in the scientific world.

Heir to a rich and illustrious family Prince Louis, later duc, de Broglie's fame burst on him at the age of thirty-seven, with the Nobel prize for physics rewarding one of the great discoveries of the twentieth century and bringing fresh lustre to the de Broglie family. A new chair at the Sorbonne was created in 1933 and occupied by him for thirty years. He was elected to the Paris Academy in the same year, by an overwhelming majority, and became permanent secretary at the age of fifty. He resigned at eighty-three, but continued in an honorary capacity. He was also one of the few scientists to be elected to the literary Académie Française. Although he seldom left France, he was elected a foreign associate of the National Academy of Sciences in Washington and a foreign member of the Royal Society of London.

In 1928 his mother died without ever having realized that her son was a genius; she took to her grave an image of him as a failure. After her death the vast family palace in the rue de Messine was sold, with most of its contents, and Prince Louis chose to live in a modest house in Neuilly-sur-Seine, his home for the remaining sixty years of his life, removed from the crowds and the noisy world. He never married but was attended by two faithful retainers. He never owned a motor car, preferring to walk everywhere or to take the metro. During the summers he stayed in Paris and never went on holiday. When his brother Maurice died in 1960, he became duc Louis de Broglie; before that he was known by his imperial title of Prince.

Nobody ever saw duc Louis angry or even heard him raise his voice. Some said that he was actually a man of strong passions and dislikes but there is no evidence for this. He was always exquisitely polite to all his visitors, no matter who they might be. At a ceremony organized for his eightieth birthday he said 'I consider that the period which followed my seventieth birthday may well have been, on the intellectual side, the most beautiful of my life.' He wrote copiously and well, rather like Poincaré, but his great work was accomplished when he was young. He came to accept Bohr's notion of complementarity and Heisenberg's indeterminancy principle. Among the scientists of his time, Einstein was probably the only one whom duc Louis admired unconditionally. It was Einstein's special relativity and his theory of light quanta that launched duc Louis on the path of his great discovery. It was Einstein's obstinacy in refusing the probabilistic implications of quantum mechanics that encouraged duc Louis in his own, solitary and in the long run fruitless, quest. If Maurice de Broglie had been for Louis a friendly and protective elder brother, one may say that Einstein had been his 'spiritual elder brother'.

For thirty-three years duc Louis lectured at the Sorbonne. He took a very exalted view of his duties as a teacher: the books that originated from his lectures, beautifully written and carefully produced, brought instruction and enlightenment. In his teaching he took great care when presenting his own ideas to explain that they were not generally accepted. However, as a lecturer in the classroom he was uninspiring. Starting scrupulously on time, he read in his high-pitched voice and in a somewhat monotonous tone from a sheaf of large sheets of paper written in long-hand. He always stopped abruptly at the end of the hour and departed immediately. He also ran a well-attended weekly seminar at which young and not-so-young theorists could expound their ideas. A member of his seminar recalled meeting him for the first time, when he was middle-aged:

> he greeted me with great courtesy and invited me to take part in the proceedings. I was terribly impressed. At the thought of shaking hands with a man who was a prince of physics and a prince by birth I was seized with an almost religious emotion. He was wearing a dark blue suit which even then seemed slightly old-fashioned, with a wing collar and a pearl in his neck-tie. He had a curiously high-pitched voice and he seldom spoke. Strange as it may seem I thought this man covered

with honours and glory was shy. The seminar did not start before he had shaken hands with all the participants, numbering about twenty, who were standing as he passed. In spite of their small number the speaker was never interrupted, remarks and questions had to wait until the end of the talk.

However, with a few exceptions the disciples who congregated around duc Louis were not of the highest calibre and perhaps not always of the highest intellectual honesty. One of the manifestations of this was the atmosphere of admiration, not to say adulation, with which they surrounded him. Although duc Louis never encouraged such behaviour in any way, he never reacted against it sufficiently strongly to put it down once and for all. Also, with advancing years, as the direction of research separated him more and more from the mainstream, he may have felt some comfort in being surrounded by disciples who agreed with his conceptions and devoutly developed them in seminars and articles. Louis de Broglie died on March 19, 1987, in his ninety-fifth year.

SATYENDRANATH BOSE (1894–1974)

With the exception of Rutherford, so far all the subjects of these profiles have been either European or American, because it seemed impossible to find

suitable subjects in other parts of the world. At the close of the nineteenth century, however, the situation began to change. Outstanding individual scientists began to emerge in other countries, and it is only to be expected that their life-stories would be quite different from the others we have been describing. I give one such example from India and another from Japan. Today the contribution to science of countries such as India and Japan is most impressive, but in many ways it is even more remarkable what was achieved by the pioneers, who worked in such a very different milieu from their European and American counterparts.

Satyendranath (or Satyendra Nath) Bose, generally known as S.N. Bose in the scientific community and affectionately called Satyen Bose in his native Bengal, was born on January 1, 1894 in Calcutta, then the capital of British India. His father, Surendranath Bose, was an accountant who held responsible posts in the executive engineering department of East Indian Railways. Surendranath had an aptitude for mathematical thinking and was interested in several branches of science; he became one of the founders of a small chemical and pharmaceutical company. He was also interested in philosophical studies, especially the Hindu scriptures such as the Bhagwag-gita; at the same time he enjoyed the dialectical speculations of Hegel and Marx. Surendranath Bose died in 1964 at the age of ninety-six, having witnessed with pride the seventieth-birthday celebrations in honour of his famous son the previous January. The mother Amodini Devi of Bose, who had died twenty-five years earlier, often had to struggle against ill health and inadequate resources to maintain a middle-class home. Although she received no more than a nominal school education, she was a woman of culture who showed considerable ability in managing domestic affairs. She possessed a remarkable fortitude of will, a warm heart and a sense of personal and family dignity.

Satyendranath was the only son and eldest child of his parents, who also had six daughters. He inherited many good qualities and noble aspirations from his parents and more than fulfilled their expectations of him. Young Satyen Bose attended a neighbourhood elementary school in Calcutta until he was thirteen years old and then went on to the excellent Hindu school. His eyesight was very weak, but his intelligence and memory were keen, and he had a great desire to study science. The headmaster and senior mathematics master gave him every encouragement.

After passing out of high school in 1909, Bose entered Presidency College, Calcutta, where he enrolled in the science courses, which were being taught by a distinguished faculty. Among his contemporaries were several notable scholars, including Meghnad Saha, who became a close friend. Whereas Bose came from a middle-class urban Bengali family, Saha came from an obscure village where educational facilities were practically non-existent. Bose came first both in the B.Sc. examination in 1913 and in the M.Sc. two years later. He also got married during his final year in college, to Ushabala Ghosh, the eleven-year-old daughter of a wealthy physician; as was customary the marriage was arranged by the parents. The couple had nine children; after two had died in infancy, two sons and five daughters were left, all of whom received a good education.

It was a time of political unrest, and in many ways the career outlook was not good. However, the Indianization of education was creating opportunities for able youngsters like Bose and Saha. The University of Calcutta, founded in 1857, was being transformed into a national institution, free of British influence. The Vice-Chancellor and moving spirit behind this development, Sir Asutosh Mookerji, was himself a mathematician. The two friends decided to devote themselves to study and research in physics. Unfortunately modern physics, such as relativity and quantum theory, was not treated in the textbooks available to them, and they had no access to scientific journals. Moreover, laboratories did not exist. Planck, Einstein and Bohr were just names to them. However, Sir Asutosh was impressed by the willingness and enthusiasm of the aspiring young scientists and agreed to help them prepare to teach postgraduate courses in physics and mathematics. He granted them scholarships and arranged facilities for procuring scientific journals and working in laboratories.

Bose and Saha had already started to learn French and German in order to be able to read the European scientific literature, but during the First World War it was not easy for them to obtain what they needed. As it happened, a visiting German possessed a good collection of advanced texts and journals in German, which he lent to them. Saha chose to study first thermodynamics, statistical mechanics and spectroscopy, while Bose decided on electromagnetism and relativity.

In 1916 a professor of mathematics was appointed by the university, who started postgraduate classes in applied mathematics and physics, assisted by Bose and Saha; the following year they were both appointed lecturers in physics. 'We took upon ourselves the task of teaching postgraduate students', Bose recalled. 'Saha taught thermodynamics and spectroscopy in the physics department and hydrostatics and lunar theory in the mathematics department. I was more ambitious, teaching both physics and applied mathematics regularly. On my shoulders fell the task of teaching general physics and giving all entrants a suitable introduction to mathematical physics. I also taught elasticity and relativity.' In 1918 a new physics professor was also appointed, a magnetic personality who attracted many students, most of whom worked in optics, his speciality.

Bose and Saha started to do research in statistical mechanics, publishing their first paper in 1918. They published, in translation, an anthology of Einstein's papers on relativity and then went on to quantum theory. Saha, who began to specialize in astrophysics, went to Germany on study leave, as Bose was hoping to do one day. Some of the other younger scientists, including Bose, decided that it was time to leave Calcutta, where the physics department was becoming overcrowded. He secured a non-tenured position, equivalent to assistant professor, at Dhaka, or Dacca, where there was an enterprising Vice-Chancellor named Hartog in charge of a newly founded university. Although it was in no sense an Islamic institution, its students came mainly from the Muslim population, who were in the majority in this part of Bengal. It was unusual in being residential in character, so the students saw much more of their teachers than at other universities. The head of department, who had been instrumental in attracting Bose to Dhaka, was an expatriate, a Cambridge graduate who seemed to have been appointed professor of physics mainly on the strength of his prowess on the football field. Bose was left to lead the postgraduate work of the department; he began by teaching thermodynamics and electromagnetism, while increasingly becoming interested in relativity and quantum theory for his own research.

For Bose Dhaka was an alien land, another world culturally from the Calcutta where he grew up, but for Saha Dhaka was his homeland. Although Dhaka is only a few hundred miles from Calcutta, the journey was an arduous one before the days of air travel. When Saha returned from Germany he came over to see Bose, and the discussions he had with Bose on this visit seem to have been the catalyst for what happened next. Around 1924 Bose wrote two important papers on the statistics of radiation. The first, called 'Planck's Law and the Light-quantum Hypothesis', he submitted to the *Philosophical Magazine*, which had published his earlier work. When he heard nothing from the editor, he took the bold step of sending a copy to Einstein. His only previous contact with Einstein had been when he wrote for his permission to translate some of the relativity papers into Bengali. He then wrote a sequel, 'Thermal Equilibrium in the Radiation Field in the Presence of Matter', and sent a copy of that to Einstein as well, asking whether he would recommend both papers for the *Zeitschrift für*

Physik, or some other German journal, and arrange for them to be translated. Einstein replied on a postcard: 'I have translated your work and sent it to the *Zeitschrift für Physik* for publication. It signifies an important step forward and I like it very much. In fact I find your objections against my work not correct . . . However, this does not matter . . . It is a beautiful step forward.'

Bose had applied to Dhaka University for two years' leave to study in Europe. His application was being considered when the postcard arrived. Such words of approval from the highest authority made all the difference. His application was promptly granted, with generous financial support, and Bose arrived in Paris in October 1924. Just why he then remained there instead of continuing to Berlin is something of a mystery. He contacted Einstein, who invited Bose to come and see him, but Bose waited a whole year before doing so. He explained afterwards: 'I wanted to go directly to Berlin, but I didn't venture to go straight on because I was not sure of my knowledge of German. I came out thinking that perhaps after a few weeks in Paris I should be able to go to Berlin to see Einstein. However, two things happened: (1) friends, (2) a letter of introduction to Langevin. My friends in Paris, who received me on arrival there, took me to this boarding house where they were staying. Then they all insisted that I should stay there. Well, I found it convenient to be among friends.' It appears that these friends were Bengali nationalists, and that Bose was becoming increasingly involved in their activities. In 1905, when Satyen was eleven years old, the autocratic viceroy Lord Curzon had declared the partition of Bengal, an act that had grave political consequences, arousing Bengalis to rebel against the British political and economic domination and spurring the nationalistic Swadeshi movement into action.

Initially Bose had planned to spend the first year of his study leave in England and the second in Germany. Hartog had written to both Rutherford and Bragg on his behalf; Bragg's answer is not known, but Rutherford replied that he could not accommodate Bose in the Cavendish. In Dhaka, it should be recalled, Bose had to teach both theoretical and experimental physics, and one of his main objectives was to improve his experimental skills. He could do this in France almost as well. Langevin suggested that he should pursue the possibility of working in Madame Curie's laboratory, perhaps with *la patronne* herself. At the interview, conducted in English, the 'great elderly lady' said that Bose should first improve his French, apparently because she had had another Indian in her laboratory who, unlike Bose, did not know much French. Later Bose spent a short while at the Radium Institute learning about experimental work in radioactivity.

Meanwhile Langevin provided him with an introduction to Maurice de Broglie, who gave him some useful experience of experimental work in X-ray spectroscopy.

It was not until October 1925 that Bose moved to Berlin. Einstein had written to say that the second paper, which he was not entirely happy with, had also appeared in the *Zeitschrift*; Bose had replied that he had written a third paper, which he believed met some of Einstein's criticisms of the second. This third paper was never published and no copies seem to have survived. Langevin read it with approval, but apparently Bose did not ask him to communicate it to a French journal. In the course of time it emerged that the ideas Bose put forward in the second paper were quite valid, but by then Bose had ceased to be active in research.

When Bose arrived in Berlin, Einstein was on his annual visit to Leiden. As soon as he returned they met and continued to do so all the time Bose was in Germany. Yet it should not be assumed that they talked about physics; it might well have been politics. Bose might have asked Einstein for his views on nationalism in India. During this period Zionism, rather than physics, was Einstein's preoccupation, and of course he was a primary target of anti-Semitic propaganda. Bose also met the other leading physicists in Berlin at that time and made a visit to Göttingen as well.

While Bose was in Europe the head of physics at Dhaka University left and Bose was encouraged to apply for the vacant chair. Einstein gave him a reference: 'the recent works of Mr S.N. Bose, especially his theory of radiative equilibrium, signify in my opinion an important and enduring progress of the physical theory. Also in personal discussion with Mr Bose, I have got the impression that he is a man of unusual gifts and depth, from whom science has much to expect. He has also at his command an extensive knowledge and certain ability in our science. As university teacher he will certainly develop a successful and prosperous activity.' Other referees, including Langevin, also supported his candidature but, even so, Bose was not offered the position until another candidate had turned it down.

So Bose became head of physics and was soon caught up in the usual round of duties that goes with such a post. He reorganized his department, to make it a research centre on the European model. As for his own research, with no proper programme, his interest kept shifting from one problem to another. He had written nothing while he was in Europe and, as he said, 'on my return to India I wrote some papers. I did something on statistics, and then something on relativity theory . . . they were not important. I was not really in science any more. I was like a comet, a comet which came once and never returned again.' Bose also kept up an interest in experimental physics, especially thermoluminescence and crystal structure. However, his creative period for research was essentially over.

After the end of the Second World War, which had come close to Dhaka, the tension between the Hindu and Muslim communities which led to the separation of East Bengal from the rest of India was disrupting the work of the university to such an extent that many of the faculty, including Bose, decided that it was time to leave. In 1945 he moved back to Calcutta, as successor to Vankata Raman as professor of physics. He wrote some research papers on his ideas for a unified field theory and had some correspondence with Einstein about it. In 1954 he was given a seat in the Rajya Sabha, the upper chamber of the national parliament, but did not play a very active role in Delhi. Two years later, after retiring from Calcutta University, he became Vice-Chancellor of the new central university of Visva-Bharati, which was closely associated with the ideas of the poet Rabindranath Tagore. Earlier Tagore had invited Bose to Santiniketan and dedicated to him his *Visva-parichaya*, a book giving an elementary account of the cosmic and microcosmic world in Bengali, in recognition of Bose's efforts to popularize science through the mother tongue. However, Bose gave up the Visva-Bharati post after two years, saying that he wanted to go back to research. In 1959 he was appointed to one of the prestigious national professorships, which left him free to work as he pleased, and he held this for the rest of his life, dying on February 4, 1974, shortly after the celebration of his eightieth birthday.

After he returned to India in 1926 it was twenty-five years before Bose went abroad again, but then he travelled extensively and was often seen at scientific conferences. A visit to Japan confirmed his belief that, even in science, university education in India should be in the mother tongue of the students, not in English. The great inspiration of Bose's life was the work and personality of Albert Einstein. To him Einstein's personality was 'beyond comparison', and he was forever grateful to Einstein for the timely encouragement he had received from him. He hoped to see Einstein again before he died but, because of his reputation as a political radical, Bose was refused an American visa.

It was given to Bose to make just one important discovery and to write a four-page paper about it. The Bose–Einstein statistics continue to have far-reaching consequences in modern physics. Einstein and Bose independently predicted that, at extremely low temperatures in a dilute, non-interacting gas, atoms would condense to the point where they fall into the same quantum state, essentially behaving like a single atom. The Bose–Einstein condensate, predicted by theory, has recently been produced experimentally, and its properties are being investigated. Among the elementary particles, the boson commemorates the name of Bose, a name that has an honoured place in the annals of physics.

PIOTR LEONIDOVICH KAPITZA (1894–1984)

Piotr Leonidovich Kapitza, later known as Peter Kapitza to the world of science, was born in Kronstadt, the island fortress on the river Neva near St Petersburg, on July 8, 1894. His father, Colonel (later General) Leonid Petrovich Kapitza, was a military engineer involved in modernizing its fortifications. The Kapitzas had been landed gentry with Polish antecedents and the family was well represented in the professions. His mother, Olga Ieronimovna, to whom he was very close until her death in 1937, was a specialist in children's literature and folklore and an important figure in the literary world of St Petersburg. Her father, General Ieronim Ivanovich Stebnitski, was a geographer of international repute, a corresponding member of the St Petersburg Academy and an ardent world traveller. Unusually for his time, he arranged for his daughters to have higher education, Olga in the humanities and Alexandra in mathematics and science.

Aunt Alexandra played an important part in the upbringing of her nephew Piotr Leonidovich and it was she who discovered that, although he was somewhat backward in other respects, he had an unusually quick grasp of arithmetic. He never overcame a certain indirectness and sloppiness of speech and never learned to spell correctly in any language. He was admitted to the classical gymnasium in 1905 but was transferred after a year to the more scientifically oriented Realschule, which was much more appropriate to his developing talents. Six years later he graduated with honours and entered the electrochemical faculty of the St Petersburg polytechnic – without Latin and Greek he could not enter the more prestigious university.

Kapitza's studies were interrupted by the First World War. After serving for two years as an ambulance driver on the Polish front, he returned to the polytechnic and graduated in 1918. After the chaos of war and the upheaval of the revolution, the economy of the country was in a disastrous state. Nevertheless, a new physico-technical institute was established under the leadership of Abraham Joffé, the grand old man of Russian physics, who recruited Kapitza to join his group of enthusiastic young scientists. There were great shortages of food and fuel and practically no scientific equipment, so experimental research could be carried out only on a 'do it yourself' basis. In spite of these difficulties, however, a surprising amount was achieved.

Following a romantic trip to Harbin in China, soon after he had been demobilized, Kapitza married Nadezhda Kyrillovna, daughter of General Chernosvitov, and their son, Ieronim, was born in 1917. A second child was expected towards the end of 1919, but disaster struck the family. Epidemics were rife in the dreadful conditions following revolution and civil war. Ieronim died from scarlet fever. Nadezhda was devastated by the loss of her first child; soon afterwards she gave birth to a daughter, only for both mother and child to succumb to the pandemic Spanish flu. Then Kapitza's father also died of this, and Kapitza himself caught it but survived. Naturally he was overwhelmed by these tragic events and unable to continue working.

Then something happened that not only distracted him from his grief but changed the course of his life. This was the setting up, on Joffé's initiative, of a commission of the Soviet Academy of Sciences for renewing scientific relations with other countries. Besides Joffé himself, an important member of the commission was the naval engineer and applied mathematician Admiral A.N. Krylov, later to become Kapitza's second father-in-law. Both Joffé and Krylov formed a high opinion of Kapitza's scientific gifts and wanted to help in his difficult personal situation; appointing him to join the commission seemed a good way to do this.

Travel abroad at that time was not easy for Soviet citizens because few other countries maintained diplomatic relations with theirs. The commission set out early in 1921 and, as soon as they had travelled as far as Berlin, began procuring scientific equipment. Kapitza wanted to go further west, but neither France nor the Netherlands was prepared to risk admitting someone who might prove to be a Communist agitator. However, Britain was more accommodating and Joffé succeeded in obtaining the necessary visas.

So Kapitza and Joffé arrived in England and early in June started a round of scientific visits, which culminated in a most significant visit to Cambridge in July. They were received very cordially by Rutherford but, when Kapitza asked whether he might work in the Cavendish for a few months, Rutherford was rather discouraging, saying that the laboratory was already very crowded and it would be difficult to find room for another person. Rutherford was rather taken aback when Kapitza replied by asking what accuracy he aimed at in his experiments. The answer to this seemingly irrelevant question was two or three per cent and Kapitza then pointed out that, since there were about thirty researchers in the Cavendish already, one more would hardly be noticed since it came within the experimental error. This ingenious approach persuaded Rutherford to admit Kapitza after all. As a postscript, it may be mentioned that a year later Kapitza asked Rutherford why he had agreed to take him on and Rutherford laughed and said 'I can't think why but I'm very glad that I did.' Later he came to regard Kapitza almost as a surrogate son.

Kapitza joined the Cavendish in July 1921 and, although the original plan was for him to stay only over the winter, he remained for thirteen years. His introduction to Cambridge life and his impressions of Rutherford and the Cavendish are vividly described in the letters he wrote at frequent intervals to his mother Olga. The usual initiation of new research students was a month or two of practical work in the Cavendish attic under Chadwick's supervision; in fact, Kapitza's skill and assiduity were such that Chadwick was satisfied after only two weeks. By early August Kapitza, at Rutherford's suggestion, was studying how the energy of the alpha particle falls off at the end of its range. This project was brought to a successful conclusion with amazing rapidity. Within nine months of the conception of the idea, he was already drafting a paper for publication and by June it was sent off. He wrote to his mother that 'Today the Crocodile summoned me twice about my manuscript . . . it will be published in the *Proceedings* of the Royal Society which is the greatest honour a piece of research can achieve here. Only now have I really entered the Crocodile's school, which is certainly the most advanced school in the world and Rutherford is the greatest physicist and organizer. It is only now that I have felt my strength. Success gives me wings and I am carried away by my work.'

Crocodile was a nickname Kapitza invented for Rutherford. As he explained: 'in Russia the crocodile is a symbol for the father of the family and is also regarded with awe and admiration because it has a stiff neck and cannot turn. It just goes forward with gaping jaws – like science, like Rutherford.' He wrote again to his mother: 'the Crocodile is taken with my idea and thinks it will succeed. He has a devilish feeling for experiment and if he thinks something will come out of it that is a very good omen. His attitude towards me gets better and better' and then, 'the preliminary experiments were completely successful. I am told that the Crocodile speaks of nothing else now and I shall remember my last conversation with Rutherford as long as I live. After a whole lot of compliments he said: "I should be very happy if I could have the possibility of creating a special laboratory for you in which you could work with your own students."'

Meanwhile Kapitza's official position in Cambridge was rapidly being consolidated. In June 1923 he received the degree of doctor of philosophy and soon afterwards he was awarded a Clerk Maxwell scholarship. Eighteen months later he was appointed assistant director of magnetic research, a university post. He was elected a research fellow of Trinity College, which enabled him to live in college up to the time of his second marriage. In the spring of 1927 Kapitza went to Paris and wrote to Rutherford: 'I am going to be married. What do you think about it? I feel you are rather angry. This is why I propose to have no honeymoon and bring my wife in a few days time after my wedding to Cambridge.' His bride was Anna Alekseyevna Krylova, daughter of the Admiral Krylov mentioned earlier. Although her father stayed in Russia after the revolution, her mother emigrated to Paris and it was there that Anna completed her training as an archaeologist. The marriage was a very happy one. Anna was not only a charming hostess to their many friends and an accomplished artist, but also a great support to her husband in the difficult times which lay ahead. Two children were born in their Cambridge period: in 1928 Sergei, who became a distinguished physicist and successful popularizer of science for Soviet television, and three years later Andrei, who became a well-known Antarctic explorer and geographer.

In 1929 Kapitza was elected to the Royal Society, at a time when it was relatively closed to foreigners, and almost simultaneously to corresponding membership of the Soviet Academy. The following year he was appointed to a Royal Society Messel professorship, and the special laboratory Rutherford had said he would like to create for him took shape as the Royal Society's Mond laboratory. At the opening of the building by Stanley Baldwin, as Chancellor of the University, a bas relief of Rutherford just inside the main entrance and another of a crocodile just outside, both sculpted by Eric Gill, were revealed. Speeches were made by Baldwin, Rutherford, Kapitza and Lord Balfour, the former prime minister. Baldwin's speech duplicated Rutherford's almost word for word, owing to his having picked up an advance copy of Rutherford's under the impression that it was his own.

Kapitza had many interests and skills outside his scientific work. One of his early enthusiasms was the internal combustion engine. Soon after his arrival he bought a motor cycle and managed to crash it, without serious injury. Before long he moved up to a car and acquired a reputation for reckless driving. One of his hobbies was repairing watches and clocks, another was conjuring. He was fond of boasting and exaggeration and had boundless self-confidence, which was usually justified. Cheerful and outgoing, a great charmer and excellent company, he loved an argument, found something interesting or amusing to say on almost any subject. He frequently dined in the college hall during his time as a fellow, but at Cambridge he missed the lively discussions of the Joffé seminars and started something similar of his own, a scientific circle that soon became known as the Kapitza Club. This continued for twelve years after he had left Cambridge, but he organized a similar one in Moscow. It is time to explain how he came to be back there.

Kapitza was rather proud of his unusual situation of directing a prestigious laboratory in Cambridge while remaining a Soviet citizen and being able to go in and out of the Soviet Union at will. From 1926 on he visited Moscow nearly every summer, taking care that his permission to return to Cambridge was underwritten by people high up in the Soviet political and scientific establishment. During these visits he gave lectures, spent time with his mother and usually managed to have a good holiday in the Caucasus or Crimea as well. Although his friends wondered whether his exclusive status could continue indefinitely, Kapitza laughed off any warnings and in 1934 made his usual summer visit, accompanied by his wife Anna. When they were preparing to return to England Kapitza was informed that his permission to leave the Soviet Union had been revoked. He thought that the problem might have arisen because the authorities had developed exaggerated ideas of the technological relevance of his work, a situation for which he was partly to blame.

However, another explanation seems more likely. The distinguished physicist George Gamov decided not to return to the Soviet Union after the Solvay conference of 1933 and when he settled in the USA he was stripped of his Soviet citizenship. His case may well have influenced the treatment of Kapitza, who was utterly devastated by his inability to return to Cambridge. At first the problem was kept secret, so that the Soviet authorities could retreat without loss of face, while all kinds of efforts were made behind the scenes. Although Kapitza had contacts at high level, these efforts led nowhere because, he suspected, it was Stalin himself who had decided that he must remain in the Soviet Union. Anna was allowed to return to the children in Cambridge and begin the difficult task of winding up their affairs there, but it was not until the end of 1935 that the family could be reunited in Moscow.

Meanwhile Kapitza, after a period of feeling very frustrated and miserable, began to cooperate in the planning of a new institute, like the Mond, of which he would be director. Life for the family became easier; they were assigned a comfortable and central flat, a good car and other privileges. The Institute for Physical Problems was in part modelled on the Mond laboratory, but it was on a larger scale; the director's office was immense, there was a spacious lecture theatre and there was ample office accommodation, essential for the paperwork generated by the formidable Soviet bureaucracy. The building was attractively situated on the Lenin hills, with gardens and parks around it. There was a magnificent house nearby for the Kapitzas to live in. The institute was provided with staff of high quality. He had to convince the authorities that his work lay primarily in pure rather than applied physics and that he could do nothing useful unless he had equipment and other facilities comparable with those he had enjoyed in Cambridge. Negotiations were begun to bring him what he required.

So Kapitza settled down to research again and within a year made his greatest discovery, the superfluidity of liquid helium. However, he lacked the freedom he had enjoyed in Cambridge. From the mid 1930s Soviet scientists found themselves increasingly cut off from their colleagues in other countries. They were barred from going abroad for any length of time and forbidden to publish outside the Soviet Union, even private correspondence was not free of risk. However, Kapitza was able to take liberties that other Soviet scientists were not. Notably, he was able to use his influence to protect at least a few of his colleagues during the purge of the late thirties, as in the case of Lev Davidovitch Landau, the head of the department of theoretical physics at his institute. Landau was arrested in 1938, like so many others during the Great Terror. He was held in the Lubyanka prison for a year, after being accused of being a German spy, and would not have survived much longer. Kapitza bravely went to the Kremlin and announced that he would resign unless Landau was released, and eventually his intervention was successful.

In 1939 Kapitza was elected a full academician, but there were signs of trouble in store. Officials kept arriving to inspect the activities of the institute, and eventually he was informed that 'P.L. Kapitza, having shown a cavalier attitude to both Soviet and foreign achievement in the technology of oxygen production, and having failed to meet the scheduled dates for introducing new installations into the metallurgical industry . . . is relieved of his duties as Director of the Institute of Physical Problems.' The underlying reason for this action was probably Kapitza's refusal to work in the organization set up under Beria, the head of the NKVD, to develop the Soviet atom bomb. Kapitza is said to have written to Stalin that Beria was 'like the conductor of an orchestra with the baton in his hand but without a score'. Beria wanted to arrest Kapitza for such insubordination, but Stalin, in his unpredictable way, perhaps because he admired Kapitza's courage, vetoed this proposal and said that dismissal would be sufficient.

Although he was no longer head of the Institute of Physical Problems, Kapitza retained his position and salary as a full academician and went to live at his country house at Nikolina Gora, where he managed to carry on scientific work while virtually under house arrest. Most of his effort went into building up a laboratory in various outhouses where, aided by his sons, particularly Sergei, he could continue experimental work, albeit only on relatively unexciting projects. While atomic physics elsewhere was moving rapidly ahead using particle accelerators and other new equipment, he was unable to contribute to this. Even after he had been reinstated in 1954 and could return to Moscow, he was still without the facilities he needed for the kind of experimental work at which he excelled. Nevertheless, he began to think about the possibility of developing a defence against atomic bombs using extremely powerful microwave emissions. Later he transferred his attention to the problem of generating energy through nuclear fusion.

For thirty years Kapitza was denied the opportunity to accept invitations to travel beyond the countries of the eastern bloc, until in 1965 he was allowed to travel to Copenhagen to receive the Niels Bohr gold medal of the Danish Engineering Society. The next year he was awarded the Rutherford medal of the Institute of Physics and returned to England after a lapse of thirty-two years. He received a particularly warm welcome in Cambridge; many friends of his youth were still there. He returned to England twice more, in 1973 to receive the Simon Memorial prize of the Institute of Physics and three years later to give the Bernal Lecture at the Royal Society. He was also able to visit other countries to receive honorary degrees and other distinctions. The culminating event was the Nobel prize in 1978 for his work in low-temperature physics, especially the discovery of superfluidity.

The Kapitzas continued to lead an active social life, enjoying the company not only of their extended family but also of a wide circle of friends. On his eightieth birthday the Soviet intelligentsia flocked to Nikolina Gora for a huge celebration. Five years afterwards he suffered a severe stroke and died on April 8, 1984 after a few days in hospital. The announcement of his death in *Pravda* was signed by all the members of the politbureau as well as all the leading Soviet scientists. A memorial meeting took place at the institute, a moving occasion with many personal recollections by close colleagues and friends.

As the years passed, Kapitza had become more and more accustomed, or at least reconciled, to life in his homeland. No matter how muted his criticism, however, he insisted on freedom to criticize. Being neither an oppositionist nor a political opportunist, he based his collaboration with the most unsavoury political authorities on the principle of compromise. He did not find it easy to establish and maintain this kind of compromise. By withholding fervour from his arguments, Kapitza would have acted against his conscience and dignity. By saying too much, he would have had to confront the powers he was trying to placate. He did not enjoy the game, but, with only a few exceptions, he succeeded in avoiding its dangers. He knew enough to restrict his comments to very specific topics of science and science policy. The authorities awarded him Stalin prizes in 1941 and 1943, the Order of Lenin in 1943, 1944 and 1945, and, amongst other marks of distinction, the Soviet Union's highest civil title: Hero of Socialist Labour.

JEAN-FRÉDÉRIC JOLIOT (1900–1958)

In dealing with husband-and-wife partnerships it seems necessary, for a satisfactory profile, to choose one or other as the principal subject. In the case of the Curies, Marie rather than Pierre was chosen; in the next generation

this is balanced by choosing Frédéric rather than Irène Joliot-Curie. Jean-Frédéric Joliot was born on March 19, 1900, in a small house in the smart sixteenth *arrondissement* of Paris. Earlier in his life his father Henri, the son of a steelworker from Lorraine, had been a soldier. He was called up during the war of 1870, at the age of twenty-seven, and took an active part in the Paris Commune, after which he was forced to leave France until an amnesty was declared. Frédéric's mother Emilie, whose maiden name was Roederer, came of an Alsatian family. Her father had been sauce-cook to Napoleon III and as such the only servant permitted to be present at the Emperor's meals and, quite frequently, converse with him. She retained the austere manner of her Protestant background but was fundamentally liberal. Her great good nature and deep sense of justice had a profound influence on her son.

Frédéric, the youngest of six children, was born when his mother was already forty-nine. Two of his brothers died young and a third was to be killed in the course of one of the first engagements of the First World War. At the age of ten Frédéric was sent to boarding school at the Lycée Lakanal in Sceaux. He was more interested in sport, particularly football, than in academic school-work. He was not a well-behaved child: 'I caused my poor mother a great deal of worry, particularly when I went shopping with her and stole the sweets and fruit which were on display.' His father owned a successful wholesale business in dry goods, which was entrusted to a manager. Frédéric often accompanied his father in shooting, fishing and other such pursuits, becoming an excellent shot and fisherman in the process. There was also music at home; his father performed on the French hunting horn.

The death of Henri Joliot and changes in the family fortunes caused the Joliot family to leave the affluent sixteenth *arrondissement* and set up house in the fourteenth, the Montparnasse quarter near the 'Lion of Belfort', that symbol of the Paris Commune. The Lycée Lakanal was replaced by a municipal school at which Frédéric prepared himself for the higher-level Ecole de Physique et de Chimie. He failed the entrance examination in 1918, at the end of the First World War; when he tried again the next year he was successful, but a major illness delayed his admission until October 1920. At the end of the first year students had to choose between physics and chemistry. The director of studies, as we know, was the charismatic physicist Langevin, whose lectures on electricity were a revelation to his pupils. Frédéric chose physics and became a life-long disciple of Paul Langevin.

All those who were taught by Langevin were in some degree affected by the experience, since he was no ordinary teacher. They tended to copy his characteristic tricks and attitudes, such as moving the hand slowly backwards and forwards over the chair, or leaving the blackboard and insistently going to make a white mark with the chalk on the radiator. He taught them that science was something living, that a teacher must convey the value and importance of what has already been achieved, while making the audience realize that everyone can make an original contribution to the edifice that is being constructed. 'When he talks about science', said Marguerite Borel, 'about literature, about philosophical theories – he understands everything and is interested in everything – his very beautiful chestnut eyes and his whole face light up.'

After graduating in 1924, Frédéric underwent six months of full-time training as an officer cadet in the artillery reserve, at the end of which he became an assistant in Marie Curie's laboratory. It was there that he began his professional career in physics and met Marie's daughter Irène. Irène was green-eyed with short cropped hair and rather awkward in her movements. She had inherited her father's noble brow and her mother's bright and limpid eyes. She also inherited the shyness of both parents as well as their abilities, and had great difficulties in greeting and dealing with strangers. Even more than her mother, remarked Einstein, the main way she expressed her feelings was by grumbling, just like a grenadier. Irène became more sure of herself as she grew up, but people found her excessively self-centred. Her imperturbable calm and her direct manner in replying to questions made her seem cold and somewhat haughty.

Marie Curie was a firm believer in the prime importance of education. In France, at that time, the quality of school education was not outstanding. It was mainly for Irène's sake that Marie organized a co-operative with a group of like-minded academics from the Sorbonne, where each gave some time teaching the ten boys and girls who belonged to it. This lasted only two years, partly because the teachers were too overworked to give it the necessary share of their attention and partly because too much of the children's time was spent travelling from place to place. Irène obtained her basic scientific education from the cooperative. After it closed she was sent to the Collège Sévigné, a private school, for two years' study towards the *baccalauréat*. When the war broke out, she was ready to enter the Sorbonne. The girls were at l'Arcouest, where they were joined by their mother. 'You and I', she told Irène, 'will try to make ourselves useful.' To start with, both girls helped to gather in the harvest. Later Irène supported her mother's radiological work and, once the war was over, she became her laboratory assistant. Eve, still just a teenager, was able to gain more of her mother's personal attention. She was more approachable than her sister, but she wrote that 'in spite of the help my mother tried to give me, my young years were not happy ones'. Both their mother and grandfather Curie were closer to Irène than to Eve. In 1921 the daughters accompanied their mother on a strenuous visit to the USA. In 1925 Irène took her doctorate with a thesis on the alpha rays of polonium, and that was the year in which she met Frédéric Joliot.

Irène's position in the laboratory was a special one, as daughter of *la patronne*; moreover, she had great knowledge of radioactivity. She was three years older than Frédéric and seemed at first to be his opposite in everything. She was as calm and serene as he was impulsive. Being by nature very reserved, she found it difficult to make friends, while he was able to make social contact with everyone. She took little interest in her appearance and dress, while he was good-looking, elegant and always a great success with the opposite sex. In argument Irène was incapable of the least deceit or artifice or of making the smallest concession. With an implacable obstinacy she would present her thesis, confronting any opposition head on. Frédéric, on the other hand, without conceding a point, had a wonderful ability to put his opponents in a frame of mind where they could accept his arguments.

As well as science, Irène enjoyed French, English and German poetry. From childhood she had also been keen on sport; she excelled chiefly at swimming and skiing. She had a love of the natural world, especially the mountains, where she several times had to stay because of pulmonary infections. Her exterior, which tended to be cold, concealed a passionate nature. It was in the course of long walks through the forest of Fontainebleau, talking of physics, art and religion, that she and Frédéric realized that they were meant for each other.

They were married on October 9, 1926. Out of respect for Irène's famous parents, they adopted the name Joliot-Curie. Although there was nothing unusual in doing this, Marie was not pleased, and both professionally and academically the young couple called themselves the Joliots most of the time. Like her mother, Irène never learned to cook, but Frédéric could be quite a competent chef on occasion. Their daughter Hélène, born in 1926, and son Pierre, born in 1932, also became physicists of distinction. Owing to pleurisy, Irène was slow to recover her strength after the birth of Hélène. Except for taking vacations in the mountains, she tried not to let poor health affect her life. She carried on with her scientific work as before, working straight through her second pregnancy, but after the birth of Pierre she always felt tired and was never able to regain her previous weight. It is generally believed that at this stage she was solely suffering from the tuberculosis which was latent in her mother.

Just like Marie and Pierre, Irène and Frédéric worked together on research. Quite soon they obtained some exciting experimental results but, because they were not so interested in the theoretical side of physics, they failed to interpret them correctly. As early as 1920 Rutherford had speculated on the existence of the neutron. What the Joliots had found essentially verified his conjecture, but they did not realize this and it was left to Rutherford's colleague Chadwick to establish the connection. This was not to be the only time when the Joliot-Curies just missed making a fundamental discovery. In 1932, they went off for a fortnight to the scientific station at the Jungfraujoch, high in the Bernese alps. Here they combined research into cosmic rays with a certain amount of skiing. Shortly afterwards American researchers announced the discovery of the positron, and, when the Joliots came to analyse more closely the results they had just obtained, they saw that they too had found evidence of its existence without realizing it.

They had been invited to give a report on their latest experimental work at the forthcoming Solvay conference in Brussels, where Langevin was to be chairman. Frédéric made the report, but in the subsequent discussion the interpretation he gave of their results was strongly attacked, particularly by Lise Meitner. It was generally believed that they must be mistaken. After returning to Paris the Joliots set about vindicating themselves. On a historic January day in 1934 they found conclusive evidence. Artificial radioactivity had been created, and a new era had begun. Almost at once the news spread around the world. For the Joliots the immediate consequence was the award of the Nobel prize in chemistry for their synthesis of new radioactive elements (the physics prize went to Chadwick). They used some of the prize money to move out of central Paris to Sceaux; and they also acquired a chalet in the Alps, where they used to go skiing. Sadly, Marie Curie was no longer alive to hear of their success.

Both the Joliots were active in left-wing politics. Irène had been influenced by the radical politics and anticlericalism of her grandfather Curie and, like her mother, she took a keen interest in the social and intellectual advancement of women. In Frédéric's case it was his liberal-minded mother who was more influential, while as a student he had been strongly influenced by the left-wing views of Langevin. The Joliots firmly supported the republican cause in the Spanish Civil War. While strongly disagreeing with its policy of non-intervention, they joined the French Socialist Party. In 1936 Irène served briefly as Under-Secretary of State for national education in Léon Blum's Popular Front government. Friends were surprised that she was prepared to do so, but the fact that this was the first time women had ever been included in a French government gave it a special significance. In France, women still did not have the right to vote in elections. She explained: 'Fred and I thought I must accept it as a sacrifice for the feminist cause in France, although it annoyed us very much.' Attendance at any sort of committee meeting was anathema to her, but the position gave her the opportunity to link science and national development. It was agreed that she would hold the office only for a short while, after which her friend Jean Perrin would take over and she could return to her scientific work, but before this could happen Léon Blum's government was forced out of office in the aftermath of severe economic distress and social unrest.

In 1935 Frédéric was made professor of nuclear chemistry at the Collège de France; two years later Irène was appointed professor at the Sorbonne. Frédéric designed the first cyclotron built in western Europe and, by the start of 1939, was using it to demonstrate fission by physical means, proving that many neutrons are produced by fission, that such reactions can be developed in explosive chains and that nuclear reactions can be controlled to release great quantities of energy. However, during the same year, fearful of Hitler's power and the misuse of the chain reaction, the Joliots suddenly ceased publishing their research.

After the Second World War had broken out, Frédéric placed the principle of the nuclear reactor on record and deposited it in a sealed envelope at the Paris Academy, where it remained until it was opened in 1949. During the war he, as a captain in the artillery, was responsible for coordinating the war-effort of various French scientific laboratories. When he heard that the Germans were trying to acquire the Norwegian stock of heavy water, the world's largest, he realized that it must be for use in atomic research. In a remarkable coup he negotiated the loan of the heavy water for the duration of the war and organized its transport to France, under the eyes of the Germans. The consignment arrived at the Collège de France, but events were moving rapidly. The German army had broken through and was about to enter Paris. The material was promptly sent on to England for safety, thus seriously impeding the German project to build a nuclear reactor. He also thwarted German efforts to seize the cyclotron and the stock of radium. In June 1940, certain French scientists were told to hurry to Bordeaux and await further orders. When they had arrived in Bordeaux they were taken to a British ship, which was ready to evacuate 150 scientists and engineers. However, some of them did not get the message in time; others came without their families and, realizing only on arrival that the order meant evacuation, did not wish to leave alone, so that only fifty sailed in the end; most of those who remained were in non-occupied France. All efforts to convince Joliot that he had better take the opportunity to leave for England were in vain, but two leading members of his research group did so and resumed their work in the Cavendish, where they played a vital role.

Unlike Irène, Frédéric joined the French Communist party, to which many of the scientific left belonged. 'I was impressed', he said, 'by the generosity, courage and hope for the future that these people in my country had. They seemed willing to do the most to give France social reform.' Until then it had been difficult to be a Communist in France because of the pact between Hitler and Stalin, but now the Germans repudiated this and invaded Russia. As a leader of the National Front, centred at the university, Frédéric stood in the forefront of the Resistance movement. Pretending to be still busy with theoretical investigations of the atom, he was actually directing the manufacture of explosives and radio equipment for the Resistance, and he also hid the staff of a clandestine newspaper in the laboratory. When the Germans became suspicious, he sent Irène and the two children to Switzerland, while he remained in Paris under an assumed name.

As we shall see, scientists in other countries had by then taken the lead in nuclear research. Ten years previously, however, it was France that led the world in research on the possible liberation and utilization of atomic energy. Only two days after the liberation of Paris Frédéric was appointed director of the National Centre for Scientific Research. He was promoted to the rank of Commander of the Legion of Honour, decorated with the Croix de Guerre and elected to the Paris Academy. Then President de Gaulle named the Joliots as members of the French Atomic Energy Commission. Frédéric, the High Commissioner, stated that 'French science does not want to have anything to do with atomic research except for peace. All our efforts are being utilized in the development of this tremendous energy for the advancement of humanity.' Irène, speaking of her mother, said that, like Pierre Curie, she had strongly believed that science should resolve human problems in the sense of making happier lives possible. 'Its use for destruction seemed to her a desecration', she continued, 'In her eyes no political consideration would have justified the use of the atomic bomb.'

In the same year of 1946 Irène was appointed director of the Radium Institute, founded by her mother and endowed by Carnegie. She was now forty-nine years old and failing in health, but nothing could keep her away from her work. Chadwick has given a picture of her at this time:

> her parents were both persons of strong and independent mind, and Mme. Joliot-Curie inherited much of her character as well as her scientific genius. She knew her mind and spoke it, sometimes perhaps with devastating frankness, but her remarks were informed with such regard for scientific truth and with such conspicuous sincerity that they commanded the greatest respect in all circumstances. In all her work, whether in the laboratory, in discussion or in committee, she set herself the highest standard and she was most conscientious in the fulfilment of any duties she undertook.

In 1949 Irène proudly watched Frédéric direct the construction and successful demonstration of France's first atomic pile. By now, however, the war-time alliance with the Soviet Union was over, and their outspoken left-wing views were creating problems. The Joliots supported uncritically the policies of Stalin, while being highly critical of the USA. On a visit to America during the Cold War, Irène was briefly detained on Ellis Island; when she applied for membership of the American Chemical Society, her application was rejected. In 1950, French science was shocked when Frédéric was removed from his post on the Atomic Energy Commission. Irène served her full term of five years but was not included in the reorganization in 1951. Time and again she denounced scientific secrecy and appealed for the free circulation of scientific ideas and discoveries. In the 1950s she had to undergo surgery repeatedly; her health was deteriorating rapidly. 'I'm getting lazy', she told her friends, but she was wasting away from too much radiation, contracted during her war work. On March 17, 1956 she died of leukaemia in the hospital named after her parents. So controversial were the Joliots that Irène, the second most famous French woman scientist, was given a state funeral only after heated debate.

Frédéric's devotion to science was accompanied by an outspoken dislike of many traditional institutions (clerical and political) and a deep concern for world peace. He was first president of the World Federation of Scientific Workers and of the World Council for Peace. He also played an active part in the formation of the Educational, Scientific and Cultural Organization and the Atomic Energy Commission of the United Nations. Frédéric succeeded Irène at the Sorbonne, while keeping his chair at the Collège de France, but on August 14, 1958 he too succumbed to radiation sickness. He was also given a state funeral and was buried at Sceaux, next to the grave of Irène. Various memorials were dedicated to his memory in France; the Soviet Union named one of its warships after him, also a mountain that had just been climbed for the first time and a crater on the far side of the moon, which had been observed by one of its satellites.

ENRICO FERMI (1901–1954)

The Fermis came from a rich agricultural region near Piacenza, in the Po valley of northern Italy. Stefano Fermi, Enrico's grandfather, was the first member of the family not to work the soil with his own hands. He had obtained a modest position in the service of the duke of Parma, one of a number of minor princes who ruled in the Po valley after Napoleon's downfall. The reactionary princelings were under strong Austrian influence, but they seem to have been close to their subjects. At his death in 1905 at the age of eighty-seven, Stefano owned a small house and some land in Caorso, near Piacenza. His wife Giulia (née Beronzi) was an example of the type of Italian countrywoman often described by nineteenth-century Italian novelists. With a large family, she worked long hours in the home and possessed all the domestic virtues. She could read but not write. She was devoted to the church and brought up her children in the Catholic faith; only Alberto, Enrico's father, left it.

Alberto Fermi, the second son of Stefano, was born at Borgonure, near Piacenza, in 1857. Probably he attended one of the technical high schools which were instituted shortly after the unification of Italy. These were intended to prepare young people of limited means for useful careers. Certain private railway companies later taken over by the state employed him in various supervisory and administrative capacities until, by the time he retired in 1921, he was a divisional chief inspector of the Italian State Railways. His wife Ida (née de Gattis) was born in the Adriatic city of Bari in 1871. She was the daughter of an army officer and had studied electrical engineering at university, after which she taught in elementary schools until she married. She was fourteen years younger than her husband, unusually intelligent and able. They were living in Rome when their three children were born in successive years: Maria in 1899, then Giulio and finally Enrico, on September 29, 1901.

As babies, all three were sent out to nurse in the country. Because of his delicate health, Enrico did not return home until he was two and a half years old. His mother taught him to read and write; soon he began to show an unusual mathematical ability and a prodigious memory. At the local high schools, first the *ginnasio* for five years, then the *liceo* for two more, he was much the best pupil of his year, with an excellent scholastic record. Because of the habits of order and discipline he had learned from his mother, he enjoyed considerable free time, most of which he seems to have devoted to scientific studies. In 1915 his parents were devastated when their eldest son Giulio died in hospital before a minor operation.

While Enrico received encouragement in his studies from members of the family, the influence of a colleague of his father's, Adolfo Amidei, was more important. Amidei lent him books on mathematics and, after becoming convinced that he was a prodigy, advised him to apply to enter the highly selective Scuola Normale Superiore in Pisa, the best place in the country for physics. Fermi spent four years in Pisa studying the classics of physics, works that he could quote almost verbatim years later. Because he knew some French and German, he could also read the latest papers on the subject. His exceptional abilities were soon recognized; he understood the old quantum theory better than anyone else in Italy at the time. He also proved competent in experimental work, although the physics laboratory of the university was in a deplorable state of neglect. In 1922 Fermi obtained his doctorate, and the following year wrote his first important research paper, showing that, in the vicinity of a world line, space-time is Euclidean. This impressed Tullio Levi-Città, the main Italian relativist.

The post-doctoral career of a young scientist in Italy at this time often began by his becoming an assistant to a professor and then obtaining the *libera docenza*, the right to give lectures, equivalent to the *venia legendi*. After a few years in this lowly state, the next stage was to apply for a professorship in a national competition called a *concorso*; usually three of the candidates were chosen and assigned to vacancies at different universities. The candidates submitted publications, but influence was important too. Subsequent promotion was usually by seniority, from one university to a better. The ultimate accolade was election to the venerable Accademia dei Lincei.

After Mussolini's march on Rome in 1922, Fermi considered the possibility of emigrating, but first he needed to establish a reputation outside Italy. He won a fellowship for post-doctoral study in natural sciences offered by the Italian Ministry of Education and used it to spend a year in Göttingen, at the Institute where Born was professor of theoretical physics. Despite receiving a warm welcome from the Borns, he did not enjoy the experience, feeling isolated and foreign. The faculty, he found, assumed an attitude of omniscience while, despite his quite adequate knowledge of German, he was not welcomed into the group around Heisenberg and Pauli. Although they became more friendly later, it meant that his year at the Georgia Augusta was not as valuable as it might have been. When he returned from Germany Fermi obtained a temporary position as assistant to Senator Orso Mario Corbino, the senior physicist at the University of Rome. He was then awarded a three-month fellowship from the International Education Board to visit Leiden and work with Ehrenfest in 1924.

From his student years onwards Fermi used to return to the family home whenever he could. His mother never really got over the loss of her eldest son; she died in 1924, and his father died three years later. That left only Enrico's sister Maria; they lived together until 1928, when Enrico married. His bride was Laura Capon, the daughter of a highly cultured and respected family of non-observant Jews; her father was an admiral in the Italian navy.

Fermi was obliged to publish in Italian, but he also published the same work in German translation, to make it more accessible. Later, after the Nazis had come to power, he published in English instead. The senior Italian physicists had agreed on a scheme to provide junior positions for their able youngsters, which resulted in Fermi going to Florence. When a chair fell vacant at the University of Cagliari in Sardinia, Fermi applied for it but was unsuccessful. However, the following year he entered a *concorso*, which resulted in him being appointed professor of theoretical physics at the University of Rome, a far more important post.

Now twenty-six, Fermi was determined to do all he could to modernize the study of physics in Italy. He gave semi-popular lectures at the annual meetings of the Italian Society for the Advancement of Science. At the university the seminar he ran attracted many young physicists. One of them was Ettore Majorana, a theoretician of outstanding ability, who later mysteriously disappeared; he was excessively modest about his research but the little he published has proved to be of great importance. Fermi also wrote an *Introduzione alla fisica atomica*, which was published in 1928, and with Corbino he set about strengthening the experimental side of physics in Rome. Unfortunately Fermi antagonized the professor in charge of that side, who had the power to block any reform. In 1929 Fermi was the only physicist appointed to the new Accademia d'Italia, created by Mussolini to replace the Lincei which he believed was too much a focus of opposition to fascism. The appointment gave Fermi the title of Excellency and put him in a better position to obtain support for his research, but the process of making university appointments remained under the control of the old guard.

Fermi crossed the Atlantic for the first time during the summer of 1930, to collaborate with Ehrenfest in teaching during a summer school at the University of Michigan. Someone who met him then described him as 'a very young and pleasant little Italian, with unending good humour, and a brilliant and clear method of presenting what he has to present in terrible English'. This was such a success that he returned twice more and increasingly came to appreciate the American way of life. On one such visit he bought a car, named it 'the flying tortoise' and, when it broke down, Fermi, who was a capable mechanic, repaired it in a garage so efficiently that the manager offered him a job on the spot. In the summer of 1934 he went on a lecture tour of South America sponsored by the Italian government, stopping in London on the way back to report on the latest research of his team in Rome at an international physics conference. They had been bombarding all kinds of elements with neutrons and had found some puzzling results in the case of uranium. The suggestion was made at the time that the uranium nucleus was being split, but other interpretations seemed much more likely. Some believe that this experimental work was the cause of the cancer which led to Fermi's death twenty years later.

By 1937 Fermi's influence in Italy was on the wane. He was unsurprised to find that he had not been chosen to succeed Corbino in Rome and began to think more and more seriously about the possibility of emigrating to the USA. The subjugation of Austria by Germany two years later was a sign that Italy was powerless to stand in Hitler's way. In 1938 Mussolini, having signed the Italian–German alliance, promulgated the *Manifesto della razza*, on the German model, which classified the Jews as aliens. The racial laws did not affect Fermi or his two children directly, but his wife's family was Jewish and it was not clear what the consequences would be for her.

Fermi had previously received offers of positions from several American universities and he now decided that the time had come to follow these up. Before long he had accepted an appointment to a chair at Columbia University in New York City. Only a few people knew that he was leaving Italy permanently; in order to avoid trouble with the fascist authorities it was necessary for him to pretend that it was just another visit to America. Towards the end of the year, he was awarded the Nobel prize in physics 'for his demonstrations of the existence of new radioactive elements produced by neutron radiation, and for his related discovery of nuclear reactions brought about by slow neutrons'. Shortly afterwards the family left for Stockholm to attend the Nobel ceremonies and then went to Copenhagen to spend a few days with Bohr, before sailing for New York. Once Fermi had made a decision he never looked back.

Like his father, Fermi's nature was that of a loyal and efficient public servant. In his last years in Italy he had served in a few high-level advisory posts in the Ministry of Education and as consultant for a few industrial firms, but he avoided controversy. His interests outside science were rather limited. Physically he was strong and had plenty of stamina. He was, after all, of peasant stock. He would challenge younger physicists to play tennis in the heat of a New York summer day and, after an exhausting hour or so under the broiling sun, would remark on their lack of vigour. When he lectured, he did so with a strong Italian accent.

In 1939 Fermi was at the midpoint of his career and about to become involved in events he could hardly have foreseen. The news of the discovery of nuclear fission, brought to America by Bohr, reached him soon after he had landed in New York. Then Germany invaded Poland and the Second World War began, although the USA was not a participant until later. At first the development of radar had top priority among leading scientific administrators in the USA; research into atomic energy received no more than the ordinary level of support. Fermi was concerned about this and, together with the *émigré* Hungarian physicists Leo Szilard and Eugene Wigner, drafted an important letter for Einstein to send to President Roosevelt pointing out the dangerous military potentialities offered by nuclear fission and warning him that Germany might be developing these weapons. This letter helped to initiate the American efforts that eventually produced the nuclear reactor and the fission bomb. Until there was a change of research priorities, a large proportion of the physicists involved in America's earliest efforts to release nuclear energy came from Europe, and Fermi was one of them.

The first step was to try to produce a chain reaction. Work on this project was concentrated in Chicago but ironically Fermi, being an enemy alien, was confined to New York, and his mail was subject to censorship. However, in 1942 Roosevelt announced that Italians would no longer be considered enemy aliens, and Fermi was able to join the other scientists working in Chicago, although his mail was still being censored. One of the other physicists involved described how Fermi played a leading part in the work:

> Fermi possessed a sure way of starting off in the right direction, of setting aside the irrelevances, of seizing all the essentials and proceeding to the core of the matter. The whole process of wresting

from nature her secrets was for Fermi an exciting sport which he entered into with supreme confidence and great zest. No task was too menial if it sped him towards his goal. He thoroughly enjoyed the whole of the enterprise. The piling of the graphite bricks, the running with the short-lived rhodium foils, and the merry clicking of the Geiger counter which effected the measurement, all was done with great energy and with obvious pleasure, but by the end of the day, in accordance with his plan, the results were neatly compiled, their significance assessed, and the progress measured, so that early in the morning of the following day the next step could begin.

It was a feature of the Fermi approach never to waste time – to keep things as simple as possible, never to construct more elaborately or to measure with more care than was required by the task at hand. In such matters his judgement was unerring. In this way, step by step, the work sped forwards until, in less than four short years, Fermi had reached his goal. A huge pile of graphite and uranium had arisen in the football stadium of the University of Chicago. When Fermi stood before that silent monster, he was its acknowledged master. Whatever he commanded it obeyed. When he called for it to come alive and pour forth its neutrons, it responded with remarkable alacrity; and when at his command it quietened down again, it had become clear to all who watched that Fermi had indeed unlocked the door to the neutron. Early in the afternoon of December 2, 1942, the chain reaction was demonstrated successfully.

Soon further research on the project was transferred to the Los Alamos National Laboratory, where Fermi became a sort of oracle to any physicist in trouble, as was the *émigré* Hungarian mathematician von Neumann. Three years later the war in Europe was over, and the race to develop the atomic bomb before the Germans could do so had been won decisively. Although some of the other physicists on the project remained at Los Alamos, Fermi decided to return to normal academic work. He accepted the offer of a professorship at the University of Chicago, which was developing a promising science programme. The teaching aspect of the position appealed to him very much, as did the experimental facilities at the Argonne National Laboratory, including the powerful cyclotron. He served as president of the American Physical Society and helped to defend Oppenheimer at the notorious security-clearance hearing, of which more later.

Fermi liked to spend the summer away from the university, but at first only within the USA. It was not until 1949 that he returned to Europe and, after attending some important conferences, went back to his homeland for the first time since 1938. The new generation of physicists who had emerged in Italy after his departure saw and heard the almost-legendary Fermi.

In the summer of 1954 he made what proved to be his last visit to Europe. His health was failing; an insidious illness had attacked him and resisted diagnosis. With great strength of will he tried to carry on as before, but, when the disease was finally diagnosed, it proved to be incurable, caused by a malignant stomach tumour. He died on November 28, 1954 in Chicago, just after his fifty-third birthday, survived by his wife Laura and their two children. There is a memorial in the Florentine church of Santa Croce, alongside those in honour of other great Italian scientists.

10 From Heisenberg to Yukawa

Our last five remarkable physicists were born in the seven years between 1901 and 1907. They came from Austria, England, Germany, Japan and America.

WERNER HEISENBERG (1901–1976)

In the great revolution of fundamental physics which started with the ideas of Planck and culminated with the impressive breakthrough of the 1920s, and in the completed framework of quantum mechanics, Heisenberg made many important contributions. Among these there were two, the basic treatment of quantum transitions and the formulation of the uncertainty principle, that were so original and impressive that they came close to the popular image, so often unrealistic, of great new concepts growing out of the thoughts of a single individual of genius. In the two years from 1925 to 1927, which saw the emergence of a new set of principles of physics, which have since then been refined and widely applied but not fundamentally changed, the steps taken by Heisenberg were large and decisive.

Werner Karl Heisenberg was born on December 5, 1901 in the elegant Würzburg suburb of Sanderau. The family moved to Munich in 1910 and he came to regard it as his home town, retaining throughout his life an intense affection for the city. His father August, the son of a master locksmith, was a domineering man of great vigour. Werner had a younger brother, Karl, who later emigrated to America and became wealthy, as well as three sisters and an elder brother, Erwin, who became an industrial chemist. August himself held the chair of mediaeval and modern Greek philology at the university and later became professor of Byzantine studies. We know little about his wife Anna (née Wecklein), except that her father, a Greek scholar, was headmaster of the renowned Maximilian Gymnasium in Munich, the school that her son Werner attended.

Disabling allergies and illnesses recurred throughout Werner's life; at the age of five he nearly died of a pulmonary infection. He was a sensitive youth, who sought security in mathematics and in other formal subjects such as grammar and science. He distinguished himself early at school, particularly in mathematics: none of the work seemed to cause him any great

effort. He showed an early interest in physics and was particularly fascinated by the possibility of applying mathematics to practical situations. When the First World War began, his father August, being a reserve officer, was called up; after a brief experience of combat at the front he was transferred to garrison duty in Munich. The gymnasium building was commandeered and the students spent much of their time on paramilitary training, leaving them with little time for schoolwork. In the spring of 1918 boys as young as sixteen were drafted for auxiliary service, and Werner was sent to work on a farm. This involved long hours of manual labour, which left him too tired to read at night the books he had brought with him, such as the philosophical works of Kant.

When the First World War ended in November 1918, conditions in Germany were chaotic, with political authority passing back and forth between different factions. At one point Heisenberg and other boys worked as messengers for a group that was trying to restore law and order in the city. The duties were not onerous and he was able to make progress with his reading, in which philosophy played an important part. More than did the farmwork, this provided many contacts and later friendships with young men of a similar age, with whom he had many earnest discussions. The youth movement in Germany included various such groups of young people who were dissatisfied with what their elders had made of the world and who were impatient with old customs and old prejudices. They felt an emotional need for a new leader who would restore Germany's greatness, to whom they would give total commitment. In his autobiographical writings Heisenberg is studiously vague about which particular group he belonged to, but it was probably the Weisser Ritter, a group that was strongly antagonistic towards science, especially physics. Thus Heisenberg had to choose between two extremes, both of which seem to have attracted him.

After graduating from the gymnasium in 1920, Heisenberg matriculated at the University of Munich. During the first two years there, he submitted four promising research papers on physics. His mathematical background was uneven; he knew a lot about number theory but had only just encountered calculus. When he came across Weyl's book *Raum–Zeit–Materie*, he was both attracted and repelled by the sophistication of the mathematical arguments and the underlying physical concepts. With the idea of becoming a mathematician, he went to ask Ferdinand von Lindemann, the old-fashioned professor of mathematics, for permission to attend his seminar, even though he was a mere freshman. As soon as he mentioned having read Weyl's book, von Lindemann turned him down.

After this rebuff Heisenberg applied to Sommerfeld, a friend of his father's, and was immediately accepted. Sommerfeld was not only a great theoretical physicist with extensive experience in all parts of the subject and an intimate knowledge of its frontiers, but also probably its greatest teacher. His seminars and colloquia attracted students and young scholars from far afield, even from America, and helped to make Munich a world centre for theoretical physics. Often before or after a colloquium Sommerfeld could be seen at the Hofgarten cafe discussing problems with members of the audience and covering the marble tables with formulae. Sometimes he would invite them to join a skiing party on the Südenfeld, two hours by train from Munich, where he was part owner of a ski-hut. In the evenings, after a simple meal, the talk would turn to mathematical physics; and this was when receptive students might learn what he was currently thinking about. The forbidding martial impression made by Sommerfeld initially very soon gave way to a feeling of benevolence and helpful authority. He trained nearly a third of Germany's professors of physics; four of his former students were awarded Nobel prizes. Theoretical physics is a subject that attracts youngsters with a philosophical mind who speculate about the highest principles without sufficient foundations. It was just this type of beginner that he knew how to handle, leading them step by step to a realization of their lack of actual knowledge and providing them with the skill necessary for fertile research.

As soon as Heisenberg became a student at Sommerfeld's institute he met the brilliant Wolfgang Pauli, another student, who was somewhat senior to him and became his mentor. Their close friendship, during which they frequently exchanged views about major mathematical, physical and philosophical problems, lasted until the death of Pauli in 1958. Despite having completed an able thesis on hydrodynamics, Heisenberg only just succeeded in passing his doctoral examination, because his strengths on the theoretical side of physics were offset by a poor performance on the experimental side.

Sommerfeld took Heisenberg to the Georgia Augusta to listen to a two-week series of seminar talks by Bohr, who made a deep impression on Heisenberg. When Sommerfeld went for the winter semester 1922/3 to the University of Wisconsin as a visiting professor, he arranged for Heisenberg to spend this period with Max Born at Göttingen. Born was enthusiastic about the young man and before long they were collaborating on research. 'Heisenberg is at least as talented as Pauli', Born reported to his friend Einstein, 'but personally more pleasant and delightful. He also plays the piano very well.' 'I have grown very fond of Heisenberg', he confided to Sommerfeld, 'he is liked and esteemed by us all. His talents are extraordinary, but his friendly, modest attitude, his good spirits, his eagerness and enthusiasm are especially pleasing.' He described Heisenberg as looking like a simple peasant boy, with short fair hair, clear bright eyes and a charming expression.

The next year, after habilitating at Göttingen at the unusually early age of twenty-two, Heisenberg spent the summer vacation with friends from the youth movement on a walking tour and then went to Copenhagen as a research associate. He already knew how deeply Bohr understood the problems of physics, and how much he could learn from him. Their discussions, often conducted on walks in the countryside, also ranged over many other fields of human life and affairs, and here Bohr's wisdom and warmth made a deep impression on his young disciple. They went on a walking tour of Sjaelland, the island on which the Danish capital is situated. When he returned the following year, Bohr wrote that 'he is as congenial as he is talented' and 'in spite of his youth he has succeeded in realizing hopes of which earlier we hardly dared dream . . . in addition his vigorous and harmonious personality makes it a daily joy to work together with him towards common goals.' Heisenberg later said that in Göttingen he learned mathematics, in Copenhagen physics. The affection and mutual respect he and Bohr felt for each other continued to deepen over the next fifteen years.

In 1925, at the age of twenty-three, Heisenberg wrote the paper that laid the foundations of quantum mechanics on which subsequent generations have built. This was not just an extension or elaboration of the work of others, but an unexpected, radical new departure, which abandoned the basic notions of the old classical physics, such as electrons moving in orbits, replacing them by a much more abstract description. Less than a year later Schrödinger, as we know, published his theory of wave mechanics, which at first appeared to be an alternative theory to Heisenberg's. However, the two theories turned out to be essentially the same. Schrödinger's mathematics is in many ways easier to handle, but both points of view are needed in order to develop a real understanding of the physical world.

In 1927 Heisenberg was offered chairs in both the Universities of Leipzig and Zürich; he chose the former and became the youngest full professor in Germany. At his first seminar he had an audience of two, but soon students and other collaborators were attracted, and frequently senior visitors attended. His duties were not light. It was then normal for the professor to give the main undergraduate lectures in theoretical physics, usually on a four- or six-semester cycle, and to set examinations on the course-work, which he had to mark. There were also the usual faculty and committee meetings, but, in spite of these demands on his time, Heisenberg was always accessible to his students. He remained as before – informal and cheerful in manner, almost boyish, with a modesty that verged on shyness. His weekly seminar was preceded by tea, and for this he would go out to a nearby bakery for some pastries. After a strenuous discussion, and during other free periods, the whole group would descend to the basement and play table tennis. Heisenberg was a very good player and could beat everyone else, until a Chinese physicist arrived, who was equally skilful, if not more so.

Problems, difficulties and new ideas in physics were debated very intensely. Heisenberg was able to help his students particularly through his powerful intuition. Usually he did not pay too much attention to the mathematical details of their work, as long as they could see where they were going, but he needed to grasp the physics of their problem himself. Once he had done so, he was usually able to guess the answer, and he was usually right. Naturally Heisenberg's own output of papers during this period did not match the pace of previous years; a further distraction was a lecture tour of the USA in 1929. Quantum mechanics was now essentially complete,

8

48

and the next task was to work out its consequences and to see how it would explain the many mysteries, paradoxes and contradictions in atomic physics. He found that it was an exhilarating experience to see how easily the solutions to the old puzzles fell into place.

When Hitler seized power in Germany and the Nazi ideology took hold in the universities, Heisenberg, like many other academics, was deeply shocked by the anti-intellectual attitude of the regime. In a book of reminiscences, published in 1969, he describes an imaginary conversation with a student who is a leader of the Hitler Youth. One feels that, while formally maintaining his refusal to have anything to do with Nazi gatherings and other activities, he can see something to admire in the ideas of his companion. The student himself deplores the anti-Semitism and other destructive features of the movement, but insists that its essential aim is to create a better world, to fight corruption and dishonesty, and to restore respect for Germany.

However, the disastrous aspects of Nazi policies began to dominate to such an extent that Heisenberg and a few colleagues soon started to talk of resignation. He went to see Planck about this; his advice was to remain. However many professors resigned, he said, it would not affect Nazi policy. Heisenberg would have to emigrate, Planck went on, and while undoubtedly he would find a position abroad, he would be taking it away from someone else who was being forced to leave Germany. The present regime was bound to end in disaster, Planck concluded, and after that happened people like Heisenberg would be needed as leaders.

Many German physicists faced the same problem as Heisenberg; other than those dismissed or expecting to be dismissed, hardly any decided to leave. Of those who remained, von Laue was outstanding for his uncompromising stand, his proud aloofness and his refusal to cooperate with the regime. To have taken that position was not in Heisenberg's character. In referring to his attraction to military service, which involved annual training as a reservist, he remarked that 'it is nice not to have to think, for a change, but only to obey'. He tried to carry on as before, to maintain the old atmosphere, in spite of the loss of his Jewish colleagues; he commented that he rather envied them, since they had no choice. This was not the only occasion when he displayed amazing insensitivity to the effects of Nazi persecution on individuals.

In 1933 Heisenberg was awarded the prestigious Max Planck medal of the German Physical Society. No Nobel prize for physics had been awarded in 1931; the 1932 prize had been deferred and now it was announced that this had been awarded to Heisenberg. Meanwhile the grip of the Nazis on German science was steadily tightening. Visits by German scientists to foreign countries required official approval, likewise visits by foreign scientists to Germany. Although the Nazis had reservations about Heisenberg's political attitudes, his prestige was such that he was seldom if ever refused permission to travel. Mainly through visits to Copenhagen, he kept in touch with the latest ideas in physics.

In 1935 Heisenberg was proposed for the chair in Munich in succession to Sommerfeld, who was retiring. This was an attractive opportunity, both because it meant succeeding his respected teacher and particularly because of his fondness for the city. However, the proposal was attacked by those who opposed relativity and quantum theory as 'Jewish physics'. The authorities ruled against his appointment in favour of a nonentity. After this personal attacks on Heisenberg became more virulent, until eventually instructions from a high party level put an end to them and Heisenberg was appointed after all.

By this time Heisenberg had found added strength and support; in January 1937 he had made the acquaintance of Elisabeth Schumacher, daughter of the great Berlin economist, and they were married three months later. Although it was a successful marriage, Heisenberg always put his career first. Twins, a son Wolfgang and a daughter Anna Maria, were born in 1938; eventually there were to be seven children, all of whom shared their parents' love of music. Of the twins, Anna Maria Hirsch became a physiologist while Wolfgang, a lawyer by training, worked for a foundation concerned with science and politics. Jochen became an experimental physicist at MIT while Martin became a professor of biogenetics. Barbara Blum married a physicist. Christine Mann became a teacher and her husband a physiologist. The youngest, Verena, became a technician in a physiology laboratory.

By the summer of 1939, when it was clear that war was inevitable, Heisenberg purchased a country house at Urfeld, in the Bavarian Alps, as a refuge for his family (there were at this stage three children) in case of need. He revisited the USA and lectured at the universities of Michigan and Chicago, where many of his old friends and colleagues tried to persuade him to leave Germany because of the impending disaster, in which his presence could not achieve anything. However, Heisenberg did not agree; above all, he believed that leaving would be disloyal to the young people in his research group, who would rely on him for guidance in keeping science going and whose responsibility it would be to rebuild science after the war. These young people could not find positions abroad as easily as he could, and he would feel that he was taking an unfair advantage.

When the war came, Heisenberg was appointed chief technical consultant for research on nuclear fission after being excused normal military service on medical grounds. For the next five and a half years this took up most of his time and energy. He developed the theory of a nuclear reactor. Experiments indicated that a system of uranium metal and heavy water of suitable size could sustain a chain reaction. A successful experimental reactor could have been constructed, but much more time would have been needed to produce anything of practical value. The type of reactor the American physicists constructed had been dismissed as impracticable because Heisenberg had miscalculated the critical mass required; since atomic research was everywhere top secret by this time, mistakes were liable to remain uncorrected. It has been suggested that Heisenberg deliberately made the production of an atomic bomb appear impracticable, because he did not want the Nazis to have such a weapon, but the evidence for this is unconvincing. As regards atomic energy, for Germany this was a long-term project and unlikely to affect the outcome of the war.

Heisenberg had reported as early as December 1939 that, although energy could be generated from ordinary uranium if it were used in conjunction with heavy water or graphite to slow down the neutrons, it would be necessary to use enriched uranium-235 to produce an explosive. These two lines of attack became the chief objectives of the German atomic-energy programme and for a year or two progress was remarkably rapid, in spite of rather lukewarm official backing. Heisenberg was at first a consultant to the Kaiser Wilhelm Institute for physics in Berlin-Dahlem, where much of the research was being done, and during that time he continued to live in Leipzig, where some unrelated work was being done at the physics institute. The Kaiser Wilhelm Institute had been placed under military control but, after reorganization, in 1942 it was returned to the Kaiser Wilhelm Gesellschaft and Heisenberg was appointed Director. He was also appointed full professor at the University of Berlin and elected to the Berlin Academy. When he moved to the German capital in the spring of 1943, his family went to live in their country house in Urfeld, but later that year, when the air raids became more intense, the laboratory was evacuated to Tailfingen.

Several reactor experiments were set up but none actually achieved criticality; indeed, the final experiment under Heisenberg's control was carried out only a few weeks before the arrival of Allied troops. It came close to being critical and it became clear that a small increase in scale would achieve that state, but time was running out; the end of the war was near and there could be no chance of practical results before then. It was not known in Germany that Fermi had already achieved a chain reaction in Chicago three years earlier.

As soon as Germany had fallen to the Allied forces, an Anglo-American team searched Heisenberg's office and other places where work on the nuclear project might have been carried out, and concluded that the Germans had got nowhere near producing a bomb. However, to avoid the risk that the German nuclear physicists might be taken to the Soviet Union, ten of them, including Hahn and Heisenberg, were rounded up and escorted to Britain, where they were interned in Farm Hall, a pleasant manor house in the Cambridgeshire village of Godmanchester. Although they were still under guard, the internees enjoyed a good deal of freedom. Their rooms were bugged and it was hoped to learn from their conversations something about secret research in Germany, especially research into atomic energy. It was there that they heard in August 1945 the news of the atomic bomb used against Japan and realized that the Americans were far ahead. The German scientists were not only amazed that this revolutionary weapon of destruction had been produced but also mystified as to how it had been possible.

Why did Heisenberg work for his country on atomic energy, and why was the total achievement of the German project so slight? As regards the first question, Heisenberg was a patriot, technically in the army, and wanted Germany to win the war. It might have been different if the work had come close to making an atomic bomb, but it did not. He wrote 'We knew then that one could in principle make atomic bombs, and knew a realizable process, but we regarded the necessary technical effort as rather greater than in fact it was.' As regards the other question, the basic reason for the slow progress was that the work was not pursued with urgency. The authorities never instructed their scientists to make an all-out effort and did not give them the support and services such an effort would have required. Even the USA, with far greater industrial resources, without serious shortages and without interruption by air raids, was not able to make atomic bombs until after the end of the war in Europe. Also, to generate electricity by atomic energy was hardly a top priority in war-time. Moreover, there were errors, misjudgements and omissions, both in the scientific work and in the organization and planning

Although the subsequent German work on atomic energy was not in the short term aimed at producing weapons, Heisenberg and his colleagues were aware of the fact that they were working on a programme that could lead to that result. For example, a nuclear reactor, intended to produce power, could be converted into an explosive device. They decided to consult Niels Bohr, and in October 1941 Heisenberg went to Copenhagen; by this time Denmark was under German occupation, and the visit was by no means welcome. Just what was said is disputed, but it seems clear that they parted with Bohr so angry that he never forgave Heisenberg.

After the war, Heisenberg cited the meeting as evidence of his reluctance to help Hitler create the ultimate weapon of mass destruction. The purpose of his visit to Copenhagen, he suggested, was to share his qualms over nuclear weapons. In Bohr's recollection, on the other hand, Heisenberg said that 'There was no need to talk about details, since you were completely familiar with them and had spent the past two years working more or less exclusively on such preparations.' He concluded that under Heisenberg everything was being done in Germany to develop atomic weapons.

Either account could be inaccurate. Bohr was known as being better at talking than listening, and he could have misunderstood what Heisenberg said to him. Of course, Bohr was a citizen of a peaceful country, occupied without provocation by the armies of a hateful regime, and Heisenberg, although an old friend and disciple, was also a citizen of the occupying power. Soon after Heisenberg's visit the Germans occupied the Bohr Institute and, as we know, the Jewish or partly Jewish members, including Bohr himself, fled the country. There was a plan to appoint Heisenberg as director and staff it with German scientists, so as to coerce the remaining Danish scientists into contributing to the German war-effort. Heisenberg succeeded in preventing this happening.

In the autumn of 1943 Heisenberg visited occupied Holland. During the visit he took a walk with an old friend and colleague from Copenhagen, in which Heisenberg began to talk about history and world politics. He explained that 'It had always been the historical mission of Germany to defend the West and its culture against the onslaught of eastern hordes and the present conflict was one more example. Neither France nor England would have been sufficiently determined and sufficiently strong to play a leading role in such a defence, and so, perhaps, a Europe under German leadership might be the lesser evil.'

At the end of the war in Europe, German scientists were dispersed, laboratories were closed and many of the cities in which they were located were severely damaged. The Allied armies which had occupied the country were trying to get life back to normal. In the British zone, at least, the policy was to encourage the leading German scientists to resume their research and teaching. With their agreement, Göttingen was chosen as the best place to serve as the centre for the rebuilding of scientific research. They gathered there, under the auspices of what was now the Max Planck Gesellschaft. At first this functioned only in the British zone, but after a time it was able to extend its activities over the whole of Western Germany.

During the Göttingen period Heisenberg devoted much of his time and energy to questions of the future organization of science in the Federal Republic. He made a major contribution to the decision to start a nuclear-reactor programme. Whereas he pleaded strongly in favour of nuclear power, he opposed, equally strongly, any suggestion that Germany should make or acquire atomic bombs. In the autumn of 1958 the Max Planck Institute for physics and astrophysics, which Heisenberg had directed in Göttingen, moved to Munich and he was able to return to the city he had always loved. He remained director of the institute until his retirement in 1970. After that most of his writings were reviews or essays of a general nature; his reputation outside Germany was clouded by his failure to explain his war-time record satisfactorily. Five years later his health began to decline, and he died of cancer on February 1, 1976, at the age of seventy-four.

Heisenberg was undoubtedly a great physicist, the creator of quantum mechanics. Nevertheless, the account he gave of his work during the war, and his justification for it, are open to question; the whole truth may never be known, but there is considerable evidence to suggest that he would have regarded it as his patriotic duty to organize the production of atomic weapons, which might have affected the outcome of the war. When it was clear that Germany was going to be defeated, he told a Swiss colleague that 'It would have been so sweet if we had won.'

Paul Dirac (1902–1984)

Paul Dirac was a theoretical physicist in the same class as the best in Europe; according to Bohr he had the purest soul of them all. He contributed as much as anyone to the establishment of the new science of quantum mechanics. Between the ages of twenty-three and thirty-one he unveiled an original and powerful formulation of the theory, a primitive but important version of quantum electrodynamics, the relativistic wave equation of the electron, the idea of antiparticles and a theory of magnetic monopoles.

The future physicist was born on August 8, 1902 in Bishopston, a suburb of the city of Bristol. His mother Florence (née Hilton), the daughter of a ship's captain, came from Liskeard in Cornwall. His father Charles was brought up in the French-speaking Valais region of Switzerland but

emigrated to England in the 1880s and supported himself by giving French lessons. They had three children, the eldest being Reginald, who was two years older than Paul, and the youngest Beatrice, four years younger. All three were registered at birth as Swiss citizens, but they and their father relinquished Swiss citizenship and became British in 1919. Although he had lost touch with his own parents, who were originally French, Charles wished his children to speak French at home and speak it correctly. 'My father made the rule that I should only speak to him in French', recalled Paul later on. 'He thought it would be good for me to learn French in that way. However, since I found I couldn't express myself in French, it was better for me to stay silent than talk in English. So I became very silent at that time – that started very early.' His mother, who had worked in the library of Bristol University before marriage, could hardly speak French at all and so she usually took meals separately with the other children.

From 1896 Charles had been teaching in the Merchant Venturers Technical College in Bristol; after this was taken over by Bristol University, part of it was hived off to become a secondary school, which was where Charles now taught French. He had the reputation of being exceptionally strict in class. He was an outstanding linguist, able to speak eight or nine languages, and an enthusiast for Esperanto. Although he cared about his children and their future careers, he succeeded in alienating his sons. Reginald, the elder, wanted to be a doctor; his father Charles made him study mechanical engineering. After graduating from Bristol University with a third-class degree, Reginald took a job as a draughtsman at an engineering works, but gave it up and three months later committed suicide at the age of twenty-four.

Paul's relations with his father were always chilly. After he grew up they had little personal contact, although Charles was proud of his son's success and tried to understand what he did. When Paul was awarded the Nobel prize and was told that he could invite his parents to attend the ceremony in Stockholm, he chose to invite only his mother. He rarely visited Switzerland, because he associated it so much with his father. As a young man Paul never had a girl-friend and seems to have had a rather platonic conception of the opposite sex for some time. He confided to a friend: 'I never saw a woman naked, either in childhood or in youth . . . The first time I saw a woman naked was in 1927, when I went to Russia with Peter Kapitza. She was a child, an adolescent. I was taken to a girls' swimming pool, and they bathed without swimming suits. I thought they looked nice.'

Paul's mathematical ability became apparent even at primary school. He went on from there to the college where his father taught, at the age of twelve. Although the academic standards there were high, the teaching had a vocational orientation. Modern languages were taught for use, metal work was in the syllabus, as was shorthand, but classics and literature were not. However, the school was particularly strong in mathematics and science. Paul was soon so far ahead of his class in mathematics that he was allowed to work largely on his own. He was recognized as a boy of exceptional intelligence; Paul's schoolfellows remember him as silent and aloof.

After leaving school in 1918, Dirac entered Bristol University, where he studied electrical engineering. In 1921, after graduating, he looked without success for work as an engineer. He won an exhibition to St John's College, Cambridge, but did not take it up because he could not afford the additional expense of being a Cambridge undergraduate. Then he was offered two years' free tuition by Bristol University to return and study mathematics, where, unlike at Cambridge, he would be able to economize by living at home. This he accepted and, after a brilliant performance in the final examination, was awarded a government grant that enabled him to go up to Cambridge as a graduate student. Two years later he was given a more generous award from the Commissioners for the 1851 Exhibition. At Cambridge his research supervisor was Rutherford's son-in-law Ralph Fowler, who recognized in Dirac a student of exceptional ability. Before long Dirac was publishing research, first on statistical mechanics and then on quantum mechanics. At that time the Bohr–Sommerfeld quantum theory was the best available theory of atomic phenomena, but it had many shortcomings and contradictions. He attempted to find ways of improving it, but without success.

Dirac first met Bohr in May 1925 when the latter gave a talk in Cambridge on the fundamental problems and difficulties of quantum theory. He said later that 'people here were pretty much spell-bound by what Bohr said . . . while I was very much impressed by him his arguments were mainly of a qualitative nature, and I was not able really to pinpoint the facts behind them. What I wanted was statements which could be expressed in terms of equations, and Bohr's work very seldom provided such statements.' In the summer of 1925 Heisenberg came to address the Kapitza Club, which had become the unofficial Cambridge forum for discussing modern physics. Afterwards Dirac said that he did not remember him talking about the new ideas in his great paper that laid the foundations of quantum mechanics; the subject of his talk was something less exciting. However, when Fowler received the proofs of the paper on quantum mechanics and showed them to Dirac, he soon realized its revolutionary significance.

Dirac's first paper on quantum mechanics, the basis for his Ph.D. thesis of 1926, paralleled much of what was being done at the same time elsewhere, but he followed it up with more innovative work that immediately attracted the attention of theoreticians everywhere. Born was about to leave on a visit to America when 'The day before I left there appeared a parcel of papers by Dirac, whose name I had never heard . . . Never have I been so astonished in my life: that a completely unknown and apparently young man could write such a perfect paper.' Fowler arranged for Dirac to spend some time first in Copenhagen and then in Göttingen and Leiden. He enjoyed the informal and friendly atmosphere of the Bohr institute and had many long conversations with the Great Dane: 'I admired Bohr very much. We had long talks together, very long talks in which Bohr did practically all of the talking.' While respecting Bohr greatly for his depth of thought, he said he did not know that Bohr had any influence on his own work because Bohr tended to argue qualitatively, while Dirac was more mathematical. At

Göttingen he got to know Robert Oppenheimer, at the time a Ph.D. student, while at Leiden he mainly worked with Ehrenfest.

In 1927 St John's elected Dirac to a fellowship, after which he lived and worked in college. In the same year he was invited to the sixth Solvay conference, where he made important contributions to the discussion and had the opportunity of meeting Einstein and Lorentz. In 1930 he was elected to the Royal Society at the unusually young age of twenty-eight. He spent much of the following year at the Institute for Advanced Study in Princeton and, soon after returning to Cambridge, he was elected Lucasian Professor of mathematics; his teacher Fowler had been elected Plumerian Professor of mathematical physics the previous year. Sadly, his extreme rationalism now led Dirac into sterile byways after his amazingly successful early years. Few of his later contributions to physics had lasting value, and none had the revolutionary character of his earlier work.

In 1933 Dirac and Schrödinger shared the Nobel prize for physics 'for the discovery of new productive forms of atomic theory'. At first Dirac was inclined to refuse it because he so hated publicity, but changed his mind when Rutherford warned him that a refusal would attract even greater publicity. At the seventh Solvay conference Dirac gave a talk on 'Structure and Properties of Atomic Nuclei'. He liked to travel, and often went to the Soviet Union to see his friend Kapitza. In the academic year 1934/5, when he was on leave from Cambridge, Dirac returned to Princeton, mainly to revise his classic textbook *The Principles of Quantum Mechanics* for a second edition. There he developed a close friendship with Eugene Wigner, whose sister Margit Balasz was visiting from Budapest at the time. Her temperament was quite unlike Dirac's, spontaneous and impulsive, with strong likes and dislikes. There was an attraction of opposites, and in January 1937 they were married in London. From then until he retired in 1969 the Diracs lived in a house in Cavendish Avenue, Cambridge. The household included two children, a son and a daughter, from Margit's first marriage, both of whom took the name of Dirac, and two daughters from the second marriage; later Dirac's mother Florence came to live with them.

There are many stories about Dirac's personality, usually related to his taciturnity, for which he was inclined to blame his father, perhaps unfairly. Heisenberg, who knew Dirac well, recalled that 'We were on the steamer from America to Japan and I liked to take part in the social life on the steamer and so, for instance, I took part in the dances in the evening. Paul, somehow, didn't like that too much but he would sit in a chair and look at the dances. Once I came back from a dance and took a chair beside him and he asked me "Heisenberg, why do you dance?" I said "Well, when there are nice girls it is a pleasure to dance." He thought for a long time about it and after about five minutes he said "Heisenberg, how do you know *beforehand* that the girls are nice?"'.

Although he was not much interested in teaching, Dirac seems to have had some success as a research supervisor. Among the doctoral students who were supervised by him, the mathematician Harish-Chandra stands out. As for Dirac's lectures, one who attended his regular course on quantum theory recalled that 'The delivery was always exceptionally clear and one was carried along in the unfolding of an argument which seemed as majestic and inevitable as the development of a Bach fugue.' Dirac was an inveterate traveller and, although not a serious mountaineer, he climbed some high mountains, notably Mount Elbruz in Turkey, which brought on an attack of altitude sickness.

Until he retired Dirac based himself in Cambridge, except for extended visits to the Institute for Advanced Study in Princeton, the Tata Institute for Fundamental Research in Bombay and Moscow State University. Although he had attractive offers from elsewhere, he chose to remain in England, so as to give a lead to young British theoreticians, until in 1972 he crossed the Atlantic and took on a new lease of life at the Tallahassee campus of Florida State University. In Cambridge he tended to work mainly at home and went to the university only for classes and seminars; in Tallahassee he was on the campus all day. He published prolifically, over sixty papers during the last twelve years of his life, although they were not research papers. Gradually his health failed and he died at Tallahassee on October 20, 1984.

Dirac refused all honorary degrees, but among the honours he accepted, apart from the Nobel prize, perhaps the Copley and Royal medals of the Royal Society and the Order of Merit should be singled out. There are several paintings of him and several portrait busts. 'Dirac was tall, gaunt, awkward and extremely taciturn', wrote a German scientist after his death, 'he has succeeded in throwing everything he has into one dominant interest. He was a man, then, of towering magnitude in one field, but with little interest and competence left for other human activities. In other words he was the prototype of the superior mathematical mind; but while in others this has coexisted with a multitude of interests, in Dirac's case everything went into the performance of his great historical mission, the establishment of the new science quantum mechanics, to which he probably contributed as much as any other man.'

J. Robert Oppenheimer (1904–1967)

In America the study of physics was slow to develop. There were outstanding individuals at times, such as Henry, Gibbs and Millikan, but it was not until after the Second World War that the USA became the world leader in physical research, with immigrants from Europe and elsewhere setting the standard for native-born Americans to attain. More than anyone else it was Oppenheimer who was responsible for raising American theoretical physics from the state of being little more than a provincial adjunct of Europe to world leadership.

J. Robert Oppenheimer was born on April 22, 1904 in New York; the letter J. in his full name may refer to his father Julius, who had come to the USA from Germany at the age of seventeen and developed a prosperous business importing textiles. Oppenheimer's mother Ella (née Friedman) came from Baltimore. She was said to have been unusually sensitive, and had studied painting in Paris. Robert had one younger brother, Frank Friedman Oppenheimer, born in 1912, who became professor of experimental physics at the University of Colorado: he also had another brother who died in infancy. The family was well-off, with a sumptuous apartment on Riverside Drive in Manhattan, furnished with post-impressionist paintings, and an estate on Long Island, where they kept a yacht. They were non-observant Jews, much interested in art and music.

In 1911 the young Oppenheimer entered the non-sectarian School of
Ethical Culture, one of the best in the city of New York, where he joined a
large number of other Jewish boys who were excluded from private schools
through the rigid quota system. He received an education in what, for those
days, were advanced liberal concepts of social justice, racial equality and
intellectual freedom. Being very shy, he was much bullied and, rather than
mixing with other boys, immersed himself in his schoolwork, in poetry and
in science, particularly physics and chemistry. When he was only five he
had started collecting mineralogical specimens, some of which came from
his grandfather in Germany. By the time he was eleven his collection was so
good and his knowledge so extensive that he was admitted to membership
of the Mineralogical Club of New York.

After graduating from school in 1921, Oppenheimer made a trip to
Europe, but became sick with colitis. He entered Harvard the next year, just
after the university had introduced its quota system for admitting Jewish
students, the intention being to restrict them to the same proportion as
obtained in the population as a whole. Originally he had intended to major
in chemistry, but soon switched to physics. It was characteristic of him not
to abandon a subject in which he had once been interested; familiarity with
chemistry was to be very useful to him later on in his career. At Harvard he
was strongly influenced by Percy Williams Bridgeman, a distinguished and
unconventional experimental physicist. Apart from this, he kept very much
to himself and devoured knowledge avidly. 'I had a real chance to learn', he
said, 'I loved it. I almost came alive. I took more courses than I was supposed
to, lived in the library stacks and just raided the place intellectually'. In
addition to studying physics and chemistry, he learned Latin and Greek and
graduated *summa cum laude* in 1925, having taken three years to complete
the normal four-year course. He retained a lifetime affection for Harvard,
serving as a member of the Board of Overseers from 1949 to 1955.

After graduation from Harvard, Oppenheimer spent four years studying at the great European centres of physics. First he spent the academic
year 1925/6 as a member of Christ's College, Cambridge. After having been
turned down by Rutherford, he was assigned a rather uninspiring research
project and altogether he was generally unimpressed by Cambridge. His parents came over to see him, having heard that he was having treatment for
depression. It was rather a disappointing year, but at least he reached the
conclusion that he preferred theoretical physics to experimental physics
and decided that he would do his doctoral work at Göttingen, under Born.
While at Cambridge he had begun a thesis on the application of quantum

theory to transitions in the continuous spectrum, which he completed so rapidly that it was ready for submission early in 1927. At the same time he was building up his knowledge of the revolutionary new developments in theoretical physics.

Next Oppenheimer was awarded a Rockefeller-funded fellowship by the American National Research Council, later the National Science Foundation. He held it first at Harvard and then at the California Institute of Technology, presided over by Millikan. In the year 1928/9 he held a fellowship from the International Education Board and used it to visit Ehrenfest in Leiden and to see Bohr's disciple Kramers in Utrecht. In the first half of 1929 he went on to the ETH in Zürich, where he worked with Pauli, another influence on his scientific development.

On his return to the USA in 1929, Oppenheimer received many offers of academic positions. He accepted two of them, becoming assistant professor of physics concurrently at the California Institute of Technology and at the Berkeley campus of the University of California. In the ensuing thirteen years he divided his time between the two institutions, spending the autumn and winter in Berkeley, the rest of the year in Pasadena. The majority of the best American theoretical physicists who matured in those years were trained by him at one stage or another in their careers. His teaching, his style and his example influenced them all.

Oppenheimer was fortunate to enter physics in 1925, just when modern quantum physics came into being. Although he was too young to take part in its formulation, he was one of the first to use it for the exploration of problems that had defeated the old quantum theory. Probably the most important ingredient he brought to his teaching was his ability to choose the most interesting problems. Although his lectures were difficult, they conveyed so well the beauty of the subject that almost every student repeated his course, which was based on the survey by Pauli in the *Handbuch der Physik*. He was interested in almost everything, but principally quantum-field theory, cosmic rays and nuclear physics.

The magnetism and force of Oppenheimer's personality was such that his students tended to copy his gestures and mannerisms; he had exquisite manners, a marvellous command of language and a ready wit. In those days students were generally short of money; he entertained them generously, to concerts, dinners and other social events. Among his many friends in the Berkeley faculty were not only scientists but also classicists, artists and so on. Most of the time he was indifferent to the events taking place around him. He never read a newspaper or listened to the radio; he never used the telephone. However, he had a passion for fast cars and particularly enjoyed racing trains when the railroad track ran alongside the highway. He was described as having titanic ambition but was tortured by self-doubt. In 1936 he was promoted to full professor at both of the institutions where he worked. The following year his father died; his mother had died six years earlier. In 1941 he was elected to the National Academy of Sciences. In 1940 he married the divorcée Katherine (Kitty) Harrison; they had one son, Peter, and one daughter, Katherine.

In 1942, after America had entered the Second World War, Oppenheimer was appointed leader of the theoretical side of the effort to design and build an atomic bomb. Research in England had made it seem very likely that the concept was viable. The chemical firm Dupont was contracted to build a production reactor, and it was time to prepare for the assembly of an atomic weapon. A permanent laboratory was needed for the work, somewhere remote because of the need for secrecy. For many years Oppenheimer and his brother had rented a ranch in the high country of the state of New Mexico, where they loved to spend their summers horseback riding. The government established its laboratory on a beautiful mesa nearby, called Los Alamos. Oppenheimer, as director of the enterprise, set about recruiting his team. Many of the best physicists were already involved in war work and disinclined to move, but he was very persuasive and those who came to join him found it an unforgettable experience.

The task they were faced with was formidable. Initially not much more was known than the fundamental theory behind a chain reaction. The details of the fission process had to be fully understood, but no fissionable material was yet available. Nuclear physicists found themselves dealing with unfamiliar subjects, like hydrodynamics, but Oppenheimer kept morale high and not only led the physicists but also argued against the tendency of the military to impose security measures the scientists regarded as excessive; for example, they had to use false names. The preparatory work was completed in 1945, at just about the time that a sufficient amount of the isotope uranium-235 became available. In 1946, at the end of it all, Oppenheimer was awarded the Medal of Merit by President Truman, 'for his great scientific experience and ability, his inexhaustible energy, his rare capacity as an organizer and executive, his initiative and resourcefulness, and his unswerving devotion to duty'.

It was obvious that a community like Los Alamos would be deeply concerned with the ominous implications of the atomic bomb. Oppenheimer was one of those most concerned, and had many discussions about this problem with Bohr amongst others. Bohr had come to the USA in 1944 as a consultant on the project, but his main interest was in talking to statesmen and trying to persuade them that international control was the only way to avoid a pernicious arms race or worse, atomic war. Bohr did not get far with the statesmen but he greatly impressed Oppenheimer and through him other scientists. In 1949, after the Soviet Union had exploded its first atomic bomb, the chances of any international agreement receded, and the scientists had little influence on the arms race which then developed.

In 1945 Oppenheimer resigned as director of the Los Alamos laboratory while remaining on a number of major national and international committees concerned with atomic energy. He returned to California briefly to resume his professorships. In 1948 he was president of the American Physical Society, and the following year he was appointed as both professor of physics and director of the Institute for Advanced Study in Princeton. Not all the permanent members of the institute supported his appointment, and he was not universally popular. He arrived with an armed guard carrying a large safe containing secret documents, and much of his time and energy was spent working in Washington.

During the Cold War a notorious witch-hunt took place in the USA, when many prominent people were accused, rightly or wrongly, of being Communists or crypto-Communists and therefore not to be trusted with secret information. Oppenheimer was a known left-winger. Although his brother and sister-in-law both joined the Communist Party, apparently he never did so himself. Nevertheless, at the end of 1953 President Eisenhower ordered that his clearance for secret government work should be suspended. The ensuing protracted security investigation became a *cause célèbre*. Many of his fellow-scientists came out in his defence, but he had made powerful enemies who testified against him. The decision was upheld, and it was not until eight years later that the American government made amends. When President Johnson, carrying out a decision made by the late President Kennedy, presented him with the prestigious Enrico Fermi award at the White House, Oppenheimer commented: 'I think it is just possible, Mr President, that it had taken some charity and some courage for you to make this award today.'

Under Oppenheimer theoretical physics at the Institute for Advanced Study was strengthened, to some extent at the expense of mathematics. The institute faculty had always included prominent physicists; Einstein as a permanent member, Bohr, Dirac and Pauli, amongst others, as visitors, but now younger physicists were increasingly encouraged to visit, as well as those who were already eminent. Although he was never quite in the top rank of researchers himself, Oppenheimer could still conduct a lively seminar and play a leading and often critical role in the vigorous discussion which usually followed the talk. His own writings and lectures, after the war, were more concerned with the problems of the human race than with physics as such; increasingly he performed the twin roles of public symbol and interpreter of modern physics. Among the many foreign honours he received were membership of the Legion of Honour and the fellowship of the Royal Society. A heavy smoker, Oppenheimer died of throat cancer in Princeton on February 18, 1967, after having taken early retirement the previous year.

Maria Goeppert-Mayer (1906–1972)

Regrettably, only three of my remarkable physicists are women. The handicaps that Marie Curie and Lise Meitner faced have already been described; those that Maria Goeppert had to deal with, several decades later, were different but equally discouraging. Maria Gertrude Kate Goeppert was the only child of well-educated upper-middle-class parents. She was born on July 28, 1906, in what was then the German city of Kattowitz in Upper Silesia, and

is now the Polish city of Katowice. Her mother Maria (née Wolff) taught French and music before her marriage. Her father, Friedrich Goeppert, who was professor of medicine at the university with a special interest in paediatrics, took special pride in being the sixth straight generation in a family of university professors; later his daughter would be proud to have continued the family tradition. Friedrich Goeppert encouraged his daughter's natural curiosity and adventurousness, and made it clear that he would like her to be more than simply a housewife. She later said that she found her scientific father more interesting than her mother, yet she was devoted to her mother, who delighted to entertain faculty members at lavish dinner parties and to provide a home filled with flowers and music for her only daughter.

Maria Goeppert was blond, with blue eyes, very earnest and unguarded in her expression, who learned the importance of duty while young and presented a reserved, somewhat aristocratic bearing as an adult. As a child she was described as active, adventurous and tense. She suffered from severe headaches and minor illnesses during childhood, perhaps exacerbated by her parents' high expectations. In 1910, when she was four, the family moved to Göttingen, where her father had been appointed professor of paediatrics at the Georgia Augusta. They had Hilbert as a neighbour, and their social circle included many of the scientists of the great university. The public schools of Göttingen had very good teachers; she excelled at languages and mathematics. To prepare for the *Abitur* she completed her school education at the *Frauenstudium*, a small private school run by suffragettes to prepare girls for higher education: unfortunately, before she could finish the three-year course it was closed due to hyperinflation. Rather than change schools again, she chose to take the *Abitur* a year earlier than normal. After passing this successfully, she entered the Georgia Augusta in 1924 to major in mathematics. Like most of the women students, she began by studying for the teaching certificate, but found the classes uninteresting. She considered medicine, but her father argued against it, fearing that she would suffer too much distress whenever a patient died.

The Georgia Augusta had long been famous for mathematics and was becoming so for theoretical physics as well. In 1927, inspired by the lectures of Max Born, Maria Goeppert switched to physics. That year her father, whom she idolized, died unexpectedly and, knowing what he would have wished, she honoured his memory by resolving to complete her doctorate, as a student of Born. She also spent a term in Cambridge, at Girton College, primarily to learn English, but she was also able to meet the Rutherfords. She was described as the prettiest girl in Göttingen, rather short and plump, with blue eyes and a fine-grained complexion. Her widowed mother, in accordance with a Göttingen tradition, began taking young people from the university as boarders in her substantial house: one was a post-doctoral student in chemistry from California named Joseph Mayer, the son of an Austrian-born engineer and an American schoolteacher. He and Maria became great friends and in 1930 they married. The same year she completed her doctoral dissertation 'On Elemental Processes with Two Quantum Jumps'. This was a theoretical treatment of two-photon processes, which she later used to determine the probability of double beta decay.

Meanwhile her husband had obtained a position as assistant professor of chemistry at Johns Hopkins University and so the couple moved to Baltimore. At that time universities were rather hesitant to hire women for academic positions. Prejudice against women scientists began to recede as the twentieth century progressed, but prejudice remained. In America many universities had 'anti-nepotism' rules, preventing husband and wife both holding academic positions in the same institution, and Johns Hopkins was one of these; there was no chance of an academic post for her there. Since no-one was working on quantum mechanics at Johns Hopkins, on Born's recommendation she collaborated in research with the physical chemist Karl Herzfeld, who was working on energy transfer and the liquid phase, and published some papers with him and with her husband. However, physical chemistry was not her chosen field and during the summers she returned to Göttingen, where she wrote several papers with Max Born on beta-ray decay.

She found it difficult to adjust to life in America and was homesick for Germany. By marrying an American she had forfeited her German nationality and become for a time stateless, which created various bureaucratic problems. In the spring of 1933 she took out American citizenship, calling herself Maria Goeppert-Mayer professionally, Maria Mayer otherwise. Her first child, Maria Ann, was born that year, the second, Peter Conrad, five years later. Although both children initially studied science in college, they did not pursue it further. The family lived in a large house, with an attractive garden, where they entertained streams of German visitors. Before their son was born her husband was appointed to a much-better-paid position as associate professor of chemistry at Columbia University and so they moved to New York, where she collaborated with him on a textbook, *Statistical Mechanics*, published in 1940, which became a classic. Unfortunately, it was assumed that her husband had done most of the work on this, so she did not receive her due credit, a common experience for married women. Until the USA entered the Second World War she was never on the payroll of an American university. Later she worked for the Manhattan project and participated actively in efforts to help the German refugee scientists. Her feelings about the war were ambivalent, due to her German roots and love of family, friends and colleagues in the land of her birth. She told her children that the war was against the Nazis, not the German people. Some of the other German scientists in exile felt differently about the homeland which had rejected them, but she continued to love Germany and the Germans and to insist that Hitler was an aberration.

At the end of 1941 Goeppert-Mayer took a part-time teaching position at Sarah Lawrence College in Bronxville, New York, where she developed a unified science course. A few months later she joined the secret project to separate the isotope uranium-235 from uranium-238, and she also worked with Edward Teller on nuclear fusion. When the first atomic bombs were dropped on Japan, with such devastating results, she was relieved that her part in their development had been very minor. After the war Joseph Mayer moved to the University of Chicago, where she was offered an associate professorship but without salary, due to anti-nepotism rules, and she joined the research group at the Argonne National Laboratory as senior physicist, the first time she held a normal scientific post. The Mayers lived in a handsome old house on the South Side of the city, ideal for entertaining. Like her mother she had a flair for this. She filled their house with flowers, often from their own garden. She specialized in cultivating orchids in a greenhouse on the top floor of their house.

Her collaboration with Teller on the origin of the chemical elements led to considerations of their relative abundance. She noticed that the most stable elements contained particular numbers of either protons or neutrons, later called the 'magic numbers'. Shell models for the nucleus had been considered and discarded earlier, but Goeppert-Mayer believed that new evidence strongly supported this concept. In 1948 she published a paper that set out the evidence but without a theory. A chance remark by Fermi, who was keenly interested in her work, triggered the insight that enabled her to solve the theoretical problem. By assuming the occurrence of spin–orbit coupling, she was able to calculate the energy levels that matched the magic numbers. Always impeccably correct in her behaviour towards others, she delayed publishing her results because she had heard that several other physicists were working on shell models in the USA. As it turned out, it was a group in Heidelberg that published a similar interpretation to hers, at about the same time as she did; later she co-authored a book on the subject with one of them, Hans Jensen, a 'dear gentle man' but rather inclined to procrastinate. It is said that through Bohr he provided information about the progress of German research into atomic weapons that was useful to the Allies.

During the war Goeppert-Mayer, an inveterate chain-smoker, began to experience health problems. In 1956 she lost most of the hearing in one ear. In 1959 both she and her husband were appointed to full professorships at the new San Diego campus of the University of California, although they were under pressure to remain in Chicago. A stroke in 1960, not long after the move to La Jolla where the campus is located, left her partly paralysed, but she continued to carry on as best she could. It was there in 1963 that the news arrived that she was to become a Nobel laureate. She and Jensen shared half the prize in theoretical physics for their discoveries concerning nuclear shell structure; the other half went to Wigner for contributing to the theory of the atomic nucleus and the elementary particles. Election to the National Academy of Sciences and several honorary doctorates soon followed, but her health gradually declined and she died of heart failure on February 20, 1972.

Goeppert-Mayer dealt with the obstacles she faced in her career partly by identifying with men at an early age and by disregarding the expectations of the society in which she lived. Like some other successful women scientists, she received much encouragement from key men in her life, beginning with her father, 'a gentle bear of a man', who wanted her to make something of her gifts. Many of her colleagues, including Born and Fermi, were supportive, but, perhaps partly because she worked so long without recognition, she was unduly modest about her work and abilities. Some of the difficulties Goeppert-Mayer experienced turned to her advantage. Unable to determine her own career path, she seized the opportunities which arose as she followed her husband's career. Although she had been educated as a mathematical physicist and was equipped with great facility in the Born–Heisenberg matrix formulation of quantum theory, she turned to physical chemistry when she went to America. The combination of theoretical and practical knowledge was important for her later work on nuclear structure. What was undertaken from expedience led to a deeper appreciation of experiment and understanding of a new field. She learned about nuclear physics, another new field to her, from Teller and from Fermi while at Columbia and Chicago. The varied strands in her background, both the planned and the unplanned, eventually converged in the studies that earned her the Nobel prize for the shell model of the nucleus. In the first century of the prizes, only two women have been awarded a Nobel prize in physics.

HIDEKI YUKAWA (1907–1981)

Hantarō Nagaoka, who studied under Boltzmann, was the first Japanese physicist to participate fully in the competitive world of theoretical and experimental physics. Some young Japanese scientists studied in Germany, others in Britain, a few in France. They encountered language and other cultural difficulties. For example, several young professors of the faculty of science of the Imperial University who went to study at the University of Berlin heard an address glorifying the progress of German science in which the dean of the faculty said 'In order to study science in Germany, there come to this country many foreigners, Americans and so on, and lately even Japanese. In future, no doubt, even apes will be coming . . .'

Until around 1935 the contribution of the Japanese nation to world physics was very limited. By the end of the Second World War, however, Japan had become a significant contributor to theoretical physics, with an impressive number of first-rate research workers. Initially by far the largest volume of Japanese research was on the most modern aspect of the subject: elementary particles and fields. The new Japanese school was essentially the school of Hideki Yukawa and his friend and contemporary Sin-itiro Tomonaga. Both were Nobel laureates in physics. Both of them wrote autobiographies that have been translated into English, but Yukawa is more informative about his early years, and for that reason I choose him as the main subject for this profile. Of the two, Tomonaga had a more open and engaging personality; he remarked that 'Yukawa's eyes look inward, mine look outward.'

The man known to the scientific world as Hideki Yukawa was born in the Azabu district of Tokyo on January 23, 1907, during the fortieth year of the Meiji era. The child was the middle one of five brothers, the fifth of seven children altogether. Their parents were Takuji Ogawa and his wife Koyuki. Until his marriage to Sumi Yukawa in 1932, when he was adopted into his wife's family, the future physicist's name was Hideki Ogawa. Hideki's father's name had been changed from Asai to Ogawa for a similar reason. For a few months after Hideki's birth his father remained on the staff of the Geological Survey Bureau in Tokyo but in 1908 he became professor of geography at the University of Kyoto. Thus Hideki regarded Kyoto, the former capital city of Japan, as his home town; he spent almost all his life there.

Both the Asai family and the Ogawa family could look back on generations of scholars steeped in the tradition of Chinese as well as Japanese culture. Hideki's grandfather was a Confucian scholar and teacher of Chinese classics. His son Takuji, brought up in the same tradition, was well-versed in Chinese religion and philosophy, collected antiques and maintained an active interest in the history of ancient China and Japan throughout his life. He was associated with the Institute of Oriental Culture, under whose auspices he frequently visited China to participate in archaeological surveys and expeditions.

Like his brothers, Hideki was exposed to ancient traditions in childhood and adolescence. He recalled that, even before he entered primary school, he had studied various Chinese classics: 'In practice this meant that I repeated aloud after my grandfather a version of the Chinese texts converted into Japanese. At first, of course, I had no idea of the sense of it at all. Yet oddly enough I gradually began to understand without being told.'

In his autobiography Ogawa, as Yukawa was called in his youth, says that he was afraid of the stormy temperament of his unpredictable father and tried to avoid him; the influence of the boy's mother and grandparents on his upbringing seems to have been dominant. We learn from Ogawa himself that as a boy he was clumsy and ill at ease in his relations with other people. 'I never did develop socially', he recalled, 'often I think human relationships are tiresome, even among Japanese. Relating to foreigners simply wears out my nerves.'

At school Ogawa consistently obtained good marks, but when asked a question would often remain silent when he knew the answer perfectly well. When the time to choose a speciality arrived, he found it difficult to make a decision. His difficulty in interacting with people was linked, so he relates, to a complete lack of interest in human problems, so that all fields within the sphere of social science in the broad sense had no appeal whatsoever. Although literature and philosophy interested him, mathematics was his favourite subject. Indeed, the latter field seemed for a time to be the most likely choice of specialization, even though he had some doubts about devoting himself to a study that seemed to be so unrelated to the natural world.

Ogawa, when young, lived in a world of his own and made his own decisions. An avid and catholic reader, at the age of thirteen he entered high school, where he discovered in its library a series of books in Japanese explaining modern physics at a level he found accessible, at least in part. Ogawa related that he found the ideas of relativity comprehensible and exciting. On the other hand, in commenting on an introduction to quantum theory, he recalled that 'however much I read it, it made no sense to me . . . so I thought it might be a good idea to study it'. He appears to have kept to this resolution with remarkable tenacity.

Ogawa relates that, while he was at high school, he took very little notice of the normal run of student life around him, feeling an aversion to the bohemian life-style of his contemporaries. He was supposed to attend a class in economics and law but relates that these lectures went in one ear and out the other. Even so, his natural ability and methodical habits seemed to carry him safely over all examination hurdles. While he was at school a well-publicized visit to Japan by Einstein in 1922 had opened up the new world of ideas in physics to the Japanese. Ogawa taught himself some German and, when browsing in a bookshop, picked up the first volume of Planck's *Einführung in die theoretische Physik*. He relates with great warmth how he found the work both intelligible and fascinating, and as a result he decided to study at the physics department of the University of Kyoto.

The year of his entry was 1926, just after the discovery of the new quantum mechanics in the west. Whatever of the older quantum theory was included in the course he was offered – and it seems that there was very little – the new ideas had certainly not penetrated far. The account he gave of his studies is entirely about his private reading: learning the older quantum theory from an English translation of Reich's *Quantentheorie* and then, just as the first news of quantum mechanics was reaching Japan, being inspired by Born's *Probleme der Atomdynamik*. From then on it was a matter of studying new papers as they became available. As with so many other workers of the day, Schrödinger's version of quantum mechanics – wave mechanics – came as a revelation. He found it so much easier to understand than the Heisenberg matrix mechanics that had preceded it. Ogawa persuaded his teacher to agree to his choosing the new theory as his main subject for graduation. It seems that the only outside stimulus during these studies was from his friendship with his classmate Tomonaga.

Like Ogawa, Sin-itiro Tomonaga came from a family of culture, renowned for its literary scholarship. He was born in 1906, the eldest son of Sanjuro Tomonaga, who in turn was of Nagasaki samurai descent. Sanjuro had studied philosophy and history of philosophy at Tokyo Imperial University and at the time of Sin-itiro's birth was professor of philosophy at Shinshu University (later known as Otani University) in Tokyo. The following year he took up the position of associate professor at Kyoto University and in 1909 went to Europe for further study, leaving his wife and child with relatives in Tokyo. When he returned after an absence of four years, the family went to live in Kyoto, where he had been appointed professor.

Also like Ogawa, Sin-itiro Tomonaga gained admission to the faculty of science at Kyoto University in 1926 and elected to specialize in physics. He was greatly disappointed at the level of the lectures, especially in physics: moreover, the laboratories provided were dark, dirty and old-fashioned. More advanced work was being done in the laboratory of Professor Kajuro Tamaki, whose speciality was fluid dynamics and whose interests extended to relativity but not quantum theory. Soon Ogawa, Tomonaga and a few other ambitious students were studying western publications together. They saw that in atomic physics one major problem after another was rapidly being resolved, often by young researchers such as Heisenberg, Dirac, Pauli and Fermi.

Ogawa realized that quite basic questions remained unanswered in the context of relativistic quantum theory, in the quantum theory of fields and in the description of the atomic nucleus. He made the bold decision to concentrate on these even newer problems. As his graduation thesis he chose to offer some investigations on the properties of Dirac's equation. While there is no record of what went into this work, it was sufficiently impressive to enable him to graduate in 1929 and, although Tamaki did not accept research students, he and Tomonaga were allowed to remain in the laboratory as unpaid assistants. That year both Heisenberg and Dirac visited Kyoto, as did also the physicist Yoshio Nishina, who had just returned from seven years in Europe, most of the time spent studying with Bohr in Copenhagen. Nishina was impressed by the two young researchers and invited Tomonaga to join him in Tokyo.

In 1932 Ogawa married the classical Japanese dancer Sumi Yukawa and, as we have seen, adopted her family name, replacing the name of Ogawa. He was also appointed instructor in physics at his *alma mater*, where he lectured on quantum mechanics, using Dirac's textbook. One of his students described his voice as being 'gentle as a lullaby, an ideal invitation to sleep'. The following year he began speculating about the nature of the proton–neutron force. Before the moment of breakthrough Yukawa was given the chance to move from the peaceful atmosphere of Kyoto to the much livelier surroundings of the industrial city of Osaka, where a new university was being built up with a strong physics department, including a large experimental group provided with modern accelerator equipment. In this stimulating environment Yukawa's ideas came to fruition and his famous first paper was published, predicting the existence of a new fundamental particle, the meson.

After 1937 Yukawa was increasingly seen as a leader of theoretical physics in Japan. Already he could attract many research students, and his appointment to a chair was not long delayed. He was happy that this should be at his *alma mater*, soon to be internationally recognized as a major centre for theoretical physics. His first visit to the west included attendance at the eighth Solvay conference in Brussels in 1939. The arrangements for this were disrupted by the outbreak of the Second World War, but Yukawa was able to take the opportunity to establish personal contact with European and American physicists.

After the war, when Yukawa's own interest was beginning to turn away from the particular topic of meson theory to more general unsolved problems, the experimental discovery of the pion further enhanced his reputation. In 1948 he accepted Oppenheimer's invitation to work at the Institute for Advanced Study in Princeton, which was followed by appointment to a chair at Columbia University in New York. He remained at Columbia until 1953, when he was offered the directorship of a new inter-university research institute in Kyoto, housed in a building to be named in his honour. Yukawa remained in the old capital for the rest of his life. After reaching the age of seventy-four he became gravely ill and died from pneumonia on

September 8, 1981, survived by his widow Sumi and two sons, Harumi and Takai. He had received the Nobel prize in 1949, and many other honours as well.

In conclusion, something should also be said about the later career of Yukawa's friend Tomonaga, a quiet person of great charm. After a period of studying nuclear theory at Leipzig University under Heisenberg from June 1937 to August 1939, Tomonaga returned to Japan and received his D.Sc. from Tokyo Imperial University. In 1941 he was appointed professor at the Tokyo University of Science and Literature. Two years later Tomonaga was mobilized with other Japanese physicists to undertake research on magnetrons and ultra-short-wave circuits at the laboratory of the Naval Research Institute at Shimada. In addition he was a part-time lecturer at Tokyo University. Because of the intense air raids on Tokyo, he sent his family to live in the country while he remained in the city. During a raid on April 13, 1945 the district to the west of the campus where many professors lived was destroyed, including the house of Tomonaga.

After the war his seminar became a Mecca for young physicists. In 1948 he was elected to the Science Council of Japan. The following year he visited the Institute for Advanced Study in Princeton. In 1965 he shared the Nobel prize in physics with Richard Feynman and Julian Schwinger for contributions to quantum electrodynamics. Tomonaga's theory of 1943 had been developed quite independently of the American one, of which he did not become aware until the war was over. Like his friend Yukawa, he was awarded the prestigious Cultural medal of Japan, and numerous other scientific honours.

Much more than Yukawa, Tomonaga took on responsibility for nuclear physics and science policy generally in Japan. Scientists had long been perceived as left-wing agitators, opposed to big business and keen to distance themselves from the bureaucracy; Tomonaga attempted to break down this stereotype. The two Nobel laureates attended the first Pugwash conference held in Canada in 1957 and put their prestige behind the movement to ban atomic and hydrogen bombs. Tomonaga's health was poor in later years, and he died on July 8, 1979. Today Japanese workers contribute significantly to many branches of physics, both experimental and theoretical, following the lead of the two pioneers.

Epilogue

In the list of contents at the front of this book the names of all the subjects of the profiles are arranged in order of date of birth. The following list consists of the same names arranged in order of date of death.

Johannes Kepler (1571–1630)
Galileo Galilei (1564–1642)
Christiaan Huygens (1629–1695)
Isaac Newton (1642–1726)
Daniel Bernoulli (1700–1782)
Roger Boscovich (1711–1787)
Benjamin Franklin (1706–1790)
Charles Augustin Coulomb (1736–1806)
Henry Cavendish (1731–1810)
Benjamin Thompson (Count Rumford) (1753–1814)
Pierre-Simon Laplace (1749–1827)
Thomas Young (1773–1829)
Jean-Baptiste Fourier (1768–1830)
André-Marie Ampère (1775–1836)
George Green (1793–1841)
Hans Christian Oersted (1777–1851)
Georg Ohm (1789–1854)
Michael Faraday (1791–1867)
Joseph Henry (1797–1878)
James Clerk Maxwell (1831–1879)
Hermann von Helmholtz (1821–1894)
Willard Gibbs (1839–1903)
Ludwig Boltzmann (1844–1906)
William Thomson (Lord Kelvin) (1824–1907)
John William Strutt (Lord Rayleigh) (1842–1919)
Wilhelm Conrad Röntgen (1845–1923)
Paul Ehrenfest (1880–1933)
Marie Curie (1867–1934)
Ernest Rutherford (Lord Rutherford) (1871–1937)
Joseph John Thomson (1856–1940)
William Henry Bragg (1862–1942)
Max Planck (1858–1947)

Robert Millikan (1868–1953)
Enrico Fermi (1901–1954)
Albert Einstein (1879–1955)
Frederick Lindemann (Lord Cherwell) (1886–1957)
Jean-Frédéric Joliot (1900–1958)
Erwin Schrödinger (1887–1961)
Niels Bohr (1885–1962)
J. Robert Oppenheimer (1904–1967)
Otto Hahn (1879–1968)
Lise Meitner (1878–1968)
Max Born (1882–1970)
Maria Goeppert-Mayer (1906–1972)
Satyendranath Bose (1894–1974)
Werner Heisenberg (1901–1976)
Hideki Yukawa (1907–1981)
Piotr Leonidovich Kapitza (1894–1984)
Paul Dirac (1902–1984)
Louis de Broglie (1892–1987)

The fifty physicists whose profiles make up this book were the sons or daughters of men who were engaged in a wide variety of occupations. Only some of them were academics: Daniel Bernoulli and William Thomson were the sons of professors of mathematics, Gibbs was the son of a professor of sacred literature, Born the son of a professor of embryology, Bohr the son of a professor of physiology, Goeppert-Mayer the daughter of a professor of paediatrics, and Yukawa the son of a professor of geography. Others were the offspring of professional men: Helmholtz and Dirac were the sons and Marie Curie was the daughter of schoolteachers, Maxwell and Planck were the sons and Meitner was the daughter of lawyers, Oersted was the son of a pharmacist, Millikan was the son of a preacher, Lindemann and Kapitza were the sons of architects or engineers, Galileo was the son of a musician, Huygens was the son of a diplomat, and Bose was the son of an accountant. Franklin, Boscovitch, Young, Ampère, Röntgen and J.J. Thompson were the sons of merchants. Newton, Rumford, Laplace and Bragg were the sons of farmers. Coulomb, Boltzmann and Fermi were the sons of public officials. Einstein, Schrödinger, Joliot and Oppenheimer were the sons of businessmen. Kepler was the son of a soldier of fortune. Maxwell and Strutt were the sons of wealthy landowners; perhaps Cavendish and de Broglie should be included in the same category. Several subjects were sons of artisans or tradesmen, in particular Fourier was the son of a tailor, Ohm and Heisenberg were the sons of locksmiths, Hahn the son of a glazier, Faraday the son of a blacksmith, Green the son of a miller, and Rutherford the son of a wheelwright. Joseph Henry's father was a day labourer, and Ehrenfest's father worked in a textile mill. At some time in their lives, usually childhood, Kepler, Franklin, Fourier, Ohm, Faraday and Marie Curie knew poverty.

A number of our subjects inherited or acquired titles of nobility, and these have been used where appropriate. Of course, the nature of such titles varied from country to country; today they survive officially in only a few countries, but they were common in Europe during most of our period. Laplace was made a marquis and Fourier a baron, while de Broglie succeeded to the title of duke. Benjamin Thompson was made a count in Bavaria; de Broglie inherited the imperial title of prince. Rayleigh succeeded to his title, while Kelvin and Rutherford received theirs for services to science. Lindemann was elevated to a viscountcy for services more political than scientific. In addition, a number of our subjects received knighthoods or similar honours.

Bearing in mind the shorter expectation of life in the eighteenth and nineteenth centuries, the great majority of our subjects lived into old age. Exceptions are Green and Maxwell, who died at forty-eight, and Ehrenfest and Fermi, who died at fifty-three. In most cases deaths were from natural causes, insofar as they are known, except that Planck died after a road accident, while Boltzmann and Ehrenfest committed suicide. Marie Curie, her daughter Irène Curie and son-in-law Frédéric Joliot essentially died from radiation sickness. Boltzmann, and probably Ehrenfest, suffered from the mood-swings characteristic of manic-depression, but retrospective diagnosis of mental disorders is notoriously difficult. People with Asperger's syndrome, a mild form of autism, seem to be attracted to physics. Newton, Cavendish, Einstein and Dirac are thought to have had the syndrome. In childhood only a few of our subjects could be described as precocious and, although several exhibited phenomenal memory, only Lindemann seems to have shown any sign of savant skills. Most subjects displayed an enthusiasm and aptitude for physics from an early age, but some only decided to become physicists relatively late.

A companion work, *Remarkable Mathematicians*, has been published. The distribution of occupations of the fathers of the mathematicians is noticeably different from that of the physicists; rather more of them were men of religion, rather fewer were professional men, and rather more of the subjects grew up in a condition of poverty. Again, for mathematicians more than for physicists there are indications that the maternal influence was the dominant one in their upbringing. Two subjects, namely Fourier and Laplace, appear in both works, since they seem to qualify equally well both as mathematicians and as physicists. Arguably this is true in some other cases.

Much more has been written about the physicists than the mathematicians, whole shelves of books about Galileo, Newton, Einstein and Faraday, for example, and a greater tendency to hagiography is apparent. The story of the development of mathematics does not run parallel to that of physics. Moreover, there are cultural differences. Increasingly, experimental physics has become a matter of teamwork whereas research in theoretical physics, as in mathematics, can still be practised successfully by individuals working in isolation. The world of physics is more intensely competitive, so a promising line of research may be kept secret in case someone else starts to compete. Moreover, progress is often achieved by discarding one theory in favour of another, after a period of controversy; there is nothing comparable in mathematics. Finally, there is the vexed question of commercial exploitation of discoveries, the possibility of which arises much more frequently in physics.

Further Reading

Many of the remarkable physicists featured in this book have been the subjects of full-scale biographies. Some that have appeared fairly recently, mainly in the English language, are listed below. To list all relevant articles would take up an excessive amount of space, but the reader who wishes for further information in a particular case will find bibliographies in such reference works as the *Dictionary of Scientific Biography* and the *Isis Cumulative Bibliography*. Journals such as the *Archive for History of Exact Sciences* often contain relevant articles. The new *Biographical Dictionary of Women Scientists* is another source of information, and there are several biographical collections, of which a few are listed, for example those by J.G. Crowther. Although they were written long ago, Crowther's informative and reliable biographies of British and American scientists make a good starting-point. Among more recent accounts of the lives and works of selected physicists, that by William Cropper, listed below, can be recommended as an excellent introduction for the general reader. More or less complete 'Collected Works', usually containing some biographical information, have been compiled for most of our subjects, and these are also listed below. In a few cases the subject's scientific correspondence has been edited and published, and this can be of particular interest.

1. Andriesse, Cornelis Dirk. *Christiaan Huyghens*. Paris: Albin Michel, 1996.
2. Appleyard, Rollo. *Pioneers of Electrical Communication*. London: Macmillan, 1930.
3. Armitage, Angus. *John Kepler*. London: Faber & Faber, 1966.
4. Bell, A.E. *Christian Huygens and the Development of Science in the Seventeenth Century*. London: Edward Arnold, 1947.
5. Berry, A.J. *Henry Cavendish; His Life and Scientific Work*. London: Hutchinson & Co., 1960.
6. Biafioli, Mario. *Galileo, Courtier*. Chicago, IL: University of Chicago Press, 1993.
7. Biquard, Pierre. *Frédéric Joliot-Curie: The Man and his Theories* (trans. Geoffrey Strachan). London: Souvenir Press, 1965.
8. Birkenhead, Earl of. *The Prof in Two Worlds*. London: Collins, 1961.
9. Blaedel, Niels. *Harmony and Unity: The Life of Niels Bohr* (trans. Geoffrey French). Madison, WI: Science Tech. Inc., 1988.
10. Born, Max. *My Life: Reflections of a Nobel Laureate*. London: Taylor and Francis, 1978.

11. Brian, Denis. *Einstein: A Life*. New York, NY: John Wiley and Sons Inc., 1996.

12. Broda, Engelbert. *Ludwig Boltzmann* (trans. Engelbert Broda and Larry Gay). Woodbridge, CN: Ox Bow Press, 1983.

13. Brown, G.I. *Count Rumford: Scientist, Soldier, Statesman, Spy: The Extraordinary Life of a Scientific Genius*. Stroud: Sutton, 1999.

14. Brown, Sanborn C. *Benjamin Thompson, Count Rumford*. Cambridge, MA: MIT Press, 1979.

15. Caban, D. *Hermann von Helmholtz and the Foundations of Nineteenth Century Science*. Berkeley, CA: University of California Press, 1993.

16. Campbell, John. *Rutherford, Scientist Supreme*. Christchurch, New Zealand: AAS Publications, 1999.

17. Cannell, D.M. *George Green: Mathematician and Physicist 1793–1841*. London: Athlone Press, 1993.

18. Cantor, Geoffrey. *Michael Faraday: Sandemanian and Scientist*. London: Macmillan, 1991.

19. Caroe, G.M. *William Henry Bragg*. Cambridge: Cambridge University Press, 1978.

20. Caspar, Max. *Kepler* (trans. and ed. by C. Doris Hellman). London and New York: Abelard-Schuman, 1959.

21. Cassidy, David C. *Uncertainty; the Life and Science of Werner Heisenberg*. New York, NY: W.H. Freeman, 1992.

22. Cercignani, Carlo. *Ludwig Boltzmann, the Man who Trusted Atoms*. Oxford: Oxford University Press, 1998.

23. Chaterjee, Santimay *et al.* (eds.). *S.N. Bose: The Man and his Work*. Calcutta: S.N. Bose National Centre for Basic Sciences, 1994.

24. Cohen, I. Bernard. *Benjamin Franklin's Science*. Cambridge, MA: Harvard University Press, 1990.

25. Cotton, Eugénie. *Les Curies*. Paris: Editions Seghers, 1963.

26. Coulson, Thomas. *Joseph Henry: His Life and Work*. Princeton, NJ: Princeton University Press, 1950.

27. Cropper, William H. *Great Physicists: The Life and Times of Leading Physicists from Galileo to Hawking*. New York, NY: Oxford University Press, 2001.

28. Crowther, J.G. *British Scientists of the Nineteenth Century*. London: Kegan Paul, Trench, Trubner, 1935.

29. Crowther, J.G. *Famous American Men of Science*. London: Secker and Warburg, 1937.

30. Crowther, J.G. *British Scientists of the Twentieth Century*. London: Routledge & Kegan Paul, 1952.

31. Curie, Eve. *Madame Curie*. London: Heinemann, 1939.

32. Curie, Marie. *Pierre Curie*. New York, NY: MacMillan, 1923.

33. Dash, Joan. *A Life of One's Own*. New York, NY: Harper & Row, 1973.

34. Davis, E.A. and Falconer, I.J. *J.J. Thomson and the Discovery of the Electron*. London: Taylor and Francis, 1997.

35. Doren, Carl Van. *Benjamin Franklin*. 1938.

36. Drake, Stillman. *Galileo Studies*. Ann Arbor, MN: University of Michigan Press, 1970.

37. Drake, Stillman. *Galileo at Work*. Chicago, IL: University of Chicago Press, 1978.

38. Drake, Stillman. *Galileo, Pioneer Scientist*. Toronto: University of Toronto Press, 1990.

39. Dunsheath, Percy. *Giants of Electricity*. New York, NY: Thomas Y. Crowell, 1967.

40. Eve, A.S. *Rutherford*. Cambridge: Cambridge University Press, 1939.

41. Everitt, C.W. *James Clerk Maxwell: Physicist and Philosopher*. New York, NY: Scribner, 1975.

42. Farah, Patricia. *Newton: The Making of a Genius*. Basingstoke: MacMillan, 2002.

43. Fermi, Laura. *Atoms in the Family*. London: George Allen and Unwin, 1955.

44. Fermi, Laura. *Illustrious Immigrants*. Chicago, IL: Chicago University Press, 1961.

45. Fölsing, Albrecht. *Albert Einstein: A Biography*. New York, NY: Viking, 1997.

46. French, A.P. and Kennedy, P.J. *Niels Bohr. A Centenary Volume*. Cambridge, MA: Harvard University Press, 1985.

47. Füchtbauer, Ritter von. *Georg Simon Ohm*. Berlin: BDF Verlag, 1939.

48. Gillispie, C.C. *et al. Pierre-Simon Laplace*. Princeton, NJ: Princeton University Press, 1997.

49. Gillmor, C. Stewart. *Coulomb and the Evolution of Physics and Engineering in Eighteenth-century France*. Princeton, NJ: Princeton University Press, 1971.

50. Giroud, Françoise. *Marie Curie. A Life* (trans. Lydia Davis). New York, NY: Holmes and Meier, 1986.

51. Glasser, Otto. *William Conrad Roentgen and the History of the Roentgen Rays*. London: John Bale & Sons, Danielsson, 1933.

52. Goading, David and James, Frank Agile. *Faraday Rediscovered*. Basingstoke: MacMillan, 1989.

53. Goldman, Martin. *The Demon in the Aether: The Story of James Clerk Maxwell*. Bristol: Adam Hilger, 1983.

54. Goldsmith, Maurice. *Frédéric Joliot-Curie*. London: Lawrence and Wishart, 1976.

55. Golino, Carlo L. *Galileo Reappraised*. Berkeley, CA: University of California Press, 1966.

56. Grattan-Guinness, Ivor. *Joseph Fourier 1768–1830*. Cambridge, MA: MIT Press, 1972.

57. Hahn, Otto. *My Life* (trans. Erns Kaiser and Eithne Wilkins). London: Macdonald, 1970.

58. Hahn, Otto. *A Scientific Autobiography* (trans. Willey Ley). MacGibbon and Kee, 1967.

59. Hamilton, James. *Faraday: The Life*. London: Harper Collins, 2002.

60. Heilbron, J.L. *The Dilemmas of an Upright Man: Max Planck as Spokesman for German Science*. Berkeley, CA: University of California Press, 1986.

61. Heisenberg, Elizabeth. *Inner Exile: Recollections of a Life with Werner Heisenberg* (trans. S. Cappellari and C. Morris). Boston, MA: Birkhäuser, 1984.

62. Heisenberg, Werner. *Physics and Beyond: Encounters and Conversations* (trans. Arnold J. Pomerans). New York, NY: Harper and Row, 1971.

63. Herivel, John. *Joseph Fourier. The Man and the Physicist*. Oxford: Clarendon Press, 1975.

64. Highfield, Robert and Carter, Paul. *The Private Lives of Albert Einstein*. London: Faber & Faber, 1993.

65. Hoffmann, Banesh, with Helen Dukas. *Albert Einstein: Creator and Rebel*. New York, NY: Viking, 1972.

66. Hoffmann, K. *Otto Hahn: Achievement and Responsibility*. Heidelberg: Springer Verlag, 2001.

67. Hofmann, James R.I. *André-Marie Ampère*. Cambridge: Cambridge University Press, 1995.

68. James, Frank A.J.L. (ed.). *The Correspondence of Michael Faraday*. London: Institution of Electrical Engineers, 1991.

69. Jungnickel, Christa and McCormmack, Russell. *Cavendish: The Experimental Life*. Philadelphia, PA: American Philosophical Society, 1999.

70. Kargon, Robert H. *The Rise of Robert Millikan*. Ithaca, NY: Cornell University Press, 1982.

71. King, Agnes Gardner. *Kelvin the Man*. London: Hodder and Stoughton, 1925.

72. Klein, Martin J. *Paul Ehrenfest*. Amsterdam: North Holland, 1970.

73. Koestler, Arthur. *The Sleepwalkers*. London: Hutchinson, 1959.

74. Königsberger, Leo. *Hermann von Helmholtz* (trans. Frances A.Welby). Oxford: Clarendon Press, 1906.

75. Kragh, Helge. *Dirac: A Scientific Biography*. Cambridge: Cambridge University Press, 1990.

76. Kursunoglu, Behram N. and Wigner, Eugene P. (eds). *Reminiscences about a Great Physicist: Paul Adrien Maurice Dirac*. Cambridge: Cambridge University Press, 1987.

77. Linsay, Robert Bruce. *Lord Rayleigh: The Man and his Work*. Oxford: Pergamon Press, 1970.

78. Machamer, Peter (ed.). *The Cambridge Companion to Galileo.* Cambridge: Cambridge University Press, 1998.

79. Manuel, Frank E. *A Portrait of Isaac Newton.* Cambridge, MA: Belknap Press, 1968.

80. Matsui, Makinsake (ed.). *Tomonaga, Sin-Itoro: The Life of a Japanese Physicist* (trans. Cheryl Dujimoto and Takako Sano). Tokyo: MYU, 1995.

81. Millikan, Robert A. *Autobiography.* London: Macdonald, 1951.

82. Moore, Ruth. *Niels Bohr: The Man and the Scientist.* London: Hodder & Stoughton, 1967.

83. Moore, Walter. *Schrödinger: Life and Thought.* Cambridge: Cambridge University Press, 1989.

84. More, Louis Trenchard. *Isaac Newton: A Biography.* New York, NY: Charles Scribner's Sons, 1934.

85. Moyer, Albert E. *Joseph Henry, the Rise of an American Scientist.* Washington, DC: Smithsonian Institution Press, 1997.

86. Nachmansohn, David. *German-Jewish Pioneers in Science, 1900–1933.* Heidelberg: Springer Verlag, 1978.

87. Nitske, W. Robert. *The Life of Wilhelm Conrad Röntgen, Discoverer of the X-Ray.* Tucson, AZ: University of Arizona Press, 1971.

88. Pais, Abraham. *Subtle is the Lord – The Science and Life of Albert Einstein.* Oxford: Clarendon Press, 1982.

89. Pais, Abraham. *Niels Bohr's Times, in Physics, Philosophy and Polity.* Oxford: Clarendon Press, 1991.

90. Pais, Abraham. *The Genius of Science.* Oxford: Oxford University Press, 2000.

91. Peacock, George. *Life of Thomas Young, M.D., F.R.S., etc.* London: John Murray, 1855.

92. Pflaum, Rosalynd. *Grand Obsession: Madame Curie and her World.* New York, NY: Doubleday, 1989.

93. Planck, Max. *Scientific Autobiography and Other Papers* (trans. F. Gaynor). New York: NY: Philosophical Library, 1949.

94. Quinn, Susan. *Marie Curie: A Life.* London: Heinemann, 1995.

95. Reston, James R. Jr. *Galileo.* London: Cassell, 1994.

96. Rife, Patricia. *Lise Meitner and the Dawn of the Nuclear Age.* Basel: Birkhäuser, 1992.

97. Rose, Paul Lawrence. *Heisenberg and the Nazi Atomic Bomb Project.* Berkeley, CA: University of California Press, 1998.

98. Rozental, Stefan (ed.). *Niels Bohr: His Life and Work as Seen by His Friends and Colleagues.* Amsterdam: North Holland, 1968.

99. Rukeyser, Muriel. *Willard Gibbs.* Woodbridge, CN: Ox Bow Press, 1988.

100. Segrè, Emilio. *Enrico Fermi; Physicist*. Chicago, IL: University of Chicago Press, 1970.

101. Sharatt, Michael. *Galileo: Decisive Innovator*. Oxford: Blackwell, 1994.

102. Sharlin, Harold I. *Lord Kelvin: The Dynamic Victorian*. Philadelphia, PA: Pennsylvania State University Press, 1979.

103. Sime, Ruth Lewin. *Lise Meitner: A Life in Physics*. Berkeley, CA: University of California Press, 1996.

104. Smith, Alice Kimball and Weiner, Charles Robert. *Oppenheimer: Letters and Recollections*. Madison, WI: University of Wisconsin Press, 1968.

105. Smith, Crosbie and Wise, M. Norton. *Energy and Empire: A Biographical Study of Lord Kelvin*. Cambridge: Cambridge University Press, 1989.

106. Sobell, Dava. *Galileo's Daughter*. London: Fourth Estate, 1999.

107. Strutt, Robert John (Fourth Baron Rayleigh). *Life of John William Strutt Third Baron Rayleigh*. Cambridge, MA: Harvard University Press, 1980.

108. Tolstoy, Ivan. *James Clerk Maxwell*. Edinburgh: Canongate, 1981.

109. Villami, R de. *Newton the Man*. London: Gordon D. Knox, 1931.

110. Westfall, Richard S. *The Life of Isaac Newton*. Cambridge: Cambridge University Press, 1993.

111. Wheeler, Lynde Phelps. *Josiah Willard Gibbs: The History of a Great Mind*. New Haven, CN: Archon Books, Yale University Press, 1970.

112. White, Michael. *Isaac Newton; the Last Sorcerer*. London: Fourth Estate, 1997.

113. Whitrow, G.J. (ed). *Einstein: The Man and his Achievement*. London: British Broadcasting Corporation, 1957.

114. Whyte, Lancelot Law (ed.). *Roger Joseph Boscovich*. London: George Allen and Unwin, 1961.

115. Williams, L. Pearce. *Michael Faraday*. London: Chapman and Hall, 1965.

116. Wilson, David. *Rutherford: Simple Genius*. London: Hodder and Stoughton, 1983.

117. Wilson, David B. *Kelvin and Stokes: A Comparative Study in Victorian Physics*. Bristol: Adam Hilger, 1987.

118. Wilson, George. *The Life of the Honourable Henry Cavendish*. London: Cavendish Society, 1851.

119. Wood, Alexander. *Thomas Young, Natural Philosopher, 1773–1829*. Cambridge: Cambridge University Press, 1954.

120. Yukawa, Hideki. *Tabibito (The Traveller)* (trans L. Brown and K. Yoshida). Singapore: World Scientific Publications, 1982.

Collections

The publications of the subjects of the profiles can generally be found most conveniently in their respective Collected Works, where these exist. In physics such collections tend to be selective rather than complete, since it is in the nature of the subject for the results of even the best research often to be rapidly superseded. In the case of Daniel Bernoulli, Ampère, Röntgen, J.J. Thomson, Bragg, Millikan, Meitner, Hahn, Lindemann, Schrödinger, de Broglie, Bose, Oppenheimer and Goeppert-Mayer no such collections exist, apparently. For the remaining subjects I have found it useful to consult the following:

Galileo Galilei: *Le opere di Galileo Galilei*. Florence, 1929–1939.

Johannes Kepler: *Gesammelte Werke*. Munich: C.H. Beck, 1937–1990.

Christiaan Huygens: *Oeuvres Complètes*. The Hague: Société Hollandaise des Sciences, Martinus Nijhoff, 1888–1950.

Isaac Newton: *The Mathematical Papers of Isaac Newton*. Cambridge: Cambridge University Press, 1967–1981.

Benjamin Franklin: *The Papers of Benjamin Franklin*. New Haven, CN: Yale University Press, 1959–1999.

Roger Boscovich: *Rogerii Josephi Boscovich opera pertinentia ad opticam et astronomiam*. Bassano: Remondini, 1785.

Henry Cavendish: *The Scientific Papers of the Honourable Henry Cavendish, FRS*. Cambridge: Cambridge University Press, 1921.

Charles Augustin de Coulomb: *Mémoires de Coulomb*. Paris: Société Française de Physique, 1884.

Pierre-Simon Laplace: *Oeuvres complètes*. Paris: Gauthier-Villars, 1878–1912.

Count Rumford: *Collected Works of Count Rumford*. Cambridge, MA: Harvard University Press, 1968.

Joseph Fourier: *Oeuvres de Fourier*. Paris: Gauthier-Villars, 1888–1890.

Thomas Young: *Miscellaneous Works*. London: John Murray, 1854.

Hans Christian Oersted: *Selected Scientific Works of Hans Christian Oersted*. Princeton, NJ: Princeton University Press, 1998.

Georg Ohm: *Gesammelte Abhandlungen*, Leipzig: Johann Ambrosius Barth, 1892.

Michael Faraday: *Experimental Researches in Electricity*. London, 1839/1855.
  Experimental Researches in Chemistry and Physics. London 1859.

George Green: *Mathematical Papers of the Late George Green*. London: MacMillan, 1871.

Joseph Henry: *The Scientific Writings of Joseph Henry*. Washington, DC: Smithsonian Institution Publications, 1886.

Hermann Helmholtz: *Wissenschaftliche Abhandlungen*. Leipzig: Johann Ambrosius Barth, 1982.

Lord Kelvin: *Mathematical and Physical Papers of Sir William Thomson/Lord Kelvin*. Cambridge: Cambridge University Press, 1882/1911.

James Clerk Maxwell: *Scientific Papers*. Cambridge: Cambridge University Press, 1890.

Willard Gibbs: *The Scientific Papers of J. Willard Gibbs*. London: Longmans Green, 1966.

Lord Rayleigh: *Scientific Papers*. Cambridge: Cambridge University Press, 1899.

Ludwig Boltzmann: *Wissenschaftliche Abhandlungen*. Leipzig: Barth, 1909.

Max Planck: *Physikalische Abhandlungen und Vorträge*. Braunschweig: Vieweg, 1958.

Marie Curie: *Oeuvres de Marie Sklodowska-Curie*. Warsaw, 1954.

Lord Rutherford: *Collected Papers of Lord Rutherford of Nelson*. London: George Allen and Unwin, 1962–1965.

Albert Einstein: *The Collected Papers of Albert Einstein*. Princeton, NJ: Princeton University Press, 1987–.

Max Born: *Ausgewählte Abhandlungen*. Göttingen: Vandenhoek & Ruprecht, 1963.

Niels Bohr: *Collected Works*. Amsterdam: North Holland, 1972.

Piotr Kapitza: *Collected Papers of P.L. Kapitza*. Oxford: Pergamon Press, 1964–1985.

Frédéric Joliot: *Oeuvres complètes scientifiques de Frédéric et Irène Joliot-Curie*. Paris: Presses Universitaires de France, 1961.

Enrico Fermi: *Collected Papers*. Chicago, IL: University of Chicago Press, 1962–1965.

Werner Heisenberg: *Gesammelte Werke*. Munich: Piper, 1984.

Paul Dirac: *Collected Works of P.A.M. Dirac: 1924–1948*. Cambridge: Cambridge University Press, 1995.

Hideki Yukawa: *Scientific Works*. Tokyo: Iwani Shoten, Publishers, 1979.

Acknowledgements

In an academic historical work it is usual to document almost every statement. Here this would be inappropriate, but historians and others who might wish to do so should have no difficulty in identifying the relevant sources. Where full-scale biographies exist for the remarkable physicists featured in this book I have, of course, made use of them; the most important and recent have already been mentioned above.

The profiles of Louis de Broglie and Piotr Kapitza are largely based on their obituaries in the Biographical Memoirs of the Royal Society. The profile of J. J. Thomson is based on the memoir by his grandson David in [34]. For the other profiles the principal sources used are listed above according to the following key:

Galileo [6, 36, 37, 38, 55, 73, 95, 101, 106], Kepler [3, 20, 73], Huygens [1, 4], Newton [42, 79, 84, 109, 110, 112]; Franklin [2, 24, 29, 39]; Boscovich [114]; Cavendish [5, 69, 118]; Coulomb [2, 39, 49]; Laplace [48]; Rumford [13, 14]; Fourier [56, 63]; Young [91]; Ampère [2, 39, 67]; Oersted [2, 39]; Ohm [2, 39, 48]; Faraday [2, 18, 52, 59, 68, 115]; Green [17]; Joseph Henry [29, 85]; Helmholtz [15, 74]; Kelvin [71, 102, 117]; Maxwell [2, 28, 39, 53, 108]; Gibbs [99, 111]; Rayleigh [77, 107]; Boltzmann [12, 22]; Röntgen [51, 87]; J.J. Thomson [34]; Planck [60, 93]; Bragg [19]; Marie Curie [25, 31, 32, 50, 92, 94]; Millikan [70, 81]; Rutherford [16, 40, 116]; Lise Meitner [96, 103]; Hahn [57, 58, 66]; Einstein [11, 45, 64, 65, 88, 113]; Ehrenfest [72]; Born [10]; Bohr [9, 46, 82, 89, 98]; Lindemann [8]; Schrödinger [83]; Bose [23], Joliot [25, 54]; Fermi [43, 44, 100]; Heisenberg [21, 61, 62, 97]; Dirac [75, 76]; Oppenheimer [104]; Goeppert-Mayer [33], Yukawa [80, 120]. The sources of the longer quotations in the text are as follows:

Franklin: 'To determine the question'/ 'the doctor having published' [24]

Laplace: 'the algebraic analysis soon'/ 'At four in the afternoon' [48]

Fourier: 'first causes are not known' [56]

Rumford: 'I shall withhold' [14]

Young: 'he was not a popular physician' [91]

Oersted: 'In my family I am' [39]

Green: 'For all his six years' [17]

Helmholtz: 'had seriously to think'/ 'We were sitting in'/ 'Professor Klein came in'/ 'In his whole personality' [74]

Maxwell: 'James Clerk Maxwell still occasioned'/ 'It was Maxwell who' [53]

Gibbs: 'Though but a few of them'/ 'Gibbs cannot be given'/ 'A little over medium height' [111]

Kelvin: 'The dear Kelvins arrived' [71]

Boltzmann: 'Erdberg has remained'/ 'He never exhibited his superiority'/ 'Boltzmann has no inhibitions' [12]

Röntgen: 'he had a well-spread nose'/ 'He had a penetrating gaze' [51]

Thomson: 'in all his theories' [34]

Planck: 'Planck loved happy' [60]

Millikan: 'I do not think' [70]

Rutherford: 'Rutherford's book has no rival'/ 'This effect though to all'/ 'The Rutherfords lived' [16]

Einstein: 'He was a very well-behaved child' [45]

Ehrenfest: 'In his usual way' [72]

Born: 'The theoretical research' [10]

Bohr: 'Father always took an interest'/ 'The discussions between' [82]

Lindemann: 'worked in Professor Nernst's laboratory'/'many pilots'/ 'I have often pondered'/ 'he came to Oxford'/ 'If we had some interesting guest' [8]

Schrödinger: 'extremely stimulating and impressive'/ 'for some years already' [83]

de Broglie: 'This little brother'/ 'Demobilized in 1919' / 'Having experienced myself'/ 'He greeted me with great courtesy'. *Louis de Broglie* by A. Abgrem, Biographical Memoirs, Royal Society, 34 (1988).

Kapitza: 'Today the crocodile' *Piotr Leonidovich Kapitza* by D. Schoenberg, Biographical Memoirs, Royal Society, 31 (1985).

Fermi: 'Fermi possessed a sure way' [100]

Most of the portraits reproduced are taken from the collected works of the subject, listed above; the sources of the remainder are as follows:

Isaac Newton: Courtesy of the National Portrait Gallery, London.

Roger Boscovich: unknown

Henry Cavendish: Bridgeman Art Gallery.

Charles Augustin Coulomb: C.C. Gillmor. *Coulomb and the Evolution of Physics and Engineering in Eighteenth-century France*. Princeton, NJ: Princeton University Press, 1971.

Pierre-Simon Laplace: Ecole Polytechnique, Paris.

Benjamin Thompson (Count Rumford): Courtesy of the National Portrait Gallery, London.

Joseph Fourier: Ecole Polytechnique, Paris.

André-Marie Ampère: J.R.I. Hofmann, *André-Marie Ampère*. Cambridge: Cambridge University Press, 1995.

Georg Ohm: Ritter von Füchtbauer, *Georg Simon Ohm*. Berlin: BDF Verlag, 1939.

Michael Faraday: Courtesy of the National Portrait Gallery, London.

Joseph Henry: Smithsonian Institution, Washington.

William Thomson (Lord Kelvin): Courtesy of the National Portrait Gallery, London.

John William Strutt (Lord Rayleigh): Courtesy of the National Portrait Gallery, London.

Wilhelm Conrad Röntgen. The Nobel Foundation.

J.J. Thomson: Courtesy of the National Portrait Gallery, London.

Max Planck: The Nobel Foundation.

William Henry Bragg: Courtesy of the National Portrait Gallery, London.

Marie Curie: The Nobel Foundation.

Robert Millikan: The Nobel Foundation.

Ernest Rutherford: Courtesy of the National Portrait Gallery, London.

Lise Meitner: Royal Society, London.

Otto Hahn: The Nobel Foundation.

Albert Einstein: Lotte Jacobi Collection, University of New Hampshire.

Frederick Lindemann: Royal Society, London.

Erwin Schrödinger: Royal Society, London.

Satyendranath Bose: Santimay Chaterjee *et al.* (eds.). *S.N. Bose: The Man and his Work*. Calcutta: S.N. Bose National Centre for Basic Sciences, 1994.

Louis de Broglie: Royal Society, London.

Maria Goeppert-Mayer: The Nobel Foundation.

Frédéric Joliot: Royal Society, London.

Paul Dirac: Courtesy of the National Portrait Gallery, London.

Robert Oppenheimer: Royal Society, London.

Hideki Yukawa: The Nobel Foundation.